Cheese Manufacturing in the Twentieth Century

The Italian Experience in an International Context

P.I.E. Peter Lang
Bruxelles · Bern · Berlin · Frankfurt am Main · New York · Oxford · Wien

Claudio Besana, Rita d'Errico
and Renato Ghezzi (eds.)

Cheese Manufacturing in the Twentieth Century

The Italian Experience in an International Context

European Food Issues
Vol. 11

This book was funded partially by the Università Cattolica del Sacro Cuore (research action 283C101/2016) as part of its program of promotion and diffusion of scientific research; partially by the Department of Business Studies of the Università Roma Tre and partially by the Department for the Study of Legal, Historical, Economic and Social Sciences of the Università "Magna Graecia" in Catanzaro.

UMG dubium sapientiae initium

UNIVERSITÀ CATTOLICA del Sacro Cuore

ROMA TRE UNIVERSITÀ DEGLI STUDI

Cover picture: Justin Bradshaw, "Pecorino romano".

The book was subject to a double blind refereeing process.

Éditions scientifiques internationales
Bruxelles, 2017
1 avenue Maurice, B-1050 Bruxelles, Belgique
www.peterlang.com ; brussels@peterlang.com

Imprimé en Allemagne

ISSN 2033-7892
ISBN 978-2-8076-0122-2
ePDF 978-2-8076-0123-9
ePub 978-2-8076-0124-6
Mobi 978-2-8076-0125-3
DOI 10.3726/b11644
D/2017/5678/44

CIP available from the British Library and from the Library of Congress, USA.
Bibliographic information published by "Die Deutsche Nationalbibliothek"
"Die Deutsche Nationalbibliothek" lists this publication in the "Deutsche Nationalbibliografie"; detailed bibliographic data is available on the Internet at <http://dnb.d-nb.de>.

Contents

Part II. *Terroir*, Typicality and Market Openness in the Dairy Sector

Part III. Forms of Enterprise in Italy in the Nineteenth and Twentieth Centuries

Foreword

The Dairy and Agri-food Industries

Francesco Chiapparino

Università Politecnica delle Marche

Food manufacturing is in many ways unlike other sectors. It was rarely at the forefront of industrial and technological innovations and, throughout the nineteenth and twentieth centuries, was represented by a loose conglomerate embracing a wide range of manufacturing styles, covering all levels of participation within the secondary sector. Purely agricultural activities were at one end of this scale of participation, with products like vegetables, legumes and fruit, that required almost no processing, while at the other end were the "modern" industries born in the eighteenth and nineteenth centuries during the age of technological innovation. The latter included manufacturers of chocolate and preserves, and the large market-dominating businesses made possible by the second industrial revolution, such as sugar manufacturers. The agricultural producers mentioned above constituted the foundation upon which a large part of the nineteenth-century food sector was built, and were themselves extremely varied in character: even pre-industrial economies like grain and oil production necessitated the use of mechanical equipment like mills and oil presses, thus involving a far greater use of processing structures than was the case for other sectors such as wine-making and dairy production, and therefore were far more dependent upon large-scale financial investments. Other processing techniques, for example those used in the manufacture of liquors, pasta, bread and pastries, should, however, be seen as urban rather than rural phenomena due to their artisan nature. We could go on expanding this list endlessly by, for example, looking at the varying influence of commercial activities within each sector of the food industry, thereby revealing their presence in the service sector along with the primary and secondary sectors, or we could expand the scope of our survey to encompass the extremely diverse world of small-scale growers whose produce was destined for the family table rather than the local market.

The vast heterogeneity of a sector comprising rural industries, artisans, small and medium sized factory producers, large oligopolies, sales-oriented companies and non-commercial domestic activities can to a certain extent explain the difficulties encountered by anyone wishing to construct a historical analysis of the food industry, and might account for the scarcity of such studies in a field that has so far been relatively unexplored by Italian and international economic historians. This is compounded by the fact that the sector has rarely been a prominent player within the manufacturing industry, or influenced its evolution. It never achieved the degree of relevance enjoyed by textile manufacturers during the medieval and early modern eras, nor has it ever held a leading position in the industrialization process of the Western world, as have steel, machinery and energy producers. Despite this, the food sector has always played an important role in the economy of Italy and other countries, and not only during the pre-industrial era when it was intimately connected to the agricultural world through which it constituted a substantial portion of all manufacturing activities. Industry surveys reveal that, during the first half of the twentieth century, its 300,000 to 400,000 employees represented around 10–12% of Italy's entire secondary sector workforce, after which it gradually declined to around half that percentage during the second half of the century, although it was still employing around 250,000 people. Predictably, its share of industry profits saw a similar reduction, falling from an estimated 17–29% during the pre-fascist era until it stabilized around the 8–11% mark at current prices, where it has remained from the 1950s up until today.[1] Along with its vast dimensions and large workforce, the importance of the food industry's role in the national economy is also due to its close (although not always harmonious) relationship with the agricultural sector. This relationship was crucial both for the agricultural and for the food industry, and changed profoundly during each phase of recent Italian history. Historically dominated by the agricultural sector's position as by far the largest economy in Italy, it changed around the 1950s-1960s, when the economic boom and mass migration from the countryside led to the success of industrial foods, perceived as being far from - and in fact often detached from - the national rural productions. The refusal of traditional diets, with their burden of poverty and malnutrition, in the name of the well-being guaranteed by modern, healthy and rational eating, went together with the crisis in large parts of the Italian agriculture, and its lack of integration with the new food industries emerged in the years of the economic miracle. In the last decades, then, this relationship has experienced a further evolution, as a result, first, of the revival of rural traditions linked to the rediscovery of the peasant world and the emergence of environmentalist concerns, and, later, of the recent economic crisis and the renewed importance of agriculture facing the crisis of the mature industrial sectors.

In the light of the situation described above, with its inherent conjectures, grey areas and unknowns, the effort to provide a reconstruction of the evolution of the dairy industry assumes a special significance. Although some grey areas remain in the evolution of Italian and international production of milk derivatives, this work provides a precious contribution to the field by

1 Ciocca, P., *Ricchi per sempre? Una storia economica d'Italia (1796–2005)*, Torino, Bollati Boringhieri, 2007, p. 40.

describing the forms, trends and procedures of a sector that until now has only partially been studied. With a workforce that varied between 20,000 and 30,000 employees according to figures compiled by twentieth-century industry surveys (excluding the 1937 survey, which is notorious for its unreliable exaggerations),[2] the milk and dairy sector represents a medium to large portion of the Italian food industry, outnumbered by the highly fragmented bread-making sector (with 30,000 to 50,000 employees between 1911 and 1981), by the milling sector during the first half of the twentieth century (around 40–50,000 employees until 1961), and by tinned vegetable manufacturing in the second half (30–45,000), while its size was comparable with that of other large sectors like pasta, olive oil and the baking industry. It was also a mid-range sector in terms of energy consumption and, in terms of average company sizes, more or less in line with the relatively small food industry businesses (on average 4–10 employees in each company, not including raw milk production, compared to 3–7 employees in the food sector as a whole between 1911 and 1981), although it should be noted that these figures represent an average value taken from a highly diversified range of businesses both in the case of milk and dairy production and in that of the food industry as a whole. Due to this relatively mid-range position, the dairy industry represents a microcosm of the food manufacturing sector as a whole and of the economic modernization of Italy between the nineteenth and twentieth centuries, and offers, not least through the studies contained in this volume, an excellent opportunity to reveal the underlying patterns of these larger systems. It may be worthwhile, then, to take a closer look at some of those patterns and to identify which parts of the dairy sector's history we might see as particularly relevant.

One of the main issues worth considering is undoubtedly that of the dairy sector's industrialization, along with the speed at which it took place and how it was managed. This was essentially a passage from a traditional rural industry that was practically a by-product of agricultural (and livestock) activities to a manufacturing system that had grown independent from the primary sector and had assumed the characteristics of a modern industry in terms of technology, management and investment. In some areas, like Lombardy, the dairy sector took part in Italy's earliest phase of industrialization during the late nineteenth century, which Luciano Cafagna described as "the first coat of paint",[3] and which was to destined to become consolidated during the Giolitti era (1901–11). Provinces like Lodi, Melzo and others saw the introduction of purpose-built mechanical equipment by dairy producers and, as a consequence, the formation of complex investment and commercial strategies. A decisive factor in the move away from the rural dairy industry, whose traditional farmhouses had until then held pride of

---

2 Chiaventi R., "I censimenti industriali italiani 1911–1951: procedimenti di standardizzazione", in *Rivista di storia economica*, 1987, IV.1, pp. 119–51.

3 Cafagna L., *Dualismo e sviluppo nella storia d'Italia*, Venezia, Marsilio, 1989, pp. 288 ff.

place within one of the world's richest agricultural sectors, was the rise of scientific institutes like the "royal experimental institutes", that contributed decisively to the modernization of agricultural technologies. Another important factor was the role played by a new class of local businessmen, many of whom had aristocratic roots. Cooperative structures also played a prominent role during this phase of modernization in the Po Valley at the turn of the twentieth century, although such structures were generally limited to milk production and were, overall, less influential than elsewhere, as in the wine-making sector, or among producers of fertilizers, who, despite not being directly involved in the food industry, were fully integrated within the agricultural world. This early phase of industrialization did not see the participation of the entire dairy sector, large parts of which continued for some time to maintain traditional dairy facilities and identify more closely with the rationale and management strategies of the livestock industry in the south of Italy and elsewhere. Once-promising traditional occupations like that of Calabrian goat herders began to experience a decline in Southern Italy and bordering regions during the inter-war period, while other specialized professions came to the fore, including the producers of Pecorino romano cheese in Latium and Sardinia, foreshadowing the strong period of growth that would take place in the post-war manufacturing industry.

Large-scale modernization of the sector took place during the economic boom of the 1950s and 1960s, when industrial manufacturing techniques were introduced in many parts of the country with the updating of pre-existing facilities and the appearance of new phenomena. This new era was exemplified by the commercially successful regions of Parma and Reggio Emilia, with unique organizational structures based on a territorial consortia system and their innovative investment in brands and advertising, and by businesses in the Tuscan Maremma, which managed to become successful in a hitherto backward and underdeveloped area. Despite the continued existence of troubled and problematic areas, not least with regard to the uncertain future of many specialized professions in Southern Italy, the modernization of this sector seems, through different timescales and methods, to have taken form. The relationship between agriculture (and livestock farming) and industry also took on a distinct form within this modernization process, and it is this relationship that allows us to understand the nature and limits of the momentous transformations that engulfed Italy's economy and its society in the third quarter of the twentieth century.[4] In areas where the pace of local industrialization went forward hand in hand with the primary sector, as in Southern Lombardy and Emilia Romagna, the 1950s and 1960s migration from the countryside was not, unlike in many other parts of the country, accompanied by a crisis in the agricultural sector.

4 Sicca L., *L'industria alimentare in Italia*, Bologna, Il Mulino, 1977, pp. 92 ff.; Zamagni V., *Dalla periferia al centro. La seconda rinascita economica dell'Italia, 1861–1990*, Bologna, Il Mulino, 1993, pp. 442 ff.

The Po Valley region benefited from the presence of rural economies that, to varying degrees, were already the wealthiest and most advanced in Italy, along with an entrepreneurial and industrial dynamism that was lacking elsewhere. Nonetheless, the aforementioned cases of Pecorino production in Latium, Sardinia and Tuscany, and that of other cheeses manufactured in mountain areas like Fontina and Asiago, demonstrate that patterns of development were not necessarily predetermined, and that the general north-south imbalance (or the disadvantages of areas traditionally backward) could in some cases be subverted.

Lastly, it is worth mentioning the contradictions that have characterized recent developments, with the globalization process of the 1990s and the last decade's long recession, which are analyzed in this volume and subjected to new and revealing studies. There seem to be at least three separate factors underlying the current situation. The first of these was the increasingly central role played by political forces in the application of those regulations, restrictions and opportunities that defined the market within which the sector found itself operating. The decisive influence of political support and protection in the primary sector had, in reality, emerged on an international scale by the 1930s and was present in many countries, although to varying degrees, since the Great Depression of the late nineteenth century. After the Second World War, and especially in Europe, this political influence solidified its dominant position through incentives, subsidies and EU legislation as well as the negotiations and lobbying activities that helped to define them, allowing it to shape the destiny of entire regional and national manufacturing compartments. From the 1990s, regulation took on an enhanced role thanks to the acceleration of the integration process of global markets: a process which is still under way, increasing the already high level of uncertainty within the food and agriculture sector, as demonstrated by the discussions that began in 2013 on the Transatlantic Trade and Investment Partnership (TTIP).

Despite the complexity of this new situation and its inherent obstacles, especially since the beginning of the financial crisis of 2007–8, Italy's food manufacturers have fared well, particularly when one considers that in the last ten years theirs is one of the few segments of the national economy to have maintained, and with time improved, its position in international markets. This success is all the more impressive when seen in the context of the difficulties faced by the Italian economy as a whole, weighed down as it was by structural inadequacies like low productivity growth and heavy financial obligations. It also disproved the dire predictions at the end of the twentieth century about the future of the food industry itself when, especially in the 1970s and 1980s, it seemed overwhelmed by the deep crisis that had hit Italy's industrial sector. In actual fact, the growth that appeared during the economic boom years was in some ways limited. The relatively wide-ranging industrialization of the food sector between the late 1950s and the 1960s was, in fact, followed by a period of difficulty during the following two decades, along with a phase of intense restructuring. A solid business and manufacturing base emerged from this chain of events, which was well equipped to satisfy domestic demand, although in the majority of cases its infrastructure was poor, limiting its readiness to meet the challenges of international competition on markets which, as already noted, were becoming increasingly interdependent. The dairy sector was fully involved in these events, and experienced a comprehensive phase of industrialization during the economic boom years while being badly affected by a change in economic trends during the 1970s, and was unable to expand into foreign markets. Quite the opposite took place towards the end of that delicate phase when, in the 1980s, many dairy brands were acquired by global multinational food companies like Kraft, Danone and Nestlé (which had owned Locatelli since 1961), in a process that has continued to evolve right down to recent times with the creation of an Italian group as part of the French multinational Lactalis. It is striking that this fate has befallen the oldest part of the dairy sector, made up of companies from Lombardy whose modernization can be traced back, as we have seen, to the decades straddling the nineteenth and twentieth centuries. Among these companies were famed Italian brands like Galbani and Invernizzi, while a similar fate affected Polenghi Lombardo, which was still Italian-owned but had changed hands during the 1990s between Federconsorzi, the Cragnotti-owned Cirio and Parmalat. The history of this latter company, Parmalat, is also significant. Although it originally operated in the packaged milk sector, it later expanded into the associated dairy products sector by acquiring Polenghi and other companies. Parmalat's growth cycle, notwithstanding the legal scandals that have plagued the company,[5] is in some ways emblematic of that experienced by Italy's food industry throughout the second half of the twentieth century. Having occurred between the 1960s and the 2003 crash, Parmalat's paradigm was representative both of the sectorial specialization some businesses adopted in their move towards multinational dimensions, for example Ferrero, Barilla and Ferruzzi, and of the limitations which in one way or another frustrated some of these attempts to compete in an international market dominated by global, multi-sector companies like Nestlé and Unilever.

The outcome of the growth cycle of the second half of the twentieth century, when the emergence of a modern and vibrant industrial sector was set against poor performance in the face of international competition and an inability to counteract acquisitions made by foreign multinationals with their own domestic investments, was met by the unexpected success of the Italian food and agriculture export sector and, in general, by the popularity of Italian foods despite the ongoing recession. Naturally, results like these cannot justify the kind of enthusiasm displayed at the 2015 Expò fair in Milan, or

---

5 Franzini G., *Il crac Parmalat. Storia del crollo dell'impero del latte*, Roma, Editori Riuniti, 2004; Sapelli G., *Giochi proibiti. Enron e Parmalat, capitalismi a confronto*, Milano, Bruno Mondadori, 2004.

that generated by promotional campaigns: we need only remind ourselves that the last decade's export figures for traditional products were unable to make up for the long-running trade deficit of Italy's food and agricultural industry. Above all, the image of close adherence to natural and traditional manufacturing methods confronts the reality that only traditional products (including wine) that are produced with the support of modern technologies, a well-structured industry and substantial investments in marketing, that is to say a series of factors that are significantly alien to concepts of nature and tradition, are able to penetrate international markets and assume true economic significance. Nonetheless, the positive performance of the food sector is undeniable, both in its own right and in comparison with the rather disappointing performance in recent decades of the country's economy as a whole. The dairy sector, on the other hand, has been involved in the majority of the processes spoken of above: it underwent widespread industrial renewal during the boom years, along with the survival of more antiquated methods and the dualism typical of Italy's economy; it took part in the failed attempt to create multinational brands during the 1980s and 1990s, which was followed by a wave of foreign acquisitions; and, finally, it was at the center of the recent successes achieved by the Italian traditional foods sector and shared its strengths as well as its ambiguities. Furthermore, and with regard to the latter, cheese (just like wine, and unlike pasta for example) represents a traditional product that is not unique to Italy and is therefore subject to comparison and competition with producers from other countries, not least with those of France, as revealed by the international studies presented in this volume. The dairy sector, then, is the best place to start when attempting to understand the recent and future evolution of Italy's food and agriculture sector and to determine whether and to what extent the current success of traditional products is a temporary phenomenon, the precursor of a new cycle of growth, or simply the next phase in the history of the food industry.

Introduction

Claudio Besana, Rita d'Errico, Renato Ghezzi*

Globally, around 20 million tons of cheese (excluding others cow's milk derivatives) are produced each year, making the cheese industry the most profitable segment of the entire dairy sector. Western Europe and North America account for a large part of this figure, with around 70% of global production (Fig. 1). Similarly, these same areas represent a high proportion of sales and per capita consumption of cheese. Experts predict that the sector will continue to grow as the demand for cheese increases in developing countries,[1] whereas they see the consumption of milk and butter as being in decline, especially in the more developed economies, following a trend that began some decades ago with changes in tastes and eating habits.

The prime position occupied by these geographical areas cannot be attributed simply to their relative abundance of raw materials. While countries like France, Germany and the USA do produce large quantities of milk, it is also true that other countries, like Italy, have historically produced much less of this raw material than their competitors, but have nonetheless achieved notable results and are among the main producers and exporters of cheese, whereas countries like India and Brazil, despite possessing an abundance of raw materials, have not yet developed a significant dairy manufacturing industry.

Access to raw materials, then, is a necessary prerequisite, but it must be accompanied by other factors in order for a dairy industry to take root, including the entrepreneurial experience gained from local traditions and techniques, the ability to innovate by improving and diversifying products and gaining access to ever greater markets, the support of institutions, and the presence of associations able to overcome the disadvantage faced by small companies in an increasingly global market. Such factors proved to be crucial around the end of the nineteenth century when those countries which

* The editors and authors wish to express their sincere gratitude to the anonymous referees and to the editorial committee for their comments and suggestions. We would also like to thank the numerous librarians and archivists who have assisted us in our research activities over the years. We are particularly grateful to Professor Franco Amatori, who encouraged this project from the beginning.

1 See the *Bulletin of the International Dairy Federation*, 2012, No. 458, p. 46; the overall growth in global consumption of cheese between 2011 and 2021 is predicted to be around 19%, of which the largest share will come from China and developing countries such as India and Brazil.

were able to take advantage of the technological developments of the second industrial revolution saw a rapid growth in their dairy sectors. This quantitative and qualitative growth allowed the sector to assume ever greater autonomy in industrialized countries and achieve autonomy from the primary sector from which it had previously been generally indistinguishable, going on to depend on it only for the supply of raw materials. This phase was accelerated by the appearance of other concurrent and related factors like improvements in land and sea transport (which removed the obstacle represented by large distances, cutting the cost and duration of intercontinental and maritime transport), and the introduction of new technological advances into the delicate dairy-making process, including advances in the cold cycle process, which led to the opening up of new and larger markets for a product that is characterized by its limited lifespan. It is, furthermore, worth noting that immigration from Europe to other continents, which intensified greatly from the mid-nineteenth century, contributed to the increase in sales of dairy products, and helped to spread knowledge of dairy traditions and techniques that had been developed in the Old World.

Figure 1. Cheese produced in 2015 (x 1,000 tons) by the largest cheese-producing nations

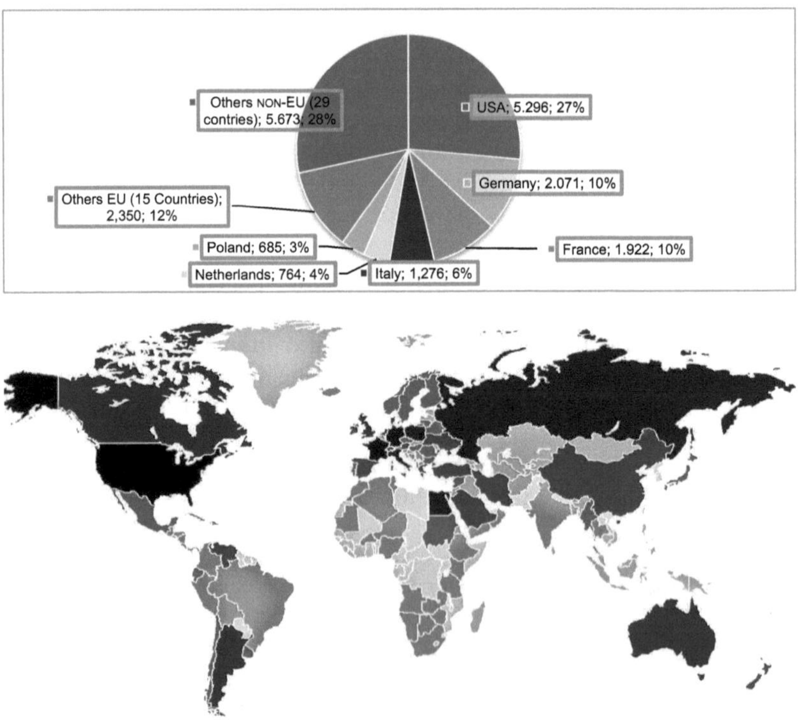

Source: our elaboration of data from <http://www.deagostinigeografia.it/wing/confmondo/confronti>.

Italy occupies a prominent position within the highly varied panorama of global cheese producers which has been taking shape since the end of the nineteenth century, being the fourth largest producer of this product despite not having, as mentioned above, a domestic supply of milk comparable with that of its competitors. It is also worth noting that the dairy industry occupies a prominent position in the multifaceted agri-food and agriculture sector of the Italian economy, with annual sales of around 15 billion euros (2013), equivalent to 11.5% of the country's overall food industry sales, which itself is the second largest category within Italy's manufacturing industry.[2]

The Italian dairy industry has achieved a position of international leadership thanks most of all to its originality and to the elevated quality and variety that are made possible by the country's special combination of environmental and geographical conditions and the diverse culinary traditions that have grown out of the many different cultural influences present across the peninsula.[3] Italy and France are the European countries with the highest number of cheeses registered as Protected Designation of Origin and Protected Geographical Indication products (Fig. 2), with more than half of Italy's milk output employed in the manufacture of PDO cheeses, of which Italy is the largest producer and exporter. For this reason Italy could perhaps be defined, in the words of Corrado Barberis, as "the land of cheese".[4]

Figure 2. Number of PDO and PGI registrations by EU countries

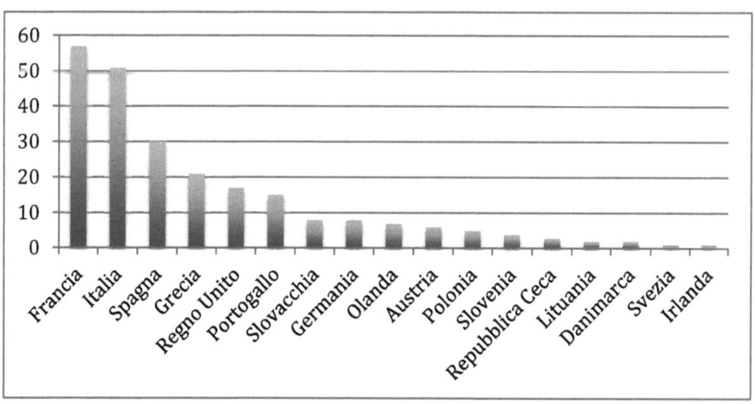

Source: "Dop. Italia prima per volumi", *Il Mondo del Latte*, LXIX, 2015, 8, pp. 40–1.

---

2 *Il mercato del latte, Rapporto 2014*, Pieri R. (ed.), Milano, Angeli, 2014, p. 213.

3 Baldrighi N., "Importanza dei DOP: fattore di qualità", in *I Georgofili*, 2014, III, pp. 19–30, Capalbo C., "L'agricoltura italiana tra biodiversità e cucine locali: una lettura di lungo periodo. Incontro con Piero Bevilacqua", in *Rivista di Storia Economica e Sociale*, I, 2015, 1–2, pp. 237–48.

4 "Introduzione", in *Atlante dei prodotti tipici. I formaggi*, Istituto Nazionale di Sociologia Rurale (ed.), Milano, 1990, p. 14.

Considering the current state of affairs in Italian dairy manufacturing, the question arises as to what its strengths and weaknesses were and which factors contributed to its present position both on a national and an international level. The chapters presented in this volume attempt to answer these questions from a variety of different viewpoints, making use of numerous interpretative approaches with the aim of highlighting the current trends in a country that can be seen as a fundamental part of a multifaceted European dairy industry that is still taking shape.

The timeline covered by the chapters in this volume stretches beyond the century referred to in the title, as far back as the last three decades of the nineteenth century and, in some cases, right up to the present day: a long twentieth century, then, that straddles three centuries and two millennia. While it may seem natural that this timeline ends where it does, the choice of where it begins is far from being a coincidence. Such a choice was born out of the consideration that, more or less from the time of the late nineteenth-century Long Depression, a series of events took place which had global consequences, bringing together different elements and leading to an acceleration in the growth of industrialized countries' dairy sectors. In view of the international nature of the phenomena examined in this volume, and how they took shape in a context of increasing globalization, it seemed appropriate to widen the territorial scope of this study beyond the borders of a single nation. The studies put forward by the chapters in question are the result of a dialogue that began in 2014 and which developed during a number of national and international conventions throughout 2014 and 2015 at which the results of the relevant research were presented and compared with the work that other scholars, some of whom have contributed to this volume, on themes related to cheese manufacturing outside of Italy. These meetings allowed us to include other countries like France and Switzerland, whose commercial and cultural ties with Italy stretch far back into history and, among the non-European countries, the USA and Canada, where Italian immigrants have played an important role in the fortunes of the dairy sector by creating a substantial demand for Italian products abroad and by bringing manufacturing traditions and know-how with them to their new homes.

Our research project did not begin with the idea of reaching a systematic and exhaustive conclusion on such a complex issue (one still largely unexplored with regard to the twentieth century) as that of the dairy industry, which finds itself at the crossroads between a number of different disciplines and research methodologies. We were, rather, interested in stimulating further interest in this subject and in inviting scholars to orient their research towards some of the underlying themes indicated in the titles of the volume's three sections.

The chapters included in the first section (The Experience of Different Nations During the Age of Globalization) compare various national scenarios in the midst of a period of transformation, when great technological advances were being made and markets were progressively becoming globalized. The chapter by Claire Delfosse ("Cheese-making in France at the beginning of the twentieth century: a spatially diffuse, complex and unique organization?") traces the contours of French dairy manufacturing during the final decades of the nineteenth century and the first twenty years of the twentieth. In a study based on a wide-ranging analysis of historical geography, the author describes the sector's spatial organization and highlights four principal systems, each characterized by different manufacturing processes and by differing company structures. Her study begins with the departments of central and western France, where Gruyère production had arisen from cooperative-style initiatives, and goes on to look at the two soft-cheese-producing areas of Normandy, the home of Camembert, and Mosa, with its Brie cheeses. These two areas were characterized by the comparatively active role played by private businessmen, as was the case with Roquefort. This latter region in the south of France was home to the Société des Caves, a large company which enjoyed a monopoly on the production of this unique goat's milk cheese and ensured its success at home, above all in Paris, and abroad. Luigi Lorenzetti ("Switzerland's dairy industry in the twentieth century: small-scale artisan production meets the market economy") takes a different approach in describing the making of cheese in the cantons of the Swiss Federation during the twentieth century. Here, too, the author analyses the different manufacturing systems that appeared throughout this European economic region to cater for the production of widely consumed cheeses, including some of the most popular dairy products on international markets. Luigi Lorenzetti focuses on the role of institutions in helping to increase the manufacture and international sales of Swiss cheese, with particular focus on the policies enacted by the federal government between 1950 and 1990, which aimed to promote their cheeses through market regulation and export subsidies. The final part of the chapter looks at the effects of a change in public policy and the reorganization process that took place within the sector from the last decade of the twentieth century. Particular mention is made of the difficulties faced by producers of Emmental, the most popular Swiss cheese internationally, and the success of local varieties and alpine cheeses made according to traditional methods by companies that also manage to operate in the global market.

Claudio Besana's chapter ("The Italian dairy industry between 1930 and 1970: production and organizational structure") focuses on Italy, and its dairy sector between the end of the 1930s and the early 1970s, which is seen to have been dominated by small and very small companies. From a geographical point of view the production of cow's milk cheeses appears to have been concentrated in the plains of eastern Piedmont, Lombardy and Emilia. That of sheep's milk cheeses, on the other hand, was centered on Latium and, above all, Sardinia. The number of varieties was extremely high due to the country's rich panorama of regional traditions. From a manufacturing point of view, the period in question was characterized by an increase in the number of cheeses produced, and by a vast assortment of varieties. Grana resisted these changes, and the production of soft cheeses grew in response to an increasing popularity among consumers. Overall, the sector continued to be characterized by the overwhelming influence of small and medium sized companies.

The changing shape of the post-Second World War dairy sector was in part connected to the expanding influence of the European Community's Common Agriculture Policy (CAP). The consequences of market integration and the evolution of the Italian dairy industry within the EU are explored by Stefanella Stranieri and Paolo Tedeschi ("The influence of European institutions and policies on the Italian dairy sector"). They concentrate particularly on the outcome of the free market policies that led to the downfall of centralized milk monopolies, and the consequences of the adoption of the milk quota system. The chapter also explores the decision made at the beginning of the 1990s to provide protection under European legislation through PDO and PGI labeling, a decision which was to prove decisive for the evolution of the Italian dairy sector over the following two decades.

Canada was the first country to experience what was an important phase in the evolution of the globalized dairy industry; the passage from a model based on an unregulated supply of milk to one regulated through a quota system, as explained in the chapter by Benoit Mario Papillon ("The Canadian dairy sector in the twentieth century: from openness to closure and self-sufficiency"). The supply management system, introduced at the beginning of the 1970s, produced unbalanced results for the categories concerned. It is, in effect, a welfare system that benefits livestock farmers at the expense of producers and consumers, who found themselves paying more for milk and milk derivatives than their foreign counterparts. A further consequence of the import barriers imposed by the quota system was Canada's withdrawal from the international dairy market, where it had previously occupied a central role both as an exporter and as an importer of cheese.

Brazil is among the developing countries currently appearing on the international stage as potential consumers and importers of dairy products, and, thanks to its vast resources in terms of land and of livestock, it could find itself in a position to challenge the supremacy of Europe and North America. Despite this, Brazil is still a net importer of milk and milk derivatives, and is struggling to achieve the required levels of quality in the manufacture of these products. This state of affairs is the result of protectionist policies and of a lack of sufficient public investment in research and technological innovation, which until recent times have impeded the sector's development (Hildete De Moraes Vodopives and Valter Galan, "Contemporary challenges to the Brazilian dairy industry").

The evolution of the dairy industry during the last century was in part dictated by the transformation of geographical areas, which in turn had either positive or negative effects on the dairy industry. Julia Lapp ("New York State cheese: nineteenth and twentieth century transformations") describes how these changes came about mostly as a consequence of population growth and urbanization. New York State, between the early nineteenth century and the First World War, was the original nucleus of the American dairy industry which emerged thanks to an efficient transport system and to a widespread use of technological innovations that favored an increase in the size of companies and of their production volumes. The pioneers of the American dairy industry included many immigrants from Italy, whose accumulation of skills and experience played an important part in the growth of the industry. Later, New York's leading position was threatened, not so much by competition from other states like Wisconsin as by population growth in urban centers and by the fact that it was more profitable for livestock farmers to sell their milk for direct consumption than for them to supply it to cheese manufacturing.

The theme that brings together the chapters contained in the second section (*Terroir*, Typicality and Market Openness) is the relationship between tradition, authenticity and innovation. Particular attention is paid to the history of a small group of cheeses that have played a prominent role in international trade, to the extent that four of these cheeses (Roquefort, Parmesan, Gorgonzola and Pecorino romano) were the first to have been recognized as deserving protection at the 1951 Stresa Convention.[5] Each country can undoubtedly boast a great many varieties of cheese – Italy alone accounts for over four hundred – that can be considered expressions of unique local conditions and of ancient traditions that cannot be reproduced elsewhere. But what is it that has led some cheeses rather than others to become international gastronomic celebrities and icons of the cultural traditions to which they belong? The pathways were many, but the fact that a cheese was the result of a unique tradition and a distinct local context was never enough to single-handedly carry it beyond its original market; behind the success stories of the most famous dairy products there has always been a skillful combination of entrepreneurial abilities and marketing strategies to ensure a cheese's fame across the globe and its reputation for consistent high quality.

A prime example of this phenomenon, that of Parmigiano-Reggiano, is the subject of Stefano Magagnoli's chapter ("Parmigiano-Reggiano cheese: the industrialization of typicality"). There were bitter clashes over the commercial use of its designation when it began to increase its foreign sales at the beginning of the twentieth century, leading producers in the provinces of Parma, Reggio Emilia, Mantua, Modena and Bologna to come together in a consortium in order to create a collective trademark promoting its

5 *Convention internationale sur l'emploi des appellations d'origine et dénomination de fromages*, 1 giugno 1951; the text of the international convention and the decree applying it to Italian law are reproduced in issue 46 of the *Gazzetta Ufficiale*, 16 February 1954 (<http://www.gazzettaufficiale.it>).

connection to the cheese's original territory of production and the traditions from which it arose. Quality and authenticity both represent a competitive advantage in a context of expanding markets where opportunities for growth increasingly depend on the ability to demonstrate the high quality of a product as well as its association with a specific territorial tradition. After the Second World War, thanks to well-managed and sophisticated promotional campaigns that made use of new forms of mass communication, Parmesan gained a foothold in domestic, European and American markets; its success in achieving the status of a household name was in part thanks to the growing popularity of pasta, an icon of Italian cooking, which was invariably topped with grated Parmesan.

Beyond Italy, too, there were examples between the two World Wars of how a connection with a specific territory could represent a successful market strategy. This often took place with the support of government institutions and through promotional campaigns and was not, therefore, a spontaneous or automatic process. The commercial success of Roquefort, a cheese produced in the village of the same name in the south of France and one of the most prominent symbols of French gastronomy around the world, owed much to the powerful organization known as Société des Caves, which worked to help Roquefort achieve international fame from the mid-1800s by participating in trade fairs and international exhibitions. One of the Societé's greatest successes was a French parliamentary law created in 1925, the first national law dedicated to the protection of a designation of origin cheese, which officialized the connection between Roquefort and its home territory by obliging producers to season the cheese in the natural caves of the village of Roquefort, the majority of which were owned by the *Société* itself (Sylvie Vabre, "Roquefort: a large company creates its territory").

Making the most out of the geographical provenance of a traditional product requires a laborious process built upon business acumen, cooperative efficiency and government support; these factors are not always in place in the original areas of production and in cases where they are not forthcoming the connection with a territory of origin can practically disappear, as in the case of Pecorino romano, a cheese which until a few decades ago was a leader in the Italian cheese export market. Pecorino romano's original manufacturing base was in the region of Latium, specifically in the outskirts of Rome, but from the first decades of the twentieth century it gradually moved towards Sardinia, where 97% of this cheese is currently produced. This happened as a result of the progressive expansion of the area of Pecorino romano production which began a century earlier in response to growing demand from the USA. Livestock farmers and producers in Latium had shown diffidence towards the formation of associations and cooperatives, and were powerless to prevent this expansion; the small efforts which they did make, furthermore, did not gain the support of government institutions or of the businesses which acted as an interface between far-away markets and the innumerable small producers of Pecorino (Rita d'Errico, "The production of

Pecorino in the Roman countryside from the end of the nineteenth century up until the 1930s").

The vicissitudes of Pecorino romano were tied in with those of the sheep's milk dairies of Sardinia and Tuscany, two regions which, together with Latium, constituted the approved production area for Pecorino romano as stipulated by the Stresa Convention in the early 1950s. Sardinia undoubtedly benefited from an influx of knowledge and manufacturing techniques from Latium, at a time when there was little interest in official protection of traditional products and larger companies found it advantageous to be able to move production to areas with lower production costs. Sardinia had, in any case, shown signs of vitality at a local level, with the birth of a strong cooperative movement, and enterprising businessmen took over from the pioneers who had come there from other regions (Giuseppe Doneddu, "Production and processing of sheep milk in Sardinia (1950–2015").

After the Second World War, many sheep owners moved to Tuscany in search of new pasture lands and new investment opportunities. It was, in part, thanks to them that the Tuscan dairy sector was able to rebuild itself in the 1960s and 1970s after a difficult post war period. The sheep population in Tuscany began to grow once more, followed by an increase in the production of sheep's milk, which in turn was used to make Pecorino toscano, a traditional cheese that achieved recognition as a PDO product in 1987. This latter cheese has, in recent decades, gained a substantial slice of the market, reaching an international consumer base thanks to the activity of a consortium set up for its protection and its ability to strengthen and promote its connection with a territory that is already well known around the world, and to raise production in terms of both quality and quantity (Renato Ghezzi, "The Tuscan ovine milk industry, 1950–2000: a quantitative analysis").

The consortia for the protection of cheeses that perform coordination and production control functions play a fundamental role in a context of market liberalization such as has taken shape since the Second World War, by promoting their cheeses at home and abroad while helping them to obtain international recognition as authentic traditional products. The importance of the role of these consortia is made clear in a chapter that examines the case of Gorgonzola, by Vanessa Pollastro ("From Milan to Novara: the production and sale of Gorgonzola in the twentieth century"). This pungent soft cheese, which was originally aged in the small Valsassina valley at the foot of the Alps, enjoyed a period of international success between the end of the nineteenth century and the first decades of the twentieth, which owed much to the ability of local businessmen like Galbani and Locatelli, who had created new markets for Gorgonzola in places like the UK, where it was appreciated as a cheaper alternative to the local cheese, Stilton. Gorgonzola's fame began to decline after the Second World War with the opening up of European markets and competition from low cost imitations produced in countries like Denmark, which had not signed the 1951 Stresa Convention on the protection of designation of origin products. The setting up of a consortium for the protection of Gorgonzola at the beginning of the 1970s was followed by a new period of ascendancy for the production and export of this cheese, thanks to the consortium's activities in the coordination and management of production, modernization of manufacturing techniques, and diversification. The revision of legislation on the protection of traditional product designation that began in 1992 (EC regulation 2081/1992) extended its provisions, which had previously been applied only in a limited number of countries, to all EU nations, and bolstered the powers of consortia to combat the phenomenon of imitation traditional products.

This new context allowed the emergence of a typical southern Italian cheese, buffalo Mozzarella, one of the main players in the growing success of Italian PDO cheeses. This product began to achieve prominence around the beginning of the 1990s, aided by the development of new packaging and refrigeration techniques which proved fundamental for a cheese with a limited shelf-life whose distribution had until then been limited to markets close to its production areas. There were undoubtedly other factors involved in the success of buffalo Mozzarella, including its position as a ingredient in the classic Italian dish, pizza, the increasing efficiency of road and air transport networks, and an evolution in consumer tastes towards delicate-tasting fresh cheeses. This cheese, then, was able in a short time to rise to become one of the most important Italian PDO cheeses (Franca Pirolo, "Buffalo Mozzarella in Campania from its origins to the twenty-first century").

The chapters included in the third section (Forms of Enterprises in Italy in the Nineteenth and Twentieth centuries) highlight the changes which took place in Italy's dairy sector during the nineteenth and twentieth centuries, by tracing the history of some of its most important companies and describing the unique structures adopted by businesses operating in this sector. The chapter by Manuel Vaquero ("The modernization of the dairy industry in Italy in the late nineteenth century: social dairies") emphasizes the importance of milk producer cooperatives in dairy modernization processes. These forms of cooperative organization appeared in the 1870s based on examples from countries like Switzerland and the USA, and became common in alpine areas, especially in the regions of the Veneto and the Friuli, often under the incentive of the local clergy. These cooperative systems arrived late in the area most intensely dedicated to dairy manufacturing, between Lombardy and Emilia, although they did eventually take hold, whereas they struggled to gain acceptance in Central and Southern Italy. It should also be noted that in alpine areas, cooperative milk producers were almost always small in scale, and were often the result of agreements made at a local level. There were, on the other hand, large cooperative milk producers in the plains, which produced a great deal of milk every day. The cooperatives that were active in the dairy sector confirmed the dualistic nature of the Italian business world at the end of the nineteenth century by combining milk production with that of butter and cheese.

The modernization of the sector was accompanied by the appearance of some large companies which had been at the center of the Italian dairy sector throughout most of the twentieth century. The chapter by Silvia A. Conca ("Enterprises, trade and industry in the Lombardy dairy sector: the origins of Locatelli and Galbani (1860–1914)") investigates the origins of two of the most important Italian food companies of the post-war years, both from the aforementioned Valsassina valley. The early manufacturing experiences of the two companies began with the seasoning and sale of soft cheeses produced in the lower alpine valleys and sold throughout Lombardy, though later their commercial pathways began to diverge; the Locatelli company saw rapid growth as a family-owned company, specializing in the large-scale seasoning and sale of Gorgonzola in Italy and abroad. In just a few years it built large seasoning warehouses and opened outlets in the United Kingdom and the Americas, while also selling large volumes of Pecorino romano, which it purchased from producers in Latium and Sardinia. Galbani, on the other hand, remained a small company managed directly by its founder until the First World War. Although Galbani, too, dedicated itself to the seasoning of soft cheeses, its founder, Egidio Galbani, chose to concentrate on the creation of new products that imitated luxury cheeses sold in the most prestigious French delicatessens. His name is connected with the invention of famous cheeses, such as 'Bel Paese', which are an integral part of the traditions of Lombardy. The company grew significantly over the following years, until in 1920 the family business became a joint stock company, no longer under the exclusive ownership of the family of its founder, an uncommon move in the context of the history of Italy's family businesses.

At the beginning of the twentieth century, the largest Italian dairy company was Polenghi Lombardo, the subject of Gianpiero Fumi's chapter, "The rise and decline of a large company: the Polenghi Lombardo". Founded in the 1870s in Codogno, a town in the province of Lodi, which had the highest concentration of milk and dairy producers in all of Italy, Polenghi Lombardo became a public limited company at the beginning of the twentieth century, and chose Lodi as the site of the largest milk manufacturing plant in the country. The company underwent an important transformative phase after the First World War when the Polenghi family lost managerial control to a group of agricultural businessmen from Lodi under the leadership of the powerful local agricultural consortium, initiating what was an unusual experience in the panorama of the Italian business world when the company was entrusted to a highly qualified manager. After the Second World War Polenghi was still one of the sector's largest companies, but it soon began to feel the weight of a difficult financial situation that was never adequately dealt with by the new controlling shareholder, Federconsorzi, leading to a series of events which ended with the latter go into liquidation at the beginning of the 1990s.

The final chapter of the volume is by Andrea Maria Locatelli ("The multinationals and the Italian agri-food industries: the dairy sector, 1974–1993"), and reflects on the theme of the gradual passage into foreign hands of the largest Italian dairy companies at a time when foreign capital was gaining ever more control over a substantial part of Italy's food and agriculture industry. These events took place over a twenty – year period during which the volume of milk and cheese imported to Italy grew significantly. The entrance of large multinationals into the sector began in the early 1960s with the Swiss company Nestlé's purchase of the family-owned business, Locatelli, at a time when it was undergoing financial difficulties. Invernizzi and Galbani, on the other hand, were in good health, having seen their market share grow during the post-war period as well as maintaining a high degree of financial stability. Their purchase by foreign multinationals was in part connected with the fact that both companies had encountered problems in finding a new generation from within the owning families to take over their management. They had also found it difficult to survive in an increasingly liberalized market, where the size of a company began to be a crucial factor in the ability to put into effect efficient promotional strategies and to renew their line of products with sufficient regularity.

As was noted at the beginning of this introduction, the volume is not intended to be a definitive study of the dairy industry in the twentieth century. The chapters it presents reveal the need to continue investigating the commercial world of Italy and the other large dairy-producing countries during the twentieth century. The actions of large multinational companies, which now control a whole series of deeply traditional brands, and the vivacity of small and medium sized companies able to perform well in a wide range of markets, deserve further attention. The question of the promotion of quality products remains largely unexamined, especially when seen in the context of the extremely topical, though not unprecedented, issue of respect for the biological processes of nature in the milk and cheese manufacturing industry and its use of new discoveries made in the field of chemistry. Finally, and of no small consequence, is the question of the dairy industries that this book has not been able to cover, including those of Holland, Spain, Germany, the Scandinavian countries, and New Zealand, a country which by the 1930s was already one of the largest dairy exporters. There is still much to be done; but what makes the research of history interesting is that, when carried out with an open mind, it always uncovers new subjects in need of study and creates new opportunities for collaboration.

Part I

The Experience of Different Nations during the Age of Globalization

Cheese-making in France at the Beginning of the Twentieth Century

A Spatially Diffuse, Complex and Unique Organization?

Claire Delfosse

Université Lyon 2

Whereas, during the "pre-industrial" period, French cheese-making presented itself as a fragmented space, new production (scientific and technical innovations, development of dairies) and trade (development of transport networks and improvement in food diets) conditions led to a process of differentiation, which modified the geography of cheese-making in France. In parallel with regional agricultural specializations and the birth of the cheese industry, successful systems for cheese production were being affirmed. Cross-referencing the 1882 and 1901[1] survey results on the dairy industry, in addition to the small amount of data available on cheeses such as Brie and Gruyère, not forgetting geographical theses and local monographs, has enabled this pattern to become clearly visible. Precursors which initiated the "industrial" production of Brie, Camembert and Roquefort and the first local agricultural cooperatives created truly successful systems, the study of which enables the analysis of the diffusion of spatial innovations.[2]

I. Emergence of cheese-making systems

These cheese-making systems may be likened to productive systems. They combine a type of dairy business, a chosen cheese, a dairy breed and a marketing mode. Four systems can be identified for cheeses: an agricultural cooperative production system, a Camembert system, a Roquefort system, and a system for butter – the cooperatives of the Charentes area (Figs 1 and 2). These spread innovation and contributed to areas of expansion, which

1 The first agricultural survey taking into account Department dairy production and its forms of development was carried out in 1882. In 1901, a Department qualitative survey devoted to "the dairy industry" was carried out by the Ministry of Agriculture (Ministry of Agriculture, Dairy Industry 1901).

2 Saint-Julien T., "Diffusion spatiale", in *Encyclopédie de géographie*, 1995, Paris, Economica, pp. 559–81.

reached their peak on the eve of the Second World War, transforming local economies. Some would contribute to the creation of a genuine regional identity; all of them would be drivers of development in the rural areas concerned, and create outward expansion from their original or "core" base areas.

Figure 1. Cheese-making system in France (1880–1920)

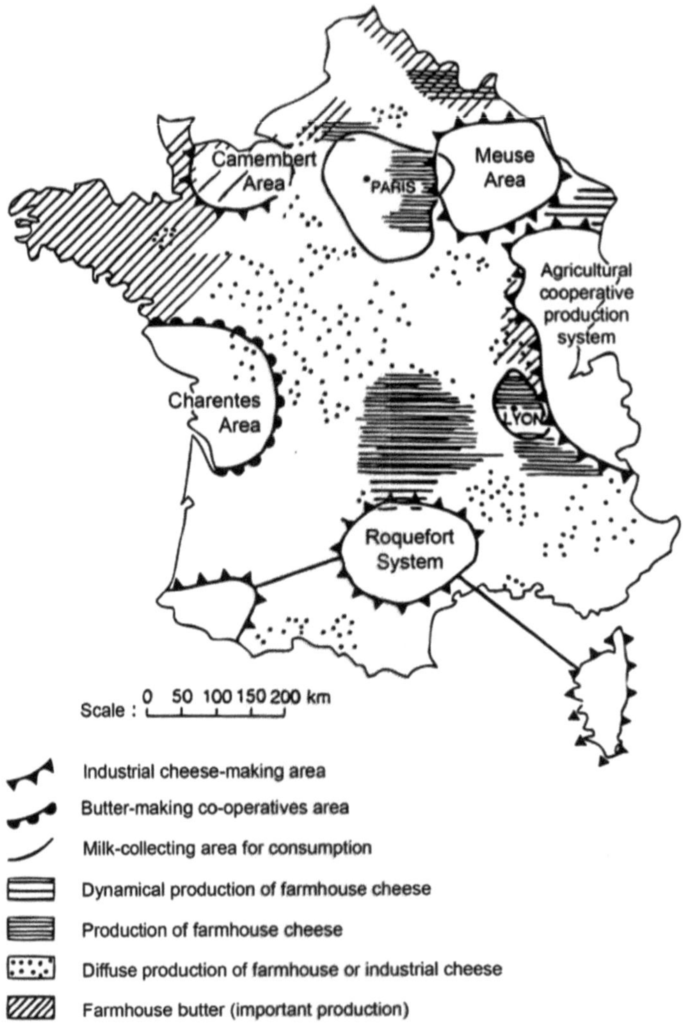

Source: *Enquête sur l'industrie laitière*, 1901; *Notice sur le commerce des produits agricoles*, 1908; *Statistiques agricoles*, 1892–1929; *Monographies agricoles départementals*, 1929 (elaborated by author).

Figure 2. The two cheese systems in the East, before the First World War

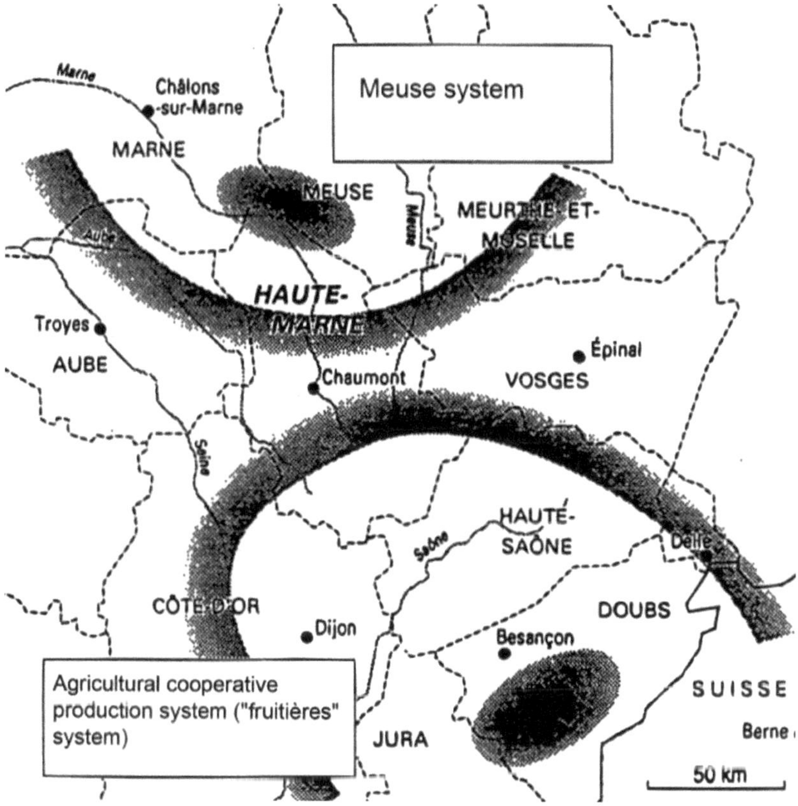

Elaborated by author.

1) Agricultural cooperatives in the East-central region

Confined to the Departments of Doubs and Jura as well as some areas in the Department of Ain, the agricultural cooperative production system, which is based on the collective processing of raw milk into Gruyère[3] within village cooperatives, called *fruitières* in French, spread widely from the last third of the nineteenth century. It spread southwards to all the mountainous regions of the Alps and the Department of Ain, northwards to the Haute-Saône plateaux south of the Haute-Marne, and westwards to the fringes of the Bresse plains.

On the Haute-Saône plateaux and south of the Haute-Marne Department, this expansion took the form of a ripple effect and was warmly

3 Very soon, a number of cheese dairies located primarily in Doubs and certain fringe areas of the agricultural cooperative system were producing Emmental.

welcomed in areas where farmers were experiencing a genuine cereal and phylloxera crisis, as they were receptive to the benefits that establishing a cooperative could provide. Cooperatives enabled dairy production to be fully developed and expanded: old vineyards and lands under wheat were converted to grassland or forage crops. Such was the case, for example, in the south of the Department of Haute-Marne. Prior to 1880, dairies and cheese-making were minor activities in this department. Cheese was made in all the cantons, according to a dairy survey in 1873.[4] These cheeses were essentially consumed by the domestic market and did not generate any cheese-making limits. The first agricultural cheese dairy was created in 1883 in the south-east of the Department, near Haute-Saône, where the Jura and Doubs cooperatives were widespread. The first cooperatives set up there replicated the process used in the neighboring Department. The initiative lay in one farmer suggesting to the other breeders in the Commune the creation of a dairy taking on the name of the cooperative. This then also happened in Haute-Marne. At the turn of the century, there were still twelve agricultural cooperatives in the Department, in addition to fourteen Gruyère cheese dairies belonging to individual farmers, which, in other sources, were likened to cooperatives. The movement for the creation of Gruyère cheese production gradually gained ground in the Vingeanne valley, then, at the beginning of the twentieth century, south of Bassigny as well as the in the Langres plateau and the Châtillonnais; these last two regions were also undergoing a crisis in sheep farming (Fig. 3).

---

4 Ministry of Trade and Agriculture, Investigation into butter and cheese production, Departmental Archives Haute-Marne.

Figure 3. The cheese cooperatives in the inter-war period

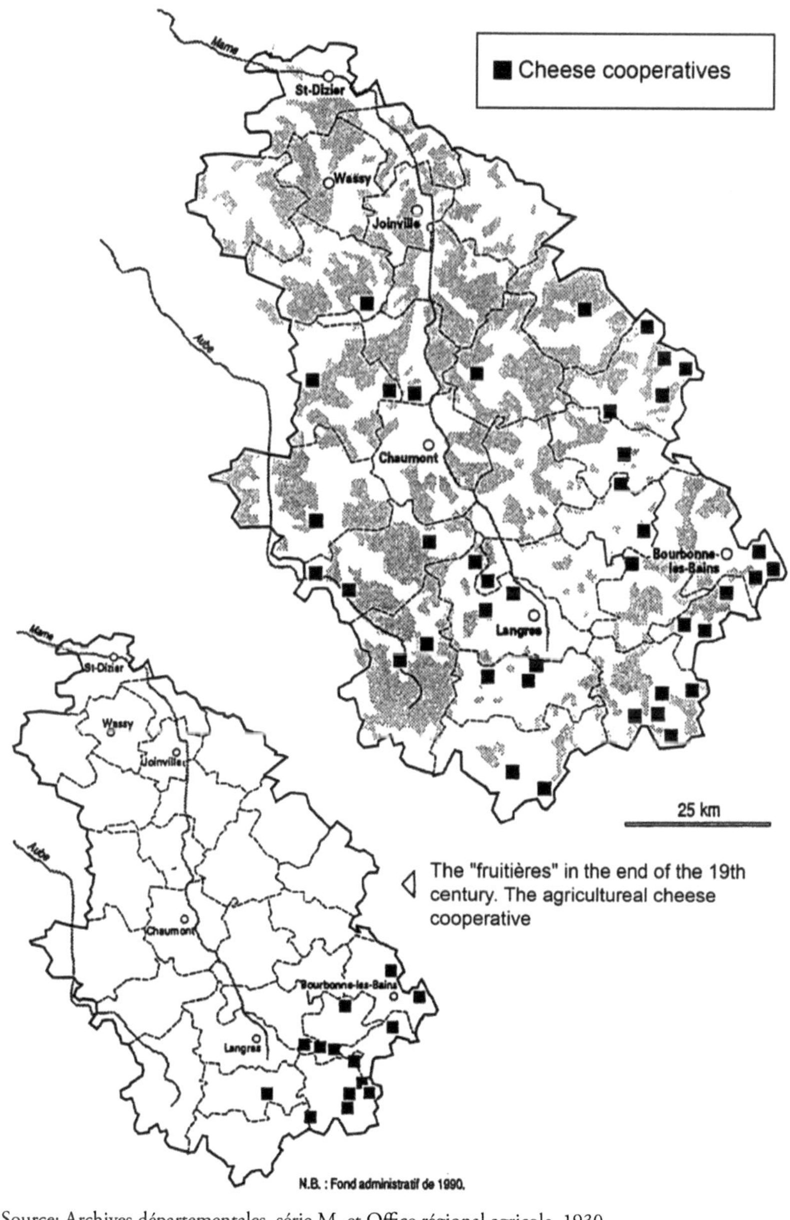

Source: Archives départementales, série M. et Office régional agricole, 1930.

This economic context, favorable to the expansion of cooperatives, generated several periods of over-production. Following the 1871 crisis, cooperatives benefited from a period of prosperity because of cheese prices that were relatively higher than those of other commodities. The number of cooperatives increased and consequently so did the production of Gruyère. This led to a crisis of over-production exacerbated by the application of new customs duties, with Switzerland favoring imports. Thus, in these difficult years, no sooner had a number of cooperatives been created than they disappeared. These crises also meant that the cooperative system expanded in fits and starts. They were stronger in the new expansions, which, however, encountered technical difficulties. Indeed, although many cheese factories were created in Doubs and Jura, and then in the Departments of Haute-Saône, Haute-Marne and Savoie, there was a lack of qualified cheese-makers. In addition, Gruyère is a large cheese requiring several hundreds of liters of milk. Making this cheese requires great attention, as it is difficult to produce and the ripening process takes several months. Therefore, a certain period of time is needed before it is possible to rectify any production errors. This explains why the issue of training cheese-makers dominated debates on cheese production throughout the second half of the nineteenth century. Given the lack of sufficiently trained specialists in France, attention turned to Swiss cheese-makers, numerous in their country. Swiss people were employed in cooperatives in the Departments of Doubs, Jura and Ain from the beginning of the nineteenth century. A century later, following the expansion of the cooperative system, they went to work in the cooperatives of Departments such as Haute-Marne or Savoie. There they maintained the cooperative structure. Their mastery in producing cooked pressed curd cheeses helped solve the problems of over-production linked to poor processing, and also contributed towards improving dairy farming.[5]

Swiss cheese-makers did not come to France purely to work for cooperatives; they also played a major role in creating cheese factories and this followed two different modes. Whereas after 1880 France adopted a resolutely protectionist policy, Swiss entrepreneurs established branches on the other side of the border, and writings of the 1920s in France characterized them as cheese barons. One of them, Eckendorf, having installed French headquarters in Delle, established a large number of cheese factories in the Department of Haute-Saône, then Haute-Marne (Fig. 4).

---

5 Delfosse C., Le savoir-faire des fromagers suisses de la France de l'Est (1850–1950), in *Etre étranger à la campagne*, *Etudes rurales* Nos. 135–6, 1995, pp. 133–44; Delfosse C., *L'impact de l'immigration des fromagers et éleveurs suisses dans l'Est central. XIXe-XXe siècles*; Ricard D. (ed.), *Les reconfigurations des filières laitières en France*, Clermont-Ferrand, Ceramac, 2013, pp. 223–48.

Figure 4. A "Cheese baron" in 1908

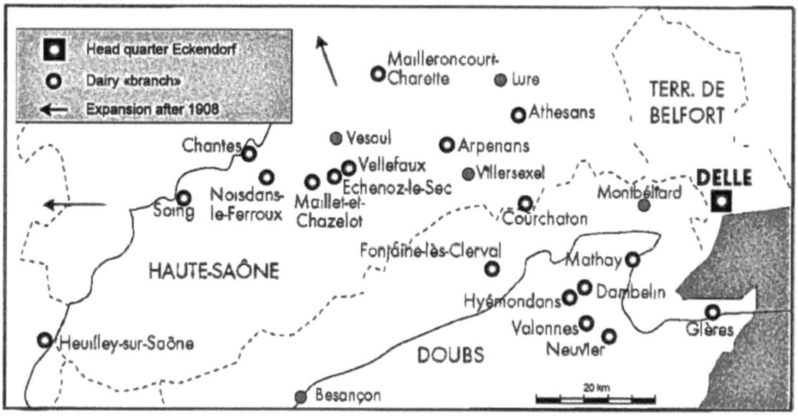

Source: *Annuaire de la crémerie*, 1908.

Many cheese-makers employed in cooperatives and by Swiss companies created their own businesses as a second step, thus contributing to the intensification and expansion of the cooperative system. In effect, from the end of the twentieth century, there were several types of cooperative: those where the milk was pooled, with the members employing a cheese-maker, known as *cooperatives in direct management*; and *cooperatives of sale or tender* whose members remained the owners of the building (cheese factory) but sold their milk all year round to a successful bidder who could be regarded as an artisan cheese-maker. This was a "developed form of the cooperative". Progress there was easier to achieve than in the direct management cooperative: the dairyman was free to act on his own to make changes he knew would be fruitful and had no hesitation in implementing them. The Swiss primarily set up this type of cooperative. Thus, they contributed not only to the spread of cooperatives, but also to the development of the system. They perfected manufacturing conditions and improved the profitability of cheese factories by adding an "industrial" piggery. They introduced, in France, the production of a new cheese which better met the expectations of consumers and traders: Emmental, which France imported in large quantities from Switzerland.[6]

Why was the cooperative model accepted so easily in the local environment? Agricultural changes alone cannot explain the adoption of this economic system, associated with a social system dominating and shaping the whole village community. As the milk was delivered twice a day by the

6 The production processes for the two cheeses are quite similar, with the ripening process differentiating the two: Comté cheese matures in a cold cellar for 4–6 months while Emmental ages in a hot (20–25°) cellar for 2–3 months. Emmental also requires more milk than Gruyère.

cooperative members, the cooperative became the center of local life, the place where people met in the morning and evening. In addition, farmers were not simple milk producers. They were also involved in operating the cooperative: they brought a quantity of wood, on a pro rata basis corresponding to their contributions of milk, to heat the tanks of Gruyère. The requirements in terms of the milk used for production were far more demanding than for other cheese dairies. In fact, making Gruyère requires milk that is highly suitable for cheese-making and which is very clean. These constraints certainly weighed on the cheese-maker, but also extended to members of cooperatives. Dairy breeds were also recommended: breeds which belonged to the "Pie-Rouge" or red-pied breeds (breeds from Franche-Comté and Switzerland, Montbéliarde, mottled, sometimes also called Simmental, more rarely Schwitz or Alpine Brown, Tarine and Abondance). Moreover, the expansion of cooperatives was often associated with the establishment of breeding unions in the village or town.

Swiss cheese-makers also played a major role in the spread of cheese dairies in the Department of Savoie. In 1889, there were 178 Swiss dairies, as against 130 in France, while the proportion was reversed ten years later. This was certainly due to a policy of education, but there were still 154 Swiss dairies, as against 223 French ones.[7] However, in Savoie, conditions governing distribution were different from those of the plateaux. In effect, cooperatives there came up against other cheese producers, in particular production linked to the transhumance herds from the high mountains.

In the Bauges mountains, the growth of cheese dairies started earlier and more intensively than in other Alpine regions. The local public authorities contributed to the process of adopting the cooperative system, the Bauges benefiting from the installation of one of the first of the Savoie cooperative schools at the chief location of the Chatelard canton and several cooperative models, such as that of Ecole. While the Bauges appeared to provide conditions favorable to the adoption of the cooperative system, other parts of the Alps, which offered some resistance, were different. In fact, while the cooperative system unified the whole of the Department of Jura, this was not the case in the Alps, where reflective barriers could be found.[8] These barriers were primarily in the Southern Alps where herds of small ruminants predominated, but also in the inland mountainous areas, such as those of Beaufort which were strongly attached to manufacture in "alpage" or summer alpine grazing in an individual manner, and the Chablais massif, which retained the domestic production of a small cheese called Reblochon and milk from "Abondance" cows on "alpage". The factors which affected the expansion of the cooperative system, which was advocated by state employees, included the physical environment or the structure of the property, but were also cultural, with a commitment to smallstock and

7 Boiret H., *Les laiteries industrielles de la Haute-Savoie*, Annecy, 1901, brochure.

8 Saint-Julien T., "Diffusion spatiale", *op. cit.*

local production. Resistance against expansion may also have been related to competition between the Alpine regions and the Jura, where the system, training framework and the dissemination of technical progress originated, and this could have impacted on the people in Savoie, before the Savoie General Councils had put in place their own policy of dissemination and training: in particular with the creation of the school at La Roche-sur-Foron.

The cooperative system therefore expanded, from the last third of the nineteenth century to the period between the two World Wars. Its development was accompanied by a radical transformation of the rural environment. In addition to the growth of livestock breeding, it also generated a certain degree of prosperity. One can distinguish a kernel, where product and production structure became real elements of identity, and where there were also structuring centers (the national schools of the dairy industry and large ripening houses). During their expansion northwards, the spread of Gruyère cheese dairies contributed to the transformation of rural areas, but the components of the system were a little different, just as in the Savoie region. In these regions, it was the cooperatives, including their "sophisticated" form (cooperatives for the sale of milk), rather than the cheese which became an identifying factor.

2) Two systems for soft cheese: the Camembert system to the west, and Brie to the east

It is possible to distinguish two systems for soft cheeses associated with two kinds of cheese: one in Normandy for Camembert, and the other in Meuse, for Brie cheeses. In both systems the dairy industry took the form of private companies producing the soft cheese family of Brie or Camembert. The cheese factories had a marked family character and often remained attached to a farm operation. However, some of them, in particular in the Department of Meuse, developed in a different way, becoming large factories. Links with agriculture and the management model of agricultural estates were more important in the Camembert system than in that of the Meuse. At first, employees were essentially family members and production was the responsibility of women, as in the case of farmhouse manufacturing. Then production became more professional and male-oriented. In both cases, cheeses were sold aged, or almost aged, to wholesalers or directly by commissioners.

The heart of the Camembert system was located in the Pays d'Auge area, which had long since specialized in breeding of livestock, both as sucklers and as dairy, with milk being processed into cheese and butter. Camembert became the fundamental cheese in this system, a cheese which, it could be said, was born with the industrialization of production. Its sales continued to grow, particularly in Paris where it won favor with gourmets and workers alike. The Camembert system made Calvados one of the leading French cheese-making Departments, with many cheese companies, which are still called "domains", being created. This system spread to the Departments of Orne and Eure, and even to the Lower Seine and Oise, and in the Pays de Bray area. There was competition in these western areas because of the expansion of the collection area of Parisian milk-for-consumption companies.

The Meuse system is located south of the river Meuse in the confines of the Department of Marne. This system spread across the Department of Meuse and reached the neighboring departments of the Marne, the Aube and, to the north, Haute-Marne. Whereas the 1901 survey showed 37 cheese factories in the Meuse Department, Mesnil records 60 there in 1912. Economic conditions were favorable for its expansion: the economic system based on industrial activities (woodworking, stone quarrying, near-surface iron ore, haulage) declined. During the 1850s, haulage was dethroned by the new Paris-Nancy-Strasbourg railway line (1850) and the canal running from the River Marne to the Rhine (1855). In the same period, the steel industry shifted to the Moselle area. Income from industry and handicrafts decreased considerably. This economic and structural change brought many villagers who, until then, had neglected farming, back to the land and livestock breeding.[9]

The history of Alfred Renard (1863–1923), in the south of the Meuse Department, sheds light on the ways in which the Meuse system[10] spread and was adopted. As a result of the crisis experienced in the region, Renard became interested in cheese production, an area previously reserved for women on family farms. At the age of 29 he set up a branch in Bailleux, a precursor of the Marne whose collection had as its starting point a neighboring village of his Commune. This encouraged him to take the first step, and in 1888 he launched himself into the "industrial" production of cheese, not local varieties, but those produced by Bailleux, the Bries and Coulommiers. To obtain a milk collection for himself, he mobilized his relations and maintained friendly relations with the families of the neighboring Commune. Assured of a guaranteed supply of milk, he sought financial support from his family, acquired a new property and turned it into a cheese-making facility: a room for receiving and heating the milk, butter, a factory building, and manufacturing, drying, ripening and packaging rooms.

This spurred imitation. Very quickly, the cheese-makers in the southern Meuse area were blocked in their expansion because of the problems of transporting milk (to a 10–15 km radius from the cheese dairy) as well as by competition from neighboring dairies. The more dynamic among them created branches further away. Thus, from expansion by contagion came a move towards a new manner of expansion, one based on leaps, by pioneering new collection points. The development of Alfred Renard's business illustrates

9 Peltre J., "Un pionnier de l'industrie fromagère: le département de la Meuse", Brunet P. (ed.), *Histoire et géographie des fromages*, Caen, Centre de publication de l'Université de Caen, 1987, pp. 79–108.

10 Henry Renard, *Alfred Renard-Gillard, my father (1863–1923)*, self-published edn.

this. At the start of the century, when his cheese dairy was expanding and there was lively competition regarding milk in the surrounding Communes, he decided to organize a collection zone further away at Liffol-le-Grand (Vosges) and at Liffol-le-Petit (Haute-Marne), whence the milk was transported by rail from Liffol-le-Grand station. In a second phase, he no longer contented himself with looking for milk in the more distant zones, but created new dairies around which he expanded the collection. To do this, he joined forces with other owners or entrepreneurs in his network of relations, whom he met while out hunting, for example. Thus, having learnt from the manager of a small steel and ironworking company in Pompierre that there was no collection in that Commune or in the surrounding area, in 1907 Renard suggested that they work together, and went with him to prospect the region. He then transformed the steel metalwork company into a cheese dairy and established a partnership (Fig. 4).

It was the first mode of expansion by association, so to speak. The "pioneers" of the Meuse system could also establish branches, provided with broad autonomy where management was entrusted to a member of the family. Large families of cheese-makers formed and spread their system by remote transfer, as much in the Meuse system (the Hutin, Bongrain and Renard families) as in the Camembert one. These large families were associated by family strategies, or sometimes by interest, when it was necessary to explore new spaces. In the first phase, these cheese-makers expanding by stages and could inspire other people to establish their own cheese-making facilities locally. However, these families very quickly began to acquire such a large technical advantage and commercial foothold that it was difficult for others to be able to come up with the capital and technical know-how required to be able to start up in business as cheese-makers.

The Meuse and Camembert system cheese-makers also competed in the increase in milk production and benefited from local developments in agriculture. The agreements with local farmers would at first glance seem to be less important than in the case of cooperatives: they were, in fact, less demanding about the quality of milk, but they had to convince neighboring farmers to deliver milk to them rather than to their competitors. Thus, no recommendation was made regarding fodder for the herd, and in the two systems no dairy breed dominated overall. Nevertheless, in the Camembert system the Auge breed did not become prominent, and the Cotentine went onto become the Norman breed *par excellence*. As for the industrialists in the east, inheritors of the farmhouse producers of Seine-et-Marne who had Norman, Flemish and Dutch breeds in their barns, these producers did not encourage any breed in particular. However, the Dutch breeds became more dominant between the two wars.[11]

Competition and imitation brought the manufacturers from each of the two systems together to create a union (*syndicat*). The Syndicat de l'Industrie

11 Delfosse C., *Histoires de bries*, Deauville, Edn La Librairie des Musées, 2008, p. 156.

fromagère de l'Est was established in Bar-le-Duc in 1895. It tried hard to resolve conflicts arising between its members, notably in relation to areas of collection and the setting of milk prices. Being unable to expand their collection area, manufacturers looked for ways of making it not only the most productive possible – through actions in favor of improvements in breeding – but also the most homogeneous possible. They signed agreements with their neighbors, officially certified before a notary and accompanied by substantial compensatory payments in gold francs with guaranties of reciprocity. These agreements permitted regularization of both supplies and prices. The union also intervened against fraud concerning milk. As the law allowed the monitor authorized by the cheese dairy to go onto the farms, to take samples of milk, and to analyze them, but did not authorize the court to prosecute on his request,[12] the union obtained the nomination of a sworn monitor, with the power to draw up official reports for any fraud discovered, which authorized the prosecution of those committing fraud. The union thus allowed manufacturers to organize themselves against the farmers, as well as against the buyers. In fact, it succeeded in obtaining a reduction in their commissions from the agents in the Halles markets (in Paris) by threatening them with a delivery strike. The creation of a group of cheese-box users also permitted more advantageous conditions to be obtained from suppliers, as well as the signing of agreements on prices.

The Syndicat des Fabricants de Véritable Camembert de Normandie (SFVCN) [Union of manufacturers of genuine Normandy Camembert], in which the Camembert system producers had become involved in 1909, played almost the same role in the regulation of internal conflicts. Aside from the question of milk, it also intervened in the question of labor. Even though, in the first phase, women continued to play a major role in cheese-making, with development by the manufacturers and the spatial expansion of the system qualified cheese-makers were required. Manufacturers competed for the best workers: not only was this competition a nuisance, but in time the factory owners feared that the workers would take advantage of this competition to obtain higher pay and weaken the employers' authority. Thus, in 1909, one of the first measures taken by this union was to forbid the poaching of workers. The following year it forbade the cheese-dairy workers, under penalty of dismissal, from seeking employment in another dairy without giving prior notice to their employer. This was the start of the exception clause (exclusivity) which still exists in the agri-food industries today. Thus, these unions sought to limit competition between the manufacturers themselves; they preferred it to be used when it concerned negotiating with a third party: farmers, box makers, workers, etc.

12 Mesnil M., "Conclusions pratiques des recherches récentes sur la fabrication des fromages à pâte molle", in *Actes du Congrès international de Paris*, 1905.

These two systems were a throwback, in terms of their operation, to the local industrial systems.[13] This was also true of the Roquefort system.

3) The Roquefort system: the "Rayon" (radius) from Corsica to the Pyrenees

The last system concerned not only the South of France, but also production of sheep's milk and large ripening houses specializing not only in the sale but more and more in the production of Roquefort. In fact, around the village of Roquefort the cheese-making industry also expanded greatly and tended to form a unique system centered on the role of ripening.[14] In 1885 the Société des Caves (company of caves) set up its first "industrial" cheese dairy. It quickly found imitators in other ripening houses, which themselves also established or sparked the creation of industrial cheese dairies. Ripening houses competed fiercely with each other for milk, and from the first decade of the twentieth century domestic manufacture tended to become negligible in relation to these cheeses from the dairy. The development of dairies allowed ripening houses to extend the supply radius of the Roquefort caves. The transition from farmer stage to industrial level in fact raised blockages linked to passing on domestic know-how: while white cheeses had been produced on farms by women, from then on they were made by men helped by women, and this know-how was then passed on to the heart of the dairies, where it concerned technical expertise. Developments in terms of the transport and storage conditions of ripened cheeses also paved the way for this expansion.

The industrialists were no longer only seeking the milk of existing sheep; they also became agents of the spatial expansion of the production of sheep's milk. Even though Larzac specialized in sheep rearing, its milk production was insufficient to meet the manufacturers' commercial ambitions and the growing success of Roquefort. Thus, the collection of sheep's milk and the manufacture of white cheese spread to Camarès, and then to southern Ségala and the Levézou areas (1880). Continuous expansion, and what is commonly called today the "Rayon" or radius, was established at the beginning of the twentieth century. In 1904, this area accounted for 275 dairies (160 of them belonging to the Société des Caves). The development of sheep breeding was

13 Courlet C., "Les systèmes productifs localisés. Un bilan de la littérature", Torre (ed.), *Le local à l'épreuve de l'économie spatiale. Agriculture, environnement, espaces ruraux*, Inra, 2002, pp. 27–40.

14 Ardouin-Dumazet V.E., in the volume of *Voyages en France* dedicated to the Roquefort region, attached great importance to this production and his testimony along with that of Emile Marre (1906), a professor of agriculture of the Aveyron Department; in the book he devoted to Roquefort cheese he describes in detail the operation of this system, its modes of diffusion and its local effects at the beginning of the 20th century. The history written by Sylvie Vabre (2015) traces this evolution around the history of the Société des Caves.

controlled by Roquefort, and even if the breeding was carried out in different agricultural conditions, the ewes, significantly, were raised in much the same way, to producers' requirements. This system also propagated a breed for which selection was organized: the Lacaune sheep.

The spatial expansion of the collection encountered other profitable agricultural productions: cattle breeding using a transhumance system towards Espalion, and breeding for mutton at Villefranche-de-Rouergue. Thus, at the end of the nineteenth century the production area of Roquefort no longer expanded in an uninterrupted manner. Sales of Roquefort were increasing steadily and manufacturers required a continuous supply of sheep's milk, so that the dairy industry had to seek out milk from other ovine production areas: firstly Corsica and, a few years later, in the Pyrenees Atlantiques.

As early as 1894, Louis Rigal, the Société's rival, conducted tests in Corsica and in 1899 established a cheese factory, in Balagne. Its success led to it being replicated by other manufacturers: in 1901, the Marie Grimal company and the Société des Caves also opened six dairies. Here, as on the Rayon, industrialists waged a fierce battle. The number of dairies rapidly increased: around 24–5 in 1909, and then a genuine explosion occurred after the First World War: 118 in 1918, 158 in 1920 and 180, for a total production of around 1,500 tons, in 1922. These dairies were located primarily in the areas of winter transhumance on the north of the island, and were able to expand inland along the railway line.[15]

Before the First World War, one of the features of this cheese appeared clearly: its ripening was strictly localized, in Roquefort, on the Combalou, while production of the white cheese did not cease expanding from the surrounding Causses plateaux to more distant regions such as Corsica or the Pyrenees, under the impetus of manufacturers which resulted from the ripening process. The elements of what we may call the Roquefort system were in place, bringing great prosperity in the Rayon, an area so renowned for its cheese production that it had a specific name. In 1900, some 70,000 people were employed in this industry: farmers, shepherds, agricultural workers and industrial merchants in the region.[16] On the other hand, expansion in Corsica and the Atlantiques Pyrenees did not experience the same level of prosperity: a lower price was paid for the milk, the Lacaune breed was not developed in Corsica, and in both cases expertise was not shared: the transformation was fully controlled by the Roquefort manufacturers. We can consider these two

15 Delfosse C., Prost J.-A., "Transmission et appropriation des savoirs fromagers: Un siècle de relations entre industriels de Roquefort et transformateurs corses", *Ruralia*, 2, 1998, pp. 23–43.

16 *Report to the Minister of Agriculture on the mission entrusted to G. Roger concerning the study of the manufacture processes of cheeses in the region of the Cevennes* (*Roquefort Region, the Forez Region*, 1900).

expansions as "associative", a term defined by Alain Reynaud[17] as spaces with limits clearly determined, dominated by an external center, but which do not totally lose their personality. The Corsican and Pyrenean expansions in fact retained breeds and the production of local cheeses – certainly transformed and jeopardized. The Corsican Brocciu, whose production decreased, remained the "National Corsican food"; the practices of island consumption were not fully transformed by the implantation of the industrialists.

These systems largely contributed to changing agriculture in the regions to which they spread, including becoming the driver of the development of dairy farming and improvements in farming practices. This spatial process generated regional specialities and, consequently, strong cheese frontiers.

II. Other forms of spatial organizations

However, the cheese-making systems did not cover the whole of France. They encountered other dairy specializations, and did not disappear completely from the areas of farmhouse production of butter and cheese which had been identified from the middle of the nineteenth century, such as in the regions of Brie near Paris, and the country regions of Bray, the Auvergne, or the Thiérache area (in the North of France).

1) Other forms of dairy specialization

Other sectors of the dairy industry also underwent a real revolution and developed in an even more spectacular manner than cheese factories. These also gave birth to regional specialities, which engendered, on the milk market, competition which was detrimental to cheese production.

Firstly, this concerned milk for consumption: the needs of cities were increasing in proportion to their population. Around these cities, genuine collection rings were forming, structured along the railway tracks which were used to route the milk as quickly as possible to the urban market. Thus, the milk from dairy farmers was centralized in some collection points near stations and transported by special trains to large cities. Major corporations with branches were created, such as the Maggi or Les Fermiers Réunis companies. Around Paris and Lyons the phenomenon took on added dimensions. Parisian consumption rose in 1901 to 800,000 liters per day and the supply area for the capital reached a radius of 200 km. Few cheese factories were created within these rings. The drinking-milk industry also competed for farm production whose manufacture moved away from the vicinity of cities, and the farmers did not hesitate to deliver their milk.

[17] Reynaud A., *Société, espace et justice: inégalités spatiales et justice socio-spatiale*, Paris, PUF, 1981, p. 263.

However, the drinking-milk industry also produced cheeses so as to use their surplus milk in certain seasons: cheeses where the technique was always better adapted to an industrial mode of production, like Camembert or the Saint-Paulins.

However, this dairy speculation, whose boundaries can be traced, did not give rise to a real system, contrary to the production of butter in the Charentes where a system was developed that was based on the processing of milk into butter.[18] The conditions were favorable for the emergence of a new agricultural specialization. In effect, phylloxera had ruined this region. Prosperity came as a result of the founding of butter-making cooperatives which knew how to take advantage of technical progress in the butter dairy, and in transport aimed at conquering the Parisian market where they competed with butter from Normandy, still deemed as unmatchable. The cooperative movement had ceased to expand, and cooperatives ended up creating a common structure: the Association centrale des laiteries Coopératives des Charentes [Central association of cooperative dairies of Charentes], in 1893. This association contracted for coal and for all the raw materials required by factories, which allowed it to obtain better rates. It played a very important role in creating and meeting the cost of the innovative collective equipment which allowed cooperatives to be at the leading edge of progress and to remain efficient. Thus, in 1897, dairy inspection, a true service for the repression of fraud, followed by a research station devoted to butter, were established after a series of conferences organized by the Association. Pierre Dornic, a scientific authority in the field of dairies and butter in particular, was appointed Director. This research station had well-equipped laboratories and was quickly accompanied by a specialized school. It was partly financed by the General Council of Charente-Maritime, a sign of the importance of this dairy product in the region. Thanks to its research infrastructure, the Association established a refrigeration cooperative in 1910, and, in 1912, a union of milk-protein cooperatives, after encouraging the enhancement of skimmed milk in the form of casein. It was a system, with its network of actors, and its structuring centers (school, control station, refrigeration, etc.), which moreover played a fundamental role in local life.

As in any milk collection business, the Charentes system inhibited the development of the cheese industry. However, these companies also produced a few cheeses with skimmed milk. They thus contributed to creating a particular spatial distribution for certain cheeses.

In other regions farmhouse production was maintained or even strengthened. This was particularly the case in the Breton Departments, which

---

18 Delfosse C., "Le savoir-faire des fromagers suisses de la France de l'Est (1850–1950)", in *Etre étranger à la campagne*, *Etudes rurales*, Nos. 135–6, 1995, pp. 133–44; Delfosse C., "L'impact de l'immigration des fromagers et éleveurs suisses dans l'Est central. XIXe-XXe siècles", Ricard D. (ed.), *Les reconfigurations des filières laitières en France*, Clermont-Ferrand, Ceramac, 2013, pp. 223–48.

remained committed, through a mixed-farming system, to the production of farmhouse butter. This was due to the importance of local consumption and to the power of the butter merchants, whose role was very often invoked in investigations; the spread of farm churns and centrifuges only increased resistance to industrialization. Thus, Brittany remained a stranger to the cheese industry, apart from a few companies around the major cities.

2) Other forms of spatial organizations of cheese factories

Islands of production of a few cheeses such as Gruyère, Brie, Camembert and Port-Salut appeared. These were cheeses appreciated on the Parisian market and whose manufacturing techniques seemed to be better known than others. They were produced both by dairy establishments in which cheese was not the main concern and in "isolated" cheese dairies not located either in a system or on its fringes. Thus, firms producing soft cheeses were created around the towns in regions which were not producers of farmhouse cheeses, such as Rennes or Bordeaux. They responded to the demands of the urban population and especially to that of workers. The Survey of 1901 also mentioned the manufacture of Brie de Meaux in the heart of the Department of Allier. This involved an "informed" farmer, who tried to encourage neighboring farmers to engage in the production of milk and deliver it to him. These industries, by exporting the production of a few cheeses outside their original area or their expansion center, also contributed to expanding the production area of some cheeses in spots.

Conversely, a few regions resisted or kept the forms of the previous period without experiencing any industry development. Such was the case with the Auvergne Departments (Cantal and Puy-de-Dome), which were among the first cheese-making Departments and which even saw their production increase.[19] The 1901 survey indicates only thirteen industrial cheese dairies in the Department of Cantal; they were still very small, and often only collected milk on a seasonal basis and primarily produced blue cheese. This is also true of the Department of Vosges, which still appeared an important cheese-making Department, but in which cheese factories were almost non-existent. There, farms which had specialized in cheese production during the previous period remained attached to farmhouse production, all the more so as the commercial structure which had been put in place did not favor the creation of industrial cheese dairies or cooperatives as the latter would reduce the role of wholesalers-ripeners.

---

19 For more details see the thesis of Brosse A.L., *Les acteurs des filières fromagères auvergnates (1881–1955)*, University of Lyon 2, 2015.

3) Areas of complex production

The areas of production which could be described as more complex appear in Fig. 1. These are, in particular, those in Lyons and the Thiérache region.

The areas surroundings Lyons, in the Department of Rhône, supplied milk both for consumption and for cheese production. Industrial cheese dairies were created in the Departments of Rhône and Isère and manufactured mainly Saint-Marcellin and Mont-d'Or-type soft cheeses. Farmhouse production did not disappear completely, at least on the outskirts. This area is so much more complex that some cheeses were made from goat's milk. But it was organized around the city of Lyons.

Another type of space allows the complexity of some spatial organizations to be illustrated. This type is represented by the Avesnois-Thiérache, a small region extending over several Departments, in the periphery of the Departments of Aisne and Nord and as far as the Belgian border. After a crisis in the meat market in the last third of the nineteenth century, farmers turned towards the production of milk, an orientation facilitated by the presence of expertise in this area in the production of both butter and cheese as well as by the presence of several major consumer markets: Lille and Roubaix, cities in full expansion, and the mining area around Valenciennes, without forgetting the nearby Belgian market. In a first phase, the grassland farmers produced mainly butter. But the French butter market was in crisis at the end of the nineteenth century and this had many more important repercussions in the Nord area, so that competition from Belgian butters was severely felt there. This crisis fostered a movement to set up private dairies and cooperatives. The cooperative model based on the production of butter was significant, as this area is in fact at the boundary of northern Europe, which at the end of the nineteenth century focused its agricultural development on dairy production, essentially processing milk into butter within cooperatives, in what was called the Danish model. The first cooperative in the Thiérache area was created following this model. But there were also companies that made cheese in the Cambrai, Avesnes and Thiérache districts, where the cheese industry was thriving.[20]

These dairies were far from using the entire milk production of the Thiérache area. In 1900, only 11,000 dairy cows out of 90,000[21] were required to supply the sixteen dairy establishments in the Thiérache area of the Department of Aisne, and barely 12,350 cows out of approximately 185,000 were sufficient to meet the needs of the fourteen Avesnes dairy establishments. Certainly, this method of evaluation is questionable, since

20 Ministry of Agriculture, Dairy Industry 1901, Malpeaux et Malotet, *L'agriculture dans la région du Nord*, Paris, Librairie des sciences agricoles, Encyclopédie de l'agriculture et des sciences agricoles, 1910.

21 Ministry of Agriculture, Dairy Industry 1901.

it does not take into account the farmers who only delivered their surplus to dairies; however, it gives an idea of the quite significant order of magnitude. And indeed, all surveys underline the importance of farm production. The production of cheese was most often localized on large farms. The Thiérache area annually exported 800,000 kg of Maroilles cheese for a value of 700,000 francs. It also sent fresh curds from butter dairies in the Nord and Belgium, where they were sold to traders, who used them for the manufacture of the cheese called "Bruxelles".

Therefore, both butter and cheese were produced, either on the farm or in private or cooperative industrial establishments. The fact remains that this activity drew the boundaries of a grassland country, which the agricultural population acknowledged.

This may serve to provide an overview of cheese-making in France. Indeed, at the end of the nineteenth century and beginning of the twentieth, when four readily identifiable cheese-making systems emerged and developed, they did not eliminate the earlier forms. They also developed in parallel with other dairy systems: that of the Charentes around the production of butter, which is similar to the cheese-making system, and systems linked to the production of milk for consumption. Some specific and complex forms were maintained and developed, such as those found in the Auvergne and the Thiérache area, for example. The organization of cheese-making in France already appears complex and original.

These cheese-making systems reached their maximum expansion in the period between the two World Wars, while diversifying. They would explode in the agro-industrial era.[22] They will, however, reappear in another form with the obtaining of the Appellation of Origin, and this happened more or less early on. The Roquefort system would continue in large part thanks to its obtaining an Appellation of Origin in 1925. The existence of the Meuse system explains the particularity of the concept of the Brie de Meaux Appellation of Origin, which was obtained in 1981.

Conclusion

This study, albeit just a brief outline, is sufficient to assert the pertinence of drawing a large historical perspective in the analysis of the geography of French cheese-making systems. Observing the very first development of dairy industries allows us to understand the diversity of those systems according to the relationships between the actors involved at a regional scale. The chronology is not exactly the same for all systems, Roquefort leading the way with the first Appellation of Origin in 1925. But one cannot tell two separate stories, one of traditional productions, developing both authenticity

22 Malassis L., *L'âge agro-industriel*, Paris, Cujas, 1997, p. 368.

and typicality, and one of industrial productions, bringing simplification and standardization. The legend of rural authenticity and dairy tradition is precisely a production of the conservative elites of the inter-war years, rewritten over and over according to the needs of the successive stages of development of the systems in question.

Actually, the context for the development of hybrid French cheese-making systems is roughly the same, taking shape in the 1920s for most of them, and so is the context for their struggle for survival in the crisis of the 1930s, with the bill on Appellations of Origin in 1935 offering a solution to some – not immediately, but, mainly, after the Second World War. The modernization process of the 1950s and 1960s created a true ordeal for those systems, torn between local and global logics, with harsh competition especially for cooked pressed curd cheeses (Gruyère, etc.). The resistance of mountainous productions has to go the way of Appellations of Origin, with the consequence of splitting the system into rival appellations and quality labels. Some other systems under lesser pressure, like Camembert, waited until the rural revival of the 1980s to take the path of legal certification of origin. But these are not always traditions that are recognized in this way, but are also true technological innovations.

What this history reveals is that there is no such thing as geographical determinism in the territorialization of cheese-making systems in the twentieth century. More convincingly, it is the result of complex technical, social and economical strategies and interactions at all levels, from the scale of the *fruitière* to that of the nation, but all recorded in history.

Switzerland's Dairy Industry in the Twentieth Century

Small-scale Artisan Production Meets the Market Economy

Luigi Lorenzetti

Università della Svizzera Italiana

1. Introduction

Cheese is one of the most iconic of Switzerland's food products. Ever since the beginning of the modern era, the dairy industry has left its mark on a large portion of Switzerland's alpine countryside, where cattle farming and alpiculture developed hand in hand with the manufacture of milk and cheese products for small-scale domestic consumption.[1] The second half of the sixteenth century saw the introduction of new hard cheeses like Emmental, Gruyère and Sbrinz,[2] whose long shelf-life and higher quality provided impetus to a rapidly growing export market.[3] By the eighteenth century, cheese had become one of Switzerland's biggest selling exports, with varieties unique to a number of different regions achieving widespread reputation,[4] and collectively contributing to the image of the '*pays des bergers*' so admired by A. von Haller, J.W. Goethe and J. Gotthelf.

1 Braun R., *Le déclin de l'Ancien Régime en Suisse. Un tableau de l'histoire économique et sociale du 18e siècle*, Lausanne, éditions d'en bas & Editions de la Maison des Sciences de l'Homme, 1988, pp. 47–67.

2 A situation witnessed by J. Stumpf, who wrote towards the middle of the 16th century that in Switzerland "there are so many animals that their care is entrusted not only to women but to men and servants, too. They are referred to as "Sennes" [mountain-dwellers], and they live and work in their chalets. No woman is involved in the making of butter or cheese". Cf. Stumpf J., *Schwytzer Chronica …*, Zürich, Christoph Froschauer, 1554.

3 Cf. Gutzwiler K., *Die Milchverarbeitung in der Schweiz und der Handel mit Milchzeugnissen. Geschichte, Betriebsformen, Markverhältnisse und volkswirtschaftliche Bedeutung*, Schaffhausen, Buchdruckerei Kühn & Co., 1923, pp. 176–206; Ruffieux R., Bodmer W., *Histoire du gruyère en Gruyère du XVIe au XXe siècle*, Fribourg, Fromage de Gruyère SA, 1972.

4 Roth A. G., *Aus der Geschichte der Schweizer Käses. Neues Quellen und Forschungen zu seiner Geschichte bis 1914*, Burgdorf, 1970; AA.VV., *Die Alpkäserei und die Geschichte des Schweizer Käses*, Bern, Schweizerische Käseunion, 1995.

Switzerland's dairy economy underwent a second major upheaval during the nineteenth century, when a substantial portion of cheese manufacturing was moved from alpine and subalpine areas into the lowlands. There followed a sharp rise in the number of rural dairies in response to an increase in domestic and foreign demand for dairy products, the need to be closer to consumer markets, improved transport networks, and the desire to continue production throughout the year. These dairies had been set up specifically to be close to the farms which in the early years of the century had begun to abandon cereal crop cultivation in favor of pasture land. At the dawn of the twentieth century, then, the Swiss dairy industry had been able to gain a strong position in international markets despite intense pressure on prices caused by the arrival of foreign competitors.

The outbreak of the First World War set the scene for yet more changes. The need to guarantee the country's food supply in a time of emergency led the federal government to promote the application of a controlled economy based on cooperation between all stakeholders in the dairy sector, from manufacturers to distributors. This policy was supported by a system of government subsidies that were shared out across the entire sector, providing crucial assistance to Switzerland's cheese export industry while at the same time leading to a sharp rise in manufacturing costs and market prices.

In the following sections we will attempt to reconstruct the evolution of the Swiss dairy industry throughout the twentieth century, focusing particularly on the official structures within which the sector operated. This will be followed by a description of the characteristics and evolution of Swiss dairy production. We aim to demonstrate how the increase in cheese production was, from the 1950s, the result of a general drop in milk consumption, and how it led to an export trade supported by a system of government subsidies which ended up causing a progressive increase in production costs and consumer prices.

2. The Swiss dairy industry: from free to "controlled" market

The structure of the Swiss dairy industry in the twentieth century is the result of a process which brought together various actors, including milk producers and milk processing factories, dairies, wholesalers, retailers and exporters, under the umbrella of a "controlled" market designed primarily to ensure the supply of milk and dairy products to the Swiss population. This system meant that prices were not dictated by supply and demand but rather, as we will demonstrate, were the result of economic policies and regulations under which the Federal Government defined the roles of those involved in the sector. The policies introduced in 1916, which remained substantially in place until the end of the 1990s, meant that the entire dairy sector was subject to regulations controlling the production, processing and consumption of dairy goods.

A period of formation and organization in the Swiss milk and dairy sector (1870–1916)

The range of policies mentioned above was the result of a long period of development that began in the 1870s when, after decades of constant growth, a worldwide economic crisis caused a sudden slowdown in the Swiss dairy industry. For the first time, the dairy sector found itself dealing with a shrinking foreign market. The crisis was made worse by a general restructuring process in the livestock sector which involved an increase in the number of cattle and the abandonment of cereal crops in favor of pasture land, and by the steadily rising price of milk which, in little more than a quarter of a century (1880 to 1908), went up by 25%.

In the 1880s and 1990s, surplus production[5] led to a crisis in the Swiss cheese export sector, which was already suffering from a certain volatility in demand. This state of affairs created tension between the different interests operating in the sector, with the value of milk and cheese becoming the main bone of contention. Up until the First World War, milk prices had been traditionally linked to the sale price of cheese destined for export. This meant, in practice, that exporters set aside a percentage of their profits in order to cover the costs of raw materials, thus guaranteeing their own profit margins. They then passed on the remaining income to cheese-makers who, after taking their own percentage, paid what was left to the milk producers.[6] At the time, however, only 15–20% of the milk that passed through dairies was destined for export, meaning that the price of milk produced in Switzerland was dictated by the demand for Swiss cheese abroad, a situation which was the source of much dissatisfaction on the part of milk producers, who felt that they suffered the most from such an arrangement.

Those involved reacted to this situation by developing strategies aimed at protecting their interests on the basis of improved business structures and the formation of guilds which were initially canton-based, but soon became national. Up until the 1970–1980s, dairy production was essentially carried out by cooperatives organized around dairies, which redistributed profits to milk producers according to how much they had contributed, or by private

---

5 Annual cheese production is estimated to have increased by 77% between 1866 and 1886, growing from 200,000 to almost 360,000 tons.

6 According to this procedure, milk producers ended up receiving around one twelfth of the final sale price of cheese. Cf. Moser P., Brodbeck B., *Portrait en images, documents et analyses de l'économie et de la politique laitière en Suisse au 20e siècle*, Baden, Hier+Jetzt, 2007, p. 94.

dairies which independently managed the production and sale of their own milk.

Thanks in part to a growing market and to the entry into force in 1881 of the Swiss Code of Obligations – which made it easier for many companies to alter their own statutes[7] – in a matter of a few years the number of cooperatives registered with the federal authorities increased substantially, from 164 in 1883 to 845 in 1890, 1,459 in 1900, 2,086 in 1910, and 2,399 in 1915.[8] At the same time these new cooperatives were joining forces under regional federations whose aim was to represent more effectively the interests of their members during negotiations with private dairies to set milk prices.

In 1887, an attempt was made to coordinate the strategies and interests of all those involved in the dairy industry (including milk producers, wholesalers and exporters of dairy products), with the creation of the Swiss Milk Industry Association [Société suisse d'industrie laitière, SSIL], although the limitations of this organization were soon revealed as a wide range of divergent aims and interests emerged among its members. As a consequence, the following years saw the formation of independent interest groups, leading to even more disaggregation within the industry. Cheese exporters and resellers were the first to organize themselves in groups, with the foundation, in 1896, of the Swiss Cheese Exporters' Association [Syndicat des Exportateurs Suisses de Fromage, SEF] which aimed at the promotion of Swiss dairy products abroad. Just over ten years later, in 1907, milk producers came together to form the Central Union for Swiss Milk Producers [Union Centrale des Producteurs Suisses de Lait, UCPL], made up of regional and canton-based milk producers' federations which in turn were composed of local milk and cheese producers' associations. In 1911, in order to avoid contractual obligations imposed by cheese retailers, the UCPL founded Emmental S.A., a company which quickly went on to become the largest cheese manufacturer in Switzerland.

The last category to unite in an association was that of the cheese producers. Since the beginning of the nineteenth century they had been affiliated with various different milk and dairy associations, but the 1880s saw the majority of these producers join new cooperatives, which began to buy milk directly from small producers and transform it into its derivative products like cheese and butter, which would then be sold locally or to wholesalers and exporters. Milk and dairy cooperatives could be divided into two main categories: those who owned their own dairy and employed a workforce (mostly in the cantons of Berne and Lucerne), and those who sold their milk to independent dairies (mainly in western and eastern parts of Switzerland). These collectives struggled to put in place an effective

7 *Ibid.*, p. 13.

8 *Annuaire Statistique de la Suisse 1933*, Berne, 1934, p. 303. The majority of these collectives sold only milk, but around one thousand of them also produced their own cheese and butter.

system to defend their interests, leaving them at the mercy of the stronger negotiating positions of cheese exporters and milk producers. The cheese producers found themselves between a rock and a hard place,[9] and in a weak position when it came to negotiating the purchase price of milk and the sale price of cheese. This weakness manifested itself in the first "milk war",[10] which broke out in the east of the country in 1908 after milk producers and dairies confronted milk federations over the constant increases in milk prices. The dairies demanded that the prices be kept in check, and threatened a boycott, but the financial support provided to milk producers by the UCPL forced the dairies to back down on the majority of their demands. The "milk war" set the stage for the struggle between milk producers and cheese retailers and exporters for control over the dairy sector through contractual supremacy. The struggle eventually led to a settlement a few years later when the Swiss Milk Buyers Union [Union Suisse des Acheteurs de Lait, USAL] was founded.

Industrial relations and the introduction of a "controlled" market (1916–1999)

During the First World War the need for a regular food supply forced the agricultural sector to increase production of cereal crops to the detriment of pasture land. This resulted in a momentary reduction in livestock numbers, consequently reducing the volume of milk produced. Another casualty was the profits from the export of dairy products, which had been suspended to avoid wartime food shortages.[11]

The Federal government took steps to avoid shortages of milk and dairy products by entrusting the UCPL with the task of guaranteeing supply to the population. This decision marked the end of the free market which had been in place until then, and opened the door to a new kind of market, in which participants were subject to regulations dictated by the political authorities. Milk producers were obliged to turn over any surpluses to milk associations, which, in turn, were subject to directives formulated by regional trade associations regarding the processing and production of dairy products (including butter, cheeses, powdered milk, etc.).[12] Hard

9 Moser P., Brodbeck B., *Du lait pour tous, op. cit.*, p. 35.

10 Between 1908 and 1913, tensions in the milk and dairy market and friction over milk prices and whose right it was to set them led to a bitter conflict between the various interest groups in the dairy sector which became known as the Milk War. Cf. *ibid.*, pp. 54–72.

11 Between 1913 and 1919 the cow population fell by 17% (from 875,000 to 729,000), and the total value of milk and dairy exports suffered a 13% reduction (from 20.2 to 17.6 million Swiss francs). Cf. Ritzmann-Blickenstorfer H. (Hg.), *Historische Statistik der Schweiz*, Zürich, Chronos, 1996; *Annuaire agricole de la Suisse 1936*, Berne, 1937, pp. 18–19.

12 Moser P., Brodbeck B., *Du lait pour tous, op. cit.*, p. 21.

cheeses (like Emmental, Gruyère and Sbrinz) were sold to retailers, who were members of the Swiss Union of Cheese Sellers [Union Suisse du Commerce de Fromage, USF],[13] on the basis of prearranged prices and quotas. The milk producers benefited from a prearranged price, which was adjusted every six months by the Federal government in collaboration with the UCPL. This meant, in practice, that the entire Swiss dairy industry was subject to a "controlled market",[14] which continued to be supported through a number of adjustments and modifications until the 1990s, when free market policies returned the milk and dairy industry to its pre-First World War structure.

This profound change in the underlying conditions that regulated dairy production was accompanied by the founding in 1917 of the USAL, whose primary aims were to promote the independence of dairies and cheese producers and ensure that they received a fair share of the profits. The new organization quickly achieved significant results. In 1920 the USF accepted the participation of the USAL in negotiations aimed at establishing the price of cheese,[15] thus strengthening the new climate of cooperation on which the Swiss milk and dairy market system was now based.

In this period the milk federations went through a continual process of transformation. After three main periods of change which led to the proliferation of a number of regional and cantonal trade associations,[16] the 1940s witnessed the merger of many of these regional federations. Of the twenty or so federations active at the beginning of the 1930s, only thirteen were left by the 1980s, nearly all of which were active at the supra-regional level.[17]

Subsidies and Export

Just like the rest of the Swiss agricultural sector, the dairy and cheese industries had, for most of the twenty-first century, benefited from a system of government subsidies which had been implemented in order to ensure a plentiful supply of food and to provide economic support to farming

13 The USF is an umbrella association that took over from the SEF as the organization which represented cheese sellers and exporters Cf. Lustenberger M., *Die Organisation des schweizerischen Käsehandels seit 1914. Die Genossenschaft schweizerischer Käseexportfirmen und die Schweizerische Käseunion*, Bern, Weinfelden, Neuenschwander, 1926.

14 Moser P., Brodbeck B., *Du lait pour tous, op. cit.*, p. 103.

15 *Ibid.*, p. 37.

16 The first two waves took place in the 1880s and 1990s, and the third during the period of the First World War in order to guarantee stability in the supply of milk. Cf. Schweizerischen Milchwirtschaftlichen Verein (Hg.), *50 Jahre Schweizerische Milchwirtschaft. Festschrift unter gefälliger Mitwirkung einer Anzahl von Fachleuten 1887–1937*, Schaffhausen, Buchdruckerei Kühn & Co., 1937.

17 Moser P., Brodbeck B., *Du lait pour tous, op. cit.*, pp. 17, 25.

communities.[18] In time it also became essential for maintaining stability in the prices of agricultural products.[19]

Subsidies granted to the milk industry went up and down between the wars, but from the 1950s they saw a gradual increase, until they reached 238 million Swiss francs in 1970, 620 million in 1980, and eventually topped 1.1 billion francs in 1990[20] (Fig. 1). The overall value of the cheese export market grew over the same period, mirroring the increase in subsidies. This market was worth between 30 and 40 million francs a year during the 1930s, and grew steadily after the Second World War, exceeding 100 million francs in 1951 and 200 million in 1965, going on to reach 400 million francs in 1974 and 550 million by 1984.[21]

Figure 1. Subsidies in thousands of Swiss francs granted to the milk sector, and total value of cheese exports 1928–1984

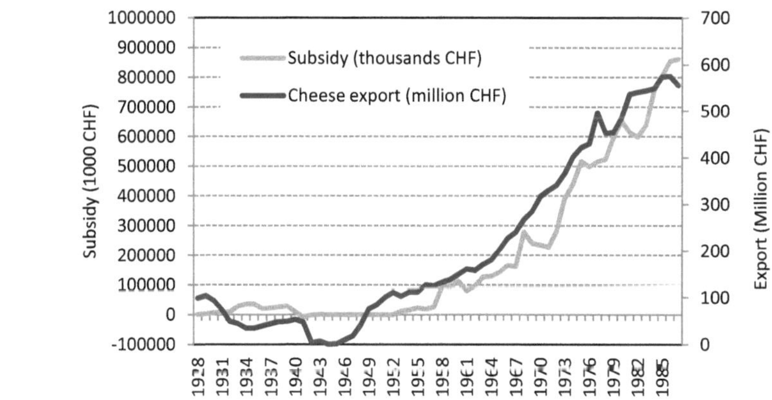

Source: Brugger H., *Agrarpolitik des Bundes seit 1914*, Frauenfeld, Huber, 1992, p. 28, 40, 49, 59, 70; <http://www.fsw.uzh.ch/hstat> (Tab. L.11b1; L13).

---

[18] Cf. Moser P., "Am Konsum orientiert, über die Produktion thematisiert. Schweizer Agrarpolitik als Ernährungspolitik 1914/18–1960", in Langthaler E., Redl Josef (Hg.), *Reguliertes Land. Agrarpolitik in Deutschland, Österreich und der Schweiz 1930–1960*, Innsbruck, Studienverlag, 2005, pp. 192–203. It has been estimated that, by the end of the 1980s and the beginning of the 1990s, around three quarters of the income of Swiss farmers came from government subsidies.

[19] In 1951, an article of the Constitution voted by referendum confirmed the special position of agriculture in the Swiss economy. This article was at the basis of federal policies providing support to the sector through stabilization of the prices of raw materials and production costs and the application of protection measures in the import market. Cf. Rieder P., Egger U., Fluckiger S., *Schweizerische Agrarmarkte*, Zurich, Vdf, 1992.

[20] At the same time, federal subsidies in favor of the cheese-producing sector also increased, from just under 100 million francs in 1970 to more than 460 million in the second half of the 1980s. Cf. Brugger H., *Agrarpolitik des Bundes seit 1914*, Frauenfeld, Huber, 1992, pp. 28, 40, 49, 59, 70.

[21] The correlation coefficient between the two figures (yearly total of federal subsidies and yearly value of cheese exports) is $r = 0.97$.

The value of exports of Swiss cheese went up steadily during the twentieth century. The two World Wars were the only exception, as foreign trade contracted because of the need to meet domestic demand.[22]

In the 1900–14 period, Germany, France and the USA together accounted for little less than two-thirds of the value of cheese exports from Switzerland.[23] This distribution gradually changed over the following decades. While the French and the US markets kept up their relative share, the German market went into decline. Thus, while in the latter half of the 1930s Germany still imported 15% of Switzerland's cheese exports, in the latter half of the 1950s this share had dwindled to a mere 6%. The Italian market, on the other hand, developed steadily to the point that, over the same period, it grew from 18% to 28% of the total export value (Tab. 1).

Tab. 1. Value of cheese exports in some countries by five-year periods (thousands CHF)

	Germany	Austria	France	Italy	Belgium	Great Brit.	USA	other
1900–04	9,895	2,988	12,079	4,519	1,134	731	5,749	5,599
1905–09	10,451	4,156	12,747	6,186	1,445	823	9,666	6,970
1910–14	11,521	6,667	13,908	6,917	1,857	993	15,496	8,006
1915–19	16,246	7,536	4,788	347	717	302	3,787	1,159
1920–24	10,855	1,544	6,086	3,872	1,856	1,495	14,774	6,832
1925–29	26,718	1,797	3,856	12,748	2,094	5,658	30,512	15,727
1930–34	6,420	1,100	12,260	7,900	2,660	3,380	16,300	9,840
1935–39	6,420	–	11,260	7,880	2,300	2,020	9,020	4,220
1940–44	11,460	–	2,240	4,620	280	0	1,680	1,900
1945–49	760	–	3,740	8,240	4,260	1,420	7,200	3,367
1950–54	1,920	–	20,640	26,560	6,780	7,980	32,020	6,080
1955–57	7,400	–	24,800	34,100	8,600	4,567	33,933	8,021

Source: *Annuaire statistique de la Suisse*, several years.

In the second post-war period, Italy became the most important export market for Swiss cheese. Towards the end of the 1980s, as much as 44% of the export value was made up by the Italian peninsula, with Germany and France each making up 15%. The USA, traditionally one of the major markets, registered a considerable drop, so much so that in 1988 its share of Swiss exports was a mere 5.8%.[24] Subsequently, however, the North American market bounced back. Indeed, in 2001, it made up 20% of the

22 Between 1910–14 and 1915–19 the volume of cheese exports shrank by 61.4% and the value dropped by 46.7%. Between 1936–40 and 1941–5, on the other hand, these quantities dropped by 84.8% while the value of exports decreased by 74.9%. During the two World War periods, the main outlets for Swiss cheese were Germany and Austria, taking in 60–70% of exports.

23 In that period (1900–14) Italy accounted for approximately 10% of cheese exports.

24 Schweizerischen Bauernsekretariat, *Statistische Erhebungen und Schätzungen über Landwirtschaft und Ernährung*, 66 Jahresheft, 1989, p. 99.

amount exported. Meanwhile, sales to Italy went from 26,565 tons in 1988 to 16,450 tons in 2001 (-38.1%); the decrease was reflected in a diminished relative value of the Italian market, which had hit 31.0%[25] in that year.

By the 1960s, government subsidies had allowed the cheese industry to reach a volume of production equivalent to 150% of domestic requirements, exceeding that of other foods (milk, butter, meat, potatoes, cereals, etc.), but they had also caused a sharp increase in manufacturing costs and sale prices. From the 1990s, market liberalization had highlighted the lack of competitiveness of the Swiss dairy export sector, and the government responded by enforcing a change in public policy in support of the agricultural sector, moving it to a system based on direct payments.

3. Swiss dairy production: industry structure and regional variations

Structural aspects of the dairy manufacturing industry

When a large part of the milk-producing sector moved from the alpine regions down to the valleys in the early decades of the nineteenth century, it led to the formation of a dense network of small-scale dairy producers. Unlike the alpine dairies, they were able to continue production throughout the year. And so, together with a more efficient infrastructure, they were able to achieve a far higher volume of production. This new infrastructure was founded on decentralization and on traditional artisan methods, and remained more or less unaltered throughout the nineteenth century and into the twentieth, not least because, until the 1960s and 1970s, milk was delivered twice a day straight after the cows had been milked in the morning and the evening in order to ensure a constant supply of milk to households and to makers of cheese and butter, etc.[26]

In 1905 there were just over 3,000 dairies factories, employing almost 8,000 workers. Over the next few years, these numbers plummeted to the point that just under 2,600 were left at the end of the 1920s, around a fifth of which (543) were located in the canton of Berne, and another fifth (565) shared between the cantons of Lucerne and St Gallen. Overall, the dairies of these three cantons employed 41% of all Swiss dairy workers, processing the milk of 38% of the country's cows.[27] This was a period in which dairies were generally small-scale artisan activities which rarely had a workforce of more

25 See Union Suisse des Paysans, *Statistique laitière de la Suisse 2003*, Travaux statistiques No. 177, Brougg, p. 45.

26 Moser P., Brodbeck B., *Du lait pour tous*, *op. cit.*, p. 140.

27 Cf. data provided by the *Annuaire agricole de la Suisse 1936*, Berne, 1937, p. 461.

than six people.[28] Things remained much the same over the following decades despite an increase in the volume of production having stretched the dairies to their limit. Proposals were put forward in the 1960s and 1970s to transfer production of hard cheeses to large-scale industrial complexes, but the dairy associations opted instead to undertake a campaign of modernization by upgrading dairy machinery (ovens, presses, etc.) and production facilities (including cellars used for ageing).[29]

The increase in the volume of milk produced and in the demand for cheese in the years following the Second World War led to a significant rise in the number of dairies and milk producers. In 1953 there were 4,856, spread right across the Jura mountain valleys and the canton of Valais (Fig. 2).[30]

In 1999, when the USF was dissolved, the dairy sector underwent a significant reorganization. Many small dairies became uncompetitive and were forced to cease production and close down; others joined forces in order to survive. The Swiss hard cheese sector has generally been able to preserve its artisan quality, with small producers still spread across much of the country.

Figure 2. Distribution of Swiss milk and dairy producers in 1953

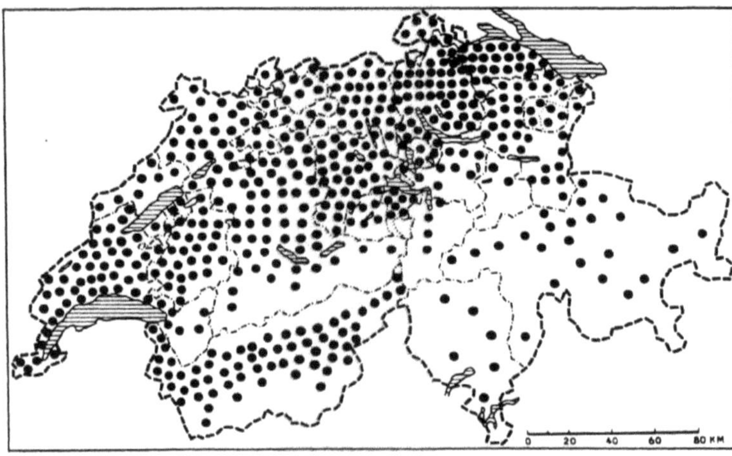

Source: Froidevaux G., "L'organisation laitière de la Suisse", in *Annales de Géographie*, t. 66, 1957, No. 354, p. 163.

Note: each dot represents 10 companies.

28 In 1929, 17% of Swiss dairies counted only one employee, while 58% employed two or three workers and 18% had four or five employees. Only 4% of all dairies employed more than six workers. Cf. *ibid.*

29 Only recently has a process of structural concentration taken place. In the case of Emmental production, for example, there were 993 small businesses in 1930 which produced 25,800 tons of cheese. In 2007 there were no more than 193 dairies, but their volume of production was 31,000 tons.

30 Froidevaux G., "L'organisation laitière de la Suisse", in *Annales de Géographie*, t. 66, 1957, No. 354, pp. 159–65.

Milk supply and dairy production

Until the early 1960s Swiss dairy production was directly linked to that of milk, which in turn was dictated by the number of dairy cattle present in the country. With the exception of the war years, when numbers fell sharply, the cow population rose constantly from 1900 to 1961, which is when it reached its peak at around 943,000 individuals. Numbers then began to fall, to under 800,000 at the beginning of the 1990s. At the same time the number of milk producers, after an initial period of expansion (again with the exception of the war years), had begun to fall in the 1960s in line with the merger of small dairies into larger companies. From 137,000 producers in 1950, the industry counted fewer than 98,500 by 1970, and just over 52,300 in 1990.[31]

Advances in animal husbandry have, nonetheless, meant an overall increase in the production of milk (on the basis of sales figures). From 2.04 million tons in the first decade of the twentieth century, average annual production reached 2.56 million tons by the second part of the 1920s. The domestic market, however, was unable to absorb such volumes, and the foreign market for cheese proved to be unpredictable and inconstant. As a result, milk prices collapsed in 1922, falling from 38 cents/kg to 22 cents/kg. Prices remained at this level until the Second World War, when the volume of milk production once more fell dramatically (-15.4% between 1936/40 and 1941/45).

The new economic outlook that followed the Second World War set off another phase of growth in the milk and dairy sector. Milk production began to increase once more, encouraged by a steady rise in the sale price of milk and cheese (Fig. 3) which grew, respectively, by 63% and 64% over the two decades between 1950 and 1970.

31 This trend continued throughout the following years. In the year 2000 there were almost 40,000, but ten years later (2010) there were only 26,100 (data from Swissmilk: <http://www.swissmilk.ch/>).

Figure 3. Average sale prices of Emmental (cents/kg) and skimmed milk (cents/l) in the cities of Berne and Zurich, 1887–1995

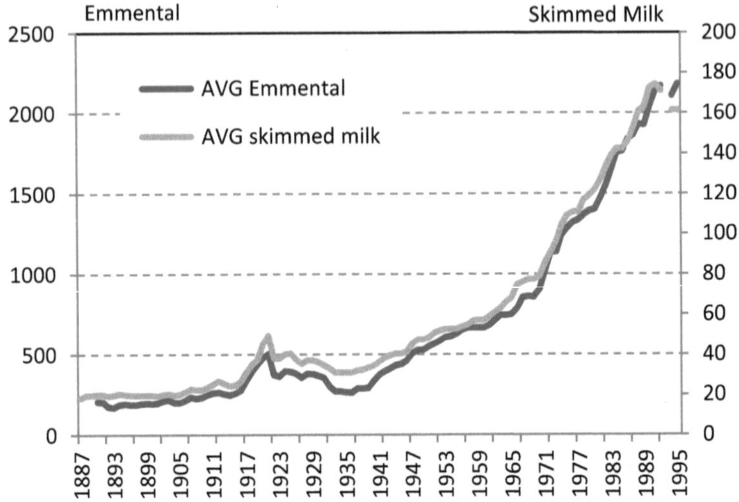

Source: <http://www.fsw.uzh.ch/histstat> (Tab. H.26a, H.27a).

Annual milk production exceeded 275,000 tons in the five-year period between 1951 and 1955, 310,000 tons in 1961/65, and 331,000 tons in 1971/75. Despite the application of milk industry quotas under a federal decree of 1977,[32] production continued to grow, reaching 378,000 tons in 1981/85 and almost 390,000 tons in 1991/95;[33] a growth that had a dramatic effect on the price of milk, which was still supported by federal subsidies.

A great deal of the increased quantities of milk produced after the 1950s and 1960s was absorbed by the growth in per capita consumption of cheese, making up for a fall in the consumption of milk.[34] Over the three decades following the Second World War there was a gradual increase in the amount of milk intended for the production of dairy products including cheese, butter, cream, etc., from 38 to 42% of all available milk[35] between the end of the 1940s and the beginning of the 1950s to 50% in 1956/60, leveling out at just over 65% from the late 1970s (Fig. 4).

32 *Feuille Fédérale*, Vol. 1, 1977, No. 5.

33 Cf. <http://www.fsw.uzh.ch/histstat> (Tab. I.23).

34 It is estimated that annual milk consumption in Switzerland fell from 233 kg per person in 1950 to 106 kg per person in 1990. Data provided by Swissmilk (<http://www.swissmilk.ch>).

35 With the exception of the First World War, when the percentage fell below 30% because of federal policies aimed at guaranteeing the supply of milk to the population.

Figure 4. Milk used to process derivatives as a percentage of overall milk production and volume of cheese produced in Switzerland (x 100 tons)

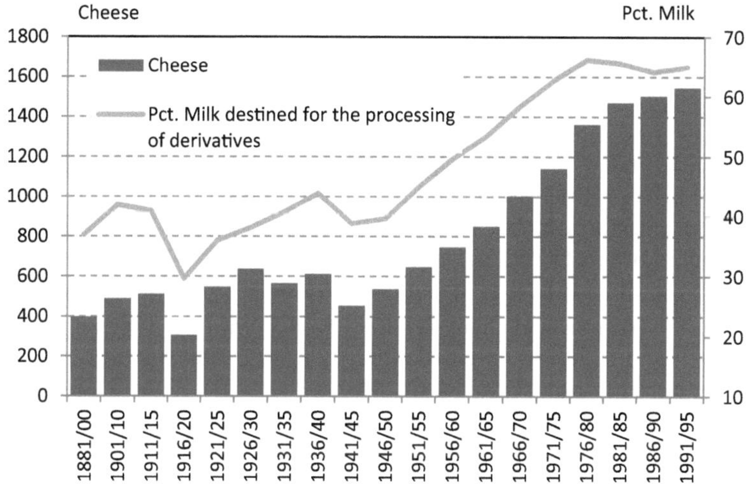

Source: <http://www.fsw.uzh.ch/histstat> (Tab. I.23).

This trend had repercussions for the volume of dairy production, which had fluctuated during the first half of the century, but began to soar rapidly from the beginning of the 1950s. Growth did not level out until the 1980s, in response to the application of production quotas as part of the federal government's new agricultural policy.

Figure 5. Evolution of the use of milk in Switzerland, 1881/1900–1991/95 (x 100 tons)

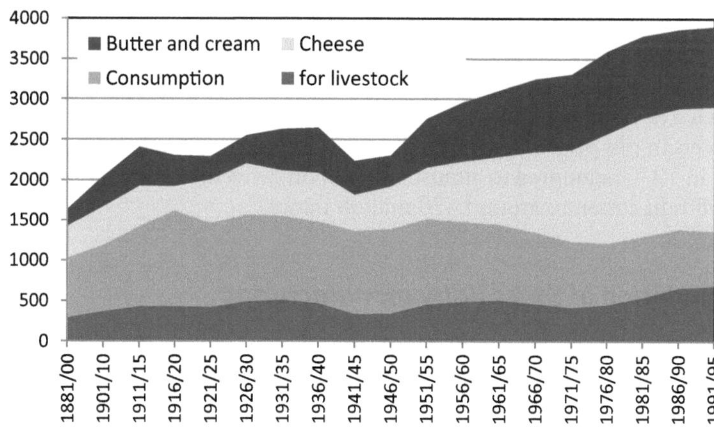

Source: Tab. I.23, <http://www.fsw.uzh.ch/histstat>

* including milk that producers set aside for animal rearing.

During the twentieth century there were significant changes in the uses to which milk was put (Fig. 5), which reflected changes in consumer behavior as well as the introduction of new regulations governing milk distribution.

With the exception of the period between 1966/70 and 1981/85,[36] the milk quota destined solely for animal consumption remained relatively unchanged, rising from 15% to 20%. The overall percentage of milk destined for direct consumption (including the milk that producers fed to animals), on the other hand, gradually fell over the course of the century. In the first decade of the twentieth century direct consumption accounted for 40% of total milk production, but in 1956/60 this portion had fallen to 33.7%, reaching 24.6% in 1971/75 and 17.4% in 1991/95.

One of the causes of this decline in milk consumption was that distribution patterns had changed over time. Up until the 1960s most milk in Switzerland had been purchased directly from milkmen who carried out daily deliveries to private homes. In 1963 a large retail group (Migros) was granted a license to sell pasteurized milk in disposable packaging, and this led to a fall in consumption and an increase in the quantity of milk set aside for the dairy processing industry.

The percentage of milk destined to become cheese had remained stable at around 20–23% until the mid-1950s, but this, too, began to rise, reaching 30% in 1966/70 and just under 40% in 1991/95.

The percentage of processed milk used in the production of fat and semi-fat cheeses changed little over the years. With a few notable exceptions (when the figure rose above 60% in 1931 and 1948), the amount of processed milk used in the production of fat and semi-fat cheeses varied between 45% and 50% over the three decades between 1930 and 1960. These included some of the principal varieties of Swiss cheese like Emmental, Gruyère and Sbrinz. Overall, cheese production grew rather moderately until the 1950s, but registered a sharp rise over the next few decades, and the annual output of 560,000 tons in the period 1931–5 (becoming 650,000 tons by the first half of the 1950s) grew by the second half of the 1960s to more than 1 million tons, going on to reach 1.47 million tons in 1981–5 and 1.54 million by 1991–5. Cheese producers in this period had enjoyed the support of generous federal subsidies, which in 1970 amounted to almost 100 million Swiss francs and by the end of the 1980s had risen to around 470 million francs.[37]

The evolution of Swiss dairy manufacturing

High fat hard cheeses are Switzerland's main dairy products, of which the most widely produced varieties are Emmental, Gruyère and Sbrinz. These

36 During this period the percentage fell to 12–14% due to a rise in the percentage destined for the production of cheese.

37 <http://www.fsw.uzh.ch/histstat> (Tab. U22b).

cheeses are made with fresh milk; therefore, for a long time the dairies that made them had to be close to where a constant supply of milk could be guaranteed, and this generally meant alpine areas. It was only in the nineteenth century that improved transport networks and better infrastructure allowed manufacturers to spread out into pre-alpine areas and the valleys. Another factor in the making of these types of cheeses is that only the milk of grass-eating cows can be used in their manufacture. Cows that are fed with products made through a process of anaerobic fermentation produce a milk which cannot be used to make hard cheeses because it contains butyric bacteria that cause swelling during the cheese-making process.[38] Hard cheeses, then, were for a long time a seasonal product, dependent on the availability of fresh grass.

The manufacture of Emmental was for a long time restricted to a specific geographic location (the Emmental valley in the canton of Berne), although it cannot be said that production throughout this area was in any way homogeneous. With the standardization of the production process it became possible to achieve such homogenization and to open up production to other parts of Central Switzerland, including the districts of Mittelland and Seeland in Berne and the cantons of Fribourg, Aargau, Lucerne, Schwyz, Solothurn and Zug to the west, and to the east, Oberland in Zürich and the cantons of Saint Gallen, Thurgau and Glarus,[39] as well as a number of other areas beyond the national borders.[40] By the beginning of the twentieth century only a small proportion of Emmental produced in Switzerland was made in the region of the same name. Local retailers also encouraged the standardization of Gruyère, which was originally produced in the districts of Gruyère and Bulle, and the manufacture of this cheese also spread beyond its original area of production to the canton of Vaud and some parts of the Canton of Jura. The three main Swiss cheeses, Emmental, Gruyère and Sbrinz, have for many years constituted a large portion of domestic cheese production (Fig. 6).

---

38 Moser P., Brodbeck B., *Du lait pour tous, op. cit.*, pp. 132–3.

39 Cf. for example, the migration of cheese-makers from Berne to the canton of Thurgau in Head-König A.-L., "Emigration bernoise, accès à la terre et insertion dans la société d'accueil en Thurgovie à la fin du XIXe siècle", in Béaur G., Dessureault Ch., Goy J. (eds.), *Familles, Terres, Marchés. Logiques économiques et stratégies dans les milieux ruraux (XVIIe-XXe siècles)*, Rennes, PUR, 2004, pp. 105–17. More generally, on the mobility of cheese-makers in Switzerland, cf. Roth Alfred G., *Talkäsereien. Zur Aufnahme des Betriebes in der Schweiz*, Burgdorf, G. Roth & Co., 1977.

40 On the migration of Swiss cheese-makers to France, cf. Delfosse C., "Le savoir-faire des fromagers suisses de la France de l'Est (1850–1950)", in *Études rurales*, Nos. 135–6, 1994, pp. 133–44 and Kronenberger S., "Des fromagers suisses en Franche-Comté: Compétences, pluriactivité et réseaux (1850–1914)", in *Histoire et sociétés rurales*, Vol. 41, 2014, No. 1, pp. 55–87. On Swiss cheese-makers in Russia, cf. Tschudin G., *Schweizer Käser im Zarenreich: zur Mentalität und Wirtschaft ausgewanderter Bauernsöhne und Bauerntöchter*, Zürich, Hans Rohr, 1990.

Figure 6. Swiss dairy production by variety of cheese, 1921–1987 (x 100 tons)

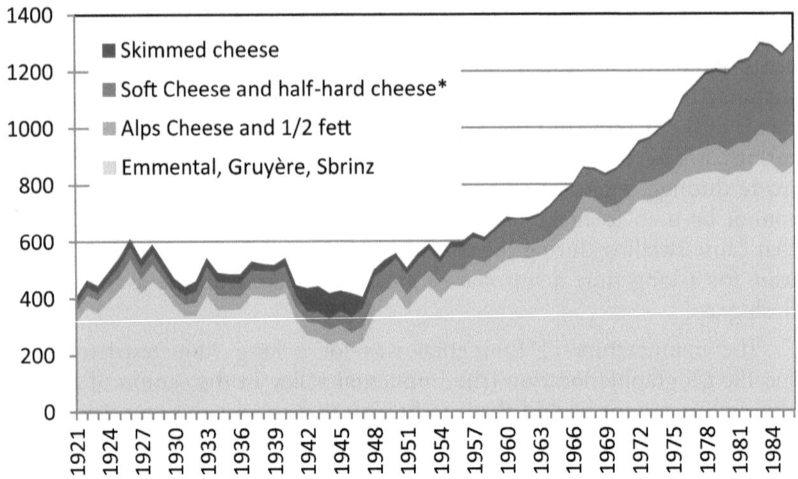

Source: *Annuaire Statistique de la Suisse*, various years.

Note: the production of Schabzieger (canton of Glarus) is not included here.

* Appenzeller, Tilsiter, Vacherin, Mutschli, Raclette Cheese.

It was a share which was well over two-thirds of the total at the beginning of the century and more than 75% by the second half of the 1920s, even reaching 80% in 1930. Wartime economics during the Second World War meant that these cheeses' share of the market fell in favor of low-fat cheeses, of which larger quantities could be produced for each liter of milk.

After the end of the war and the return of economic prosperity, production of Emmental, Gruyère and Sbrinz returned to its previous strong position, even reaching 88% of total cheese production in 1967/68. Although the volume of production continued to grow, this high point was followed by a phase in which their proportion of the market as a whole began to contract. They accounted for merely 66% of all Swiss cheese produced by the end of the 1980s, and this proportion continued to fall over the following years until, at the end of the century, the three main Swiss cheeses accounted for less than half of the total market.[41]

At the same time, cheeses that were produced in mountain areas achieved a growing slice of the market from the 1970s, thanks to an increase in demand and improved management of alpine pasture land. Their proportion of total production had until then hovered around the 6–9% mark, still below the peaks of above 10% during the Second World War when alpiculture was prioritized. In recent years, changes in consumer taste and the intervention of federal policies in the dairy sector have led to

[41] Cf. *Annexe à la statistique annuelle du marché du lait – année 2002.*

an increase in the production of soft cheeses and of fresh milk. Between 2002 and 2011 their volume of production went from 36.5 to 47.0 tons, an increase of 28.7%, while share of the overall cheese market grew over the same period from 23% to 26%.[42]

4. Conclusion

The deregulation of international markets in the 1990s brought an end to the economic model that characterized the Swiss dairy industry throughout most of the twentieth century. The end of the "controlled" market system and federal subsidies to the export sector in favor of direct payments gave way to a process of reorganization in the dairy sector which was most heavily felt in the export industry. Although the volume of cheese exports grew by 22.1% between 2002 and 2012 (from 54.8 to 66.9 tons), the share occupied by Emmental, traditionally the most important Swiss cheese export, fell by 34% (from 27.2 to 17.9 tons).[43]

At the same time, deregulation meant that the manufacturing process of the principal varieties (Emmental and Gruyère) underwent a streamlining process, while the production of butter, soft cheese and other dairy products became more concentrated. Nonetheless, it has not yet undermined the production structure itself, which continues to be based on traditional "artisan" techniques, which in some cases have been revitalized by the rediscovery and renewed promotion of local varieties and alpine cheeses produced with traditional methods.

42 Cf. TSM Fiduciaire, *Statistiques laitières* (<http://www.tsm-gmbh.ch>).

43 The lost share of the market was taken over by new semi-hard cheese varieties (growing from 345 to 9,200 tons) and soft cheeses (from 80 to 993 tons). Cf. *ibid.*

The Italian Dairy Industry between 1930 and 1970

Production and Organizational Structure

Claudio Besana

Università Cattolica del Sacro Cuore, Milan

1. The production of cheese in Italy in the 1937 Production Census

The *1937 Production Census*, which was supervised by the Central Institute of Statistics of the Italian Kingdom, concerned only some sectors of the production. Among the census data is *The processing industry of milk and its by-products.*[1] We have, therefore, a long series of data that allow us to outline the situation of the sector seventy years after the industrializing processes of the sector began. These processes led to the first separation between production of butter and cheese and the other activities carried out in farms.[2]

A butter production of 447,000 quintals emerges from the census. This is a very different amount from that recorded in the countries most involved in the production of this milk product.[3] The manufacture of butter was effectively concentrated in Northern Italy (437,000 quintals), in particular in Lombardy (181,000 quintals) and Emilia (104,000 quintals). With regard to cheese production, the 1937 census surveys led to an estimated amount of 2,229,000 quintals. Such a share set Italy at third place among

1 For the criteria used for the data collecting and their elaboration, see *Censimento industriale 1937-XV. Industria della lavorazione del latte e dei prodotti derivati. Censimento del 25 maggio 1937-XV*, by the Central Institute of Statistics of the Italian Kingdom, Roma, Tip. Failli, 1939, pp. 1–7. As regards the general features of the 1937–40 surveys, there are many instructions in the monograph concerning *L'industria dello zucchero*.

2 For these processes of first modernization of the sector concerning Lombardy see Besana C., *Tra agricoltura e industria. Il settore caseario nella Lombardia dell'Ottocento*, Milano, Vita e Pensiero, 2012, pp. 157–241.

3 In 1939, nearly 8 million quintals of butter were produced in the USA; in 1938 a production of 5 million quintals in Germany and one of 2 million quintals in France were surveyed (Savini E., *Caseificio. Il latte e la sua produzione*, Milano, Hoepli, 1946, p. 656, table CXLVII).

the world's cheese producers, behind the USA and Germany, but ahead of France.[4]

Just like that of butter, the production of cheese was concentrated in Northern Italy, where around 75% of the cheese manufactured in Italy was produced. In those years the production of cow's milk cheese was particularly significant in Lombardy (823,000 quintals), in Emilia (382,000 quintals) and in Veneto (239,000 quintals);[5] the production of sheep's milk cheeses was, instead, concentrated in Latium and in the two biggest islands; for these cheeses the contribution of Sardinia amounted to 176,000 quintals, that of Sicily to 77,000 quintals and that of Latium to 69,000 quintals.

Figure 1. Main types of cheese produced in Italy in 1937

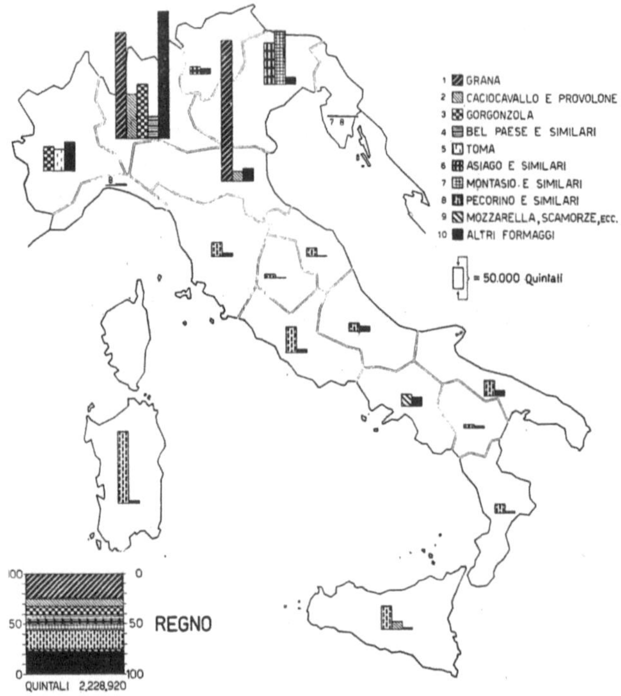

Source: *Censimento industriale 1937-XV. Industria della lavorazione del latte, op. cit.*, p. 103.

4 At the end of the 1930s, Italy was third on the list of the most important world producers of cheeses; in first place were the USA with around 3 million quintals of cheese produced in 1939. These countries were followed by Germany accounting for 2.8 million quintals (1937 borderlines), France accounting for 2.17 million quintals (1937), and the small Netherlands accounting for 1.2 million quintals (1939); see Savini E., *Caseificio. Il latte e la sua produzione, op. cit.*, table CXLVIII.

5 Note that, according to the census of March 1930, there were 2,388,000 dairy cows in Italy, 2,104,000 of which were registered in the farms of the north of Italy; see Savini E., *Produzione lattiera e casearia italiana*, Rome, Ramo editorial degli agricoltori, 1934, p. 4.

The global production of cheese was estimated to amount to 2,230,000 quintals (Tab. 1);[6] that obtained from cow's milk was assessed to amount to 1,750,000 quintals and was the outcome of transforming around 23,354,000 hectolitres of milk.[7] By using sheep's milk, around 465,000 quintals of cheese were obtained,[8] whereas by using goat's milk and buffalo milk around 13,000 quintals of cheese were obtained. The supply was characterized by a high degree of variety, because there were, among Italian cheeses, "one hundred of very different kinds as regards technique, organoleptic characteristics, duration of ripening and size".[9] It must be remembered, however, that many "dairy kinds" were produced in limited amounts and only in specific areas of Italy. The varieties that allowed production of more than 1,500 tons a year were fewer than twenty; even fewer were those bolstering significant sales flows in international markets. Among these cheeses, the most important were Grana (585,000 quintals),[10] followed by kneaded-curd products such as Caciocavallo and Provolone (161,000 quintals), Gorgonzola (184,000 quintals),[11] Asiago (131,000 quintals), Montasio (133,000 quintals), and "Swiss-kind" cheeses such as Emmental and Sbrinz (around 129,400 quintals). In fact, the production of hard and medium-hard cheeses was dominant, to which the ripe soft ones were added. The production of fresh and short-ripening soft cheeses was affected by the shortcomings of refrigerators in the chain of distribution and by the lack of fridges in Italians' houses.

6 It must be remembered that the data concerning the different productions have to be accepted with caution; they represent approximate figures that changed considerably year by year, depending on different factors such as the quantity of milk to be processed in dairies or the price trend of the single products. In another publication of 1937 the production of Grana, for example, was estimated to amount to 850,000 quintals and that of Gorgonzola to be next to 350,000 quintals; see Fotticchia N., *La disciplina della produzione dei formaggi di pecora*, Confederazione fascista degli agricoltori. Comitato nazionale lana (eds.), *Congresso nazionale armamentario*, Roma 13–14 febbraio 1937, with no typographical notes, p. 4.

7 *Censimento industriale 1937-XV. Industria della lavorazione del latte e dei prodotti derivati*, pp. 160–1.

8 Processed sheep's milk amounted to 2,580,000 hectolitres.

9 Savini E., *Produzione lattiera e casearia italiana*, *op. cit.*, p. 10.

10 Among the kinds of Grana, Parmigiano-Reggiano had a great majority amounting to around 396,000 quintals (cf. Table 1). In Fascetti's opinion, the production of Grana tripled between the end of the 19th century and the 1930s, see Fascetti G., *Enciclopedia del caseificio*, II, *Burro e formaggio*, Milano, Hoepli, 1935, p. 634.

11 The production of Gorgonzola, in another expert's opinion, was even larger than Fotticchia pointed out; in Savini's opinion, it reached 400,000 quintals. We must add that the producers of this cheese were the most damaged by the fall in exports after 1931. In that year 403,000 quintals of cheese were exported, with foreign sales of Gorgonzola amounting to 107,000 quintals; in 1937 243,000 quintals of cheeses were exported in all and 51,000 quintals of Gorgonzola were sold abroad, see Savini E., "Formaggi "molli" e autarchia", in *Latte, latticini e conserve animali*, 1938, 1, p. 15.

Table 1. Italy. 1937 census. Cheeses produced

Kinds of cheese	Quantities produced in quintals
Grana	584,788
of which Parmigiano-Reggiano	396,132
Reggiano kind	165,622
Pecorino and similar cheeses	464,684
of which Pecorino named proper	133,575
Pecorino romano	121,438
Caciocavallo, provolone and similar cheeses	160,515
Gorgonzola	183,523
Montasio and similar cheeses	132,548
Asiago and similar cheeses	131,162
Emmental and similar cheeses	68,035
Sbrinz and similar cheeses	61,363
Bel Paese and similar cheeses	53,110
Toma and similar cheeses	51,105
Toma and Toma of Lanzo	50,664
Crescenza and similar cheeses	49,843
Quartirolo	45,556
Mozzarella and similar cheeses	40,880
Vacchino	21,322
Scamorza and Provolone	19,242
Fontina	18,307
Bitto	15,698
Other cheeses	97,877
Total	2,228,920

Source: *Censimento industriale 1937-XV. Industria della lavorazione del latte, op. cit.*, p. 102, table 61.

The comprehensively undertaken census[12] offers a considerable series of data representing a stage of recovery of a sector that was badly damaged by the 1929 crisis[13] and revived by the increase in domestic consumption.[14]

12 "The survey has been widened to the overall very broad range of milk processing: from the pasteurization to the production of powdered milk; from the production of butter and cheese to that of butter and casein; from the production of cream and skimmed milk to that of butter, cheese, sugars and salts; from milk itself, in short, whose bacterial status has been nevertheless modified, to milk divided into its components. The survey was complete also in the sense that it meant to consider any amount of processed milk, regardless of the equipment of the activity" (*Censimento industriale 1937-XV. Industria della lavorazione del latte e dei prodotti derivati, op. cit.*, p. 2). The statistical surveys were carried out in different places of Italy on 25 May 1937.

13 If we consider the price of cow's milk on the market of Lodi, one of the biggest dairy townships in Italy, we will find out that in 1926 it amounted to 94.6 lire/hectolitre; in 1934, in the same place, it decreased to 33 lire; in 1937 milk would be paid 67.6 lire/hectolitre again on the market of Lodi. If we look, instead, at the price of Pecorino cheese on the market of Rome, we discover this price trend: 1,828 lire in 1926 for a quintal of ripe and choice cheese, 656 lire in 1934 and 954 in 1937 (*Censimento industriale 1937-XV. Industria della lavorazione del latte e dei prodotti derivati, op. cit.*, pp. 8–9).

14 The average per-head annual consumption of cheese in the 1930s appears to exceed by 25% that estimated for the 1920s, 5 kg versus 4 kg; see Zamagni V., "L'evoluzione dei

It gives an account of the processing carried out in the farmers' kitchens, the productions of cheese performed in makeshift buildings by transhumant farms, and the production of butter and cheese in small businesses and in "impressive" factories, which were provided with suitable machinery to enable every kind of transformation of this unique commodity. As the aim of the survey was to set "a corporative policy of the direction of production, rules and regulations governing trade, consumption control", those in charge of the statistical survey strove not to ignore any activity. With regard to the division of the production according to different "manufacturing places", we may note that, among cheeses produced using cow's milk (1,750,000 quintals), the proportion of so-called industrial plants[15] amounted to 1,555,000 quintals, with 175,000 quintals derived from home productions, whereas 19,000 quintals of cheese were obtained in transhumant farms.[16]

As regards the number of industrial plants processing milk, it is obvious that their number was high and that they were concentrated in Northern Italy. The figure was taken from a census of around 16,600 production units in total. These were concentrated in Lombardy, where more than 4,800 "dairies" were registered, in Piedmont, in Veneto and in Emilia (Tab. 2). In Lombardy, in particular in the provinces of Milan, Cremona and Pavia, there were a little less than half of the plants operating all over the year. The importance, from the statistical point of view, of Northern Italy may be explained by the fact that the fixed structures processing only milk transformed cow's milk, at almost all times. This was produced in a very much small quantities in the central and southern regions of Italy. As regards to sheep dairy products, processed in "industrial" structures, they were quite significant in Tuscany and, most of all, in Sardinia (Tab. 2).

consumi tra tradizione e innovazione", in Capatti A., De Bernardi A., Varni A. (eds.), *Storia d'Italia. Annali*, 13, *L'alimentazione*, Torino, Einaudi, 1998, p. 183, table 3.

15 In the census, according to the term "industrial business", processes performed in buildings or in rooms only used for processing milk and equipped on purpose are identified, without distinguishing whether they were rooms belonging to real firms or to big, or small, farms (*Censimento industriale 1937-XV. Industria della lavorazione del latte e dei prodotti derivati, op. cit.*, p. 6); the data we are discussing do not include businesses salting down cheeses, ripening them, or exclusively or mostly producing food milk.

16 *Ibid.*, pp. 160–1.

Table 2. Italy. 1937 census. Industrial businesses producing by-products of milk in the year of the survey

Divisions of the national territory		Number of businesses	Businesses operating all over the year
Piedmont		**3,472**	**1,194**
of which	Aosta	1,314	100
	Cuneo	343	213
	Novara	481	274
	Turin	229	563
	Vercelli	272	658
Liguria		**48**	**27**
Lombardy		**4,829**	**2,858**
of which	Bergamo	772	235
	Brescia	653	331
	Cremona	278	237
	Mantua	703	183
	Milan	943	865
	Pavia	684	673
Tridentine Venice		**1,023**	**212**
of which	Trento	974	169
Veneto		**3,170**	**1,432**
of which	Belluno	541	137
	Friuli	879	510
	Verona	548	157
	Vicenza	770	340
Venice Julia and Zara		**250**	**165**
Emilia		**2,469**	**458**
of which	Modena	752	12
	Parma	550	234
	Reggio Emilia	820	10
Northern Italy		**15,261**	**6,346**
Central Italy		**219**	143
of which	Tuscany	115	87
Southern Italy		**544**	**323**
of which	Campania	250	209
The Italian islands		**609**	**615**
of which	Sardinia	564	51
Kingdom		**16,633**	**6,857**

Remarks: businesses salting down cheeses, ripening them, exclusively or mostly producing food milk are not included. Source: *Censimento industriale 1937-XV. Industria della lavorazione del latte e dei prodotti derivati, op. cit.*, p. 6, table I.

In toto, 18,678 factories were registered in the whole milk processing and milk products sector. They were dependent on 17,704 firms. There was much sole proprietorship (almost 12,000 units), whereas cooperatives numbered around 4,400 and only 183 joint-stock non-cooperatives (Tab. 3). The complexity of the sector and the different kinds of firm may also be seen in the choices made by those in charge of the firms when they joined the corporate structures organizing production activities. Of the registered firms 30.4% joined the Confederation of Farmers and 31.8% that of Manufacturers; a quarter of the firms in the sector were represented by the Board of Cooperatives, whereas "the remaining 12.6% depended, from the trade-union point of view, on the Confederation of Traders".[17]

Table 3. Italy. 1937 census. Features of the registered "industrial" firms

	Kingdom	Northern Italy	Lombardy	Piedmont	Veneto	Emilia
Juridical kinds of firm						
registered firms	17,704	15,693	5,058	3,499	3,269	2,567
of which						
sole proprietorship	11,824	10,119	3,676	2,948	1,396	1,680
cooperatives	4,407	4,316	790	367	1,719	627
joint-stock non-cooperatives	183	154	72	19	28	28
Trade-union organizations representing firms						
Confederation of Farmers	5,388	5,024	1,606	1,849	1,037	171
Confederation of Manufacturers	5,627	4,838	1,963	764	391	1,570
Confederation of Traders	2,224	1,477	589	505	150	202
National Board of Cooperatives	4,452	4,344	897	381	1,691	618

Source: *Censimento industriale 1937-XV. Industria della lavorazione del latte e dei prodotti derivati, op. cit.*, pp. 20–1, tables 4 and 5.

Table 4. Italy. 1937 census. Production features of the registered firms

firms ordered by number of plants	firms		number of plants		staff		installed power	
	n.	%	production	ripening	n.	%	hp	%
with 1 plant	17,165	97	15,831	1,334	39,353	82.1	22,864	74.3
with 2 plants	368	2	631	105	2,955	6.2	2,616	8.6
with 3 plants	102	0.6	252	54	1,154	2.4	768	2.3
with 4 and more plants	69	0.4	400	71	4,478	9.3	4,268	14.8
total	**17,704**		**17,114**	**1,564**	**47,940**	**100**	**30,516**	**100**

Source: *Censimento industriale 1937-XV. Industria della lavorazione del latte e dei prodotti derivati, op. cit.*, p. 222, table 6.

It must be remembered that, at the date of the census, the main firm ran twenty production plants and four ripening plants, and gave work to 886 people; moreover, it had installed engines whose power was estimated around 2,000 horsepower (6.6% of the registered total). A census was taken

[17] *Ibid.*, p. 21.

of a second firm too having 36 production plants and 9 storehouses where cheese ripened, was taken too; in this case, there were only 259 workers and the installed engines had an overall power of 13 horsepower.

Table 5. Italy. 1937 census. Production features of the registered firms

firms ordered by number of plants	production plants		processed milk		produced butter		produced cheese	
	n.	%	hectolitres	%	quintals	%	quintals	%
with 1 plant	15,831	92.5	19,189,940	79.3	468,811	81.5	1,275,928	75.4
with 2 plants	631	3.7	1,883,895	7.8	40,286	7.0	121,788	7.2
with 3 plants	252	1.5	630,356	2.6	14,292	2.5	51,943	3.1
with 4 and more plants	400	2.3	2,491,139	10.3	51,472	9	243,576	14.3
total	**17,114**	**100**	**24,195,330**	**100**	**574,861**	**100**	**1,693,235**	**100**

Source: *Censimento industriale 1937-XV. Industria della lavorazione del latte e dei prodotti derivati, op. cit.*, p. 222, table 6.

The three firms with the highest number of plants (23, 24, 45) produced 22,624 quintals of butter and 59,624 quintals of cheese.

Table 6. Italy. 1937 census. Global data from the census

Kinds of business	businesses		processed milk		produced butter[1]		produced cheese	
	registered	operating	hectolitres	%	quintals	%	quintals	%
industrial businesses for producing by-products[2]	16,983	16,633	21,545,444	86.4	386,126	86.4	1,690,495	75.9
of which mostly producing food milk	60	60	166,657	0.6	3,423	0.8	8,867	0.4
Transhumant zootechnical firms	18,303	18,303	822,788	2.9	1,847	0.4	134,093	6.0
home processes	633,758	633,758	3,843,777	13.3	59,098	13.2	401,592	18.1
businesses for ripening[3]	1,564	1,537	–	–	–	–	–	–
businesses for producing food milk[4]	191	191	2,816,543	9.7	3,423	0.8	8,867	0.4

Kinds of business	businesses		processed milk		produced butter		produced cheese	
	registered	operating	hectolitres	%	quintals	%	quintals	%
of which with the production of by-products too	60	60	802,119	2.8	3,423	0.8	8,867	0.4
Total	**670,739**	**670,362**	**28,861,895**	**100**	**447,071**	**100**	**2,226,180**	**100**

Remarks: 1. After reuses; 2. These businesses dealt with ripening 80,648 quintals of cheese; 3. These businesses dealt with ripening 1,110,626 quintals of cheese; 4. These businesses produced 2,637,186 quintals of food milk.

Source: *Censimento industriale 1937-XV. Industria della lavorazione del latte e dei prodotti derivati, op. cit.*, pp. 30–1, table 18.

As emerges from the data, the production, especially of some cheeses, continued being carried out in production units corresponding with the farms, although some very interesting differences appeared. As regards Parmigiano-Reggiano, 48.1% of its production was based in structures that were part of farms, and 51.9% in dairies labeled as industrial in the census. Nearly all the cheese registered in generic terms as Pecorino (134,000 quintals) was produced directly by shepherds transforming the milk of their flocks,[18] but 31% of Pecorino romano (total production, 121,000 quintals) was produced in dairies that were not directly run by shepherds. This percentage increased to 84.5% for Sardinian Pecorino romano (total production, 56,000 quintals). The 87.3% of Green Gorgonzola (total production, 179,000 quintals) was the outcome of industrial processes, whereas 86% of the Asiago put on the market (total production, 119,000 quintals) was the product of farms that were gathered together in cooperatives.

As producers were numerous, producers putting onto the market small quantities of cheese were definitely predominant.[19] With regard to Grana and Parmigiano-Reggiano, for example, a census of 2,818 production structures belonging to 2,677 firms was carried out. In toto, the output of these plants was estimated in 396,000 quintals; half of the Parmigiano-Reggiano produced was, therefore, obtained in production units manufacturing every year a quantity of cheese between 100 and 200 quintals. Only three firms produced more than 1,000 quintals of Grana a year using different production units. The production of Gorgonzola was a little more concentrated, the volume of production was 183,000 quintals and the business units were 917, belonging to 862 firms. Among producers of Gorgonzola the biggest category consisted

18 See also the chapter by Rita d'Errico in this volume.

19 Fragmentation was typical of the whole agricultural and food sector, see Chiapparino F., *Tra polverizzazione e concentrazione. L'industria alimentare dall'Unità al periodo tra le due guerre*, in Capatti A., De Bernardi A., Varni A. (eds.), *Storia d'Italia. Annali*, 13, *L'alimentazione*, Torino, Einaudi, 1998, pp. 209–29.

of the factories producing a quantity of cheese between 200 and 500 quintals (275 production units and 81,000 quintals of manufactured Gorgonzola). Three firms, which used to produce an amount of Gorgonzola between 2,000 and 5,000 quintals a year were also operating.

The fact that this productive "dispersal" persisted was yet linked to unsolved technical problems, such as milk freight. This productive dispersal seemed to be the legacy of the old nineteenth-century organization, which was based on small producers manufacturing their cheeses for wholesalers involved in ripening and marketing these milk products. Still in 1946 some workers in the sector remembered that transformation factories processed the milk collected in farms that stood at a maximum distance of around 15 km.[20] If it was still difficult to transport the commodity, which could not be transported by train, it was easier to move the processed products, that is, the cheese that had just been manufactured. This is why many small producers of Gorgonzola worked for larger firms which picked up the product a few days after it had been processed for the first time. They selected the parcels of goods and completed the manufacturing and ripening processes. In the same way, some large Lombard firms had large ripening storehouses in Emilia, where the cheese produced by small private dairies was stored.

The set-up of the sector was actually changing, however, at the speed of a sector still strongly linked to the primary one. The transformations in progress were radical, even more than those occurring in other segments of the agricultural and food sector. The increases and improvements in cattle-breeding had made a larger and larger quantity of milk available.[21] The role of medium and large sized firms equipped with high-technology systems, involved in elaborate processes while producing different kinds of cheese, had become more important. The manufacturing activity regarding milk products, which had previously been an important part of the management of the rural business, was gradually converging in firms that did not produce milk, but purchased it in order to transform it. In the area of Italy that most focused on dairy farming, around Lodi, where by the mid-1930s 50,000 quintals of butter and 150,000 quintals of cheese were being produced every year, 450 dairies were still operating. There were, nevertheless, around forty large dairies among those processing, at least, 50 quintals of milk every day.

---

20 Interrogation of Mr. Ettore Piani. Director of the sales department of the joint-stock company Galbani in Melzo (Milan), in Ministry for Constituent Assembly (ed.), *Rapporto della Commissione economica presentato all'Assemblea Costituente*, II, *Industria*, II, *Appendice alla relazione (interrogatori)*, Roma, Istituto Poligrafico dello Stato, 1946, p. 235.

21 According to the evaluations by Carlo Besana, director of Lodi's dairy station, in 1914 38 million hectolitres of milk were produced in Italy, of which 31.2 million were hectolitres of cow's milk; his successor, Elia Savini, estimated, for 1934, a production of milk amounting to 58 million hectolitres, with a production of cow's milk amounting to 50 million hectolitres. At this date the production of buffalo milk was still insignificant, as it amounted to 60,000 hectolitres, see Savini E., *Produzione lattiera e casearia italiana*, *op. cit.*, p. 7.

In total they were able to process as many as 900 quintals.[22] A decisive role in the processes of modernization was played by cooperative structures, and cooperative dairies, which had been increasing the amount of technical equipment used in their production sites since the beginning of the twentieth century.[23]

2. Trends in the sector between 1945 and 1970

During the twenty-five years following the end of the Second World War the sector showed some important innovations from demand point of view. The huge economic transformation that took place in the 1950s and the consequent improvements in the population's living conditions caused a significant increase in the diary products consumption, in particular cheese.[24] Just in the 1956–66 decade, the per-head consumption of cheese increased by 28%; in 1966, each Italian used to eat, every year, around nine kilograms of cheese. It was an amount that, in Europe, was exceeded only in Denmark and France (11.2 kg/inhabitant).[25] This increase in consumption went with an increasing recourse to the market, as the intense exodus from the countryside experienced during this period caused an inevitable fall in private consumption, which was typical for farmers' families. There was an increase in the demand for fresh cheeses, especially after the mid-1950s, when fridges were introduced in many Italians' houses. These cheeses were enjoyed because of their mild taste and appreciated thanks to their being definitely

22 Note that a small traditional dairy that stood in a farm transformed, as a rule, 6 hectolitres of milk every day, and it produced butter and Grana cheese as a result. For data from the 1930s, see Savini E., *Produzione e industria lattiera del Lodigiano*, Lodi, Tip. Biancardi, 1934, pp. 4–5.

23 *Censimento industriale 1937-XV. Industria della lavorazione del latte e dei prodotti derivati*, *op. cit.*, pp. 6–7. A study of the role of cooperation in the development of Italian dairy in the 20th century is not included in the aims of this work. For remarks on the role of cooperatives, especially in the area of Parmigiano-reggiano and in Sardinia, in the first twenty-five years of the 20th century, see Bardellini S., "Latterie cooperative", in *L'Italia agricola*, 63, 1926, 12, pp. 652–60. The vicissitudes of the Dairy of Soresina are certainly interesting; see Fumi G., "Una grande impresa cooperativa nell'industria alimentare: la Latteria Soresinese (1900–1940)", in Battilani P., Bigatti G. (eds.), *Oro bianco. Il settore lattiero-caseario in Val Padana tra Ottocento e Novecento*, Lodi, Giona, 2002, pp. 428–45.

24 These changes became part of a process of general increase and switch-over in consumption starting in Italy in the period after the Second World War; see on this matter Zamagni V., *L'evoluzione dei consumi tra tradizione e innovazione*, *op. cit.*, pp. 175 and ff.

25 If there is no indication to the contrary, the data afterwards reported are quoted from Ferrari P., *L'industria del latte in Italia*, Piacenza, Camera di commercio di Piacenza, no date [1971]. In 1970 the per-head consumption of cheeses in Italy was estimated to amount to 10.2 kg; see "Assemblea annuale del Consorzio del formaggio Parmigiano reggiano, Reggio Emilia, 30 ottobre 1971. Sintesi della relazione del presidente avv. Giampaolo Mora", in *Il mondo del latte*, 1971, 11, p. 767.

cheaper than hard cheeses, which had to undergo a ripening process, causing their costs to rise.[26]

Changes also happened from the point of view of supply. The commodity's supply conditions gradually changed, because of the introduction of technical means that gradually enabled an increase in the distances that milk could travel without undergoing deterioration. Techniques for pasteurizing milk and its heat treatment became widespread. At the same time, people reverted more and more to cultures of selected milk enzymes in order to insert the specific microflora for different kinds of cheese into the milk.[27] Processing and heating tanks for the commodity were replaced; old copper boilers having the capacity of a few hundred liters were abandoned. Stainless steel containers, where even 10,000 liters of milk could be processed each time, were introduced. There was a big change also in the organization of the distribution system, which continued to be characterized by a superabundance of retailers' shops where consumers used to go to buy

26 The increase in the consumption of fresh products and low-price cheeses contributed to the success of some big national producers, but in the meantime it was also to foreign producers' advantage, as outlined below. Among the national businesses that best understood consumers' needs were two Lombard firms: Invernizzi and Galbani. For information about their position in the Italian food industry hierarchy see the chapter by Gianpiero Fumi in this volume. Little is known about Invernizzi. As for Galbani – which in the second post-war period remained an unlisted family-owned private company – data are more numerous. We know, for example, that in 1962 it was the most important business in the Italian dairy sector and that it operated in meat processing too; see also the chapter by Silvia Conca in this volume. At that time Galbani had approximately 4,000 employees and 142 warehouses. Its turnover kept increasing and in the financial year 1960/61 had reached 46 billion lire. Dairy products accounted for 34 billion lire, meat for 12 billion lire. Buildings and machines were worth 10.5 billion lire, while stocks exceeded 7 billion lire; pigs to be fattened up were almost 35,000. Galbani as a whole reached around 2,050,000 hectolitres of milk collected in one year, at a cost of 11 billion lire. In order to develop its activities Galbani bought dairy products for 10.5 billion lire, pigs for 3.1 billion lire, cold meats and meats for 3.8 billion lire, feeds for 850 million lire. The company was in debt to banks for 2.9 billion lire. During the financial year ended 30 April 1961 Galbani had sold 510,000 quintals of cheese (approximately 10% of the Italian output) and 200,000 quintals of cold meats. According to the management of Banca Commerciale, "while the official balance sheet showed a net profit of 230.8 million lire, the real one amounted to 2.2 billion lire". This information had been collected by consulting the minutes of the Banca Commerciale Italiana's board of directors. As a matter of fact, when the Bank renewed loans to companies – and Galbani was one of its most important clients – the managers often reported back on the situation of the firms. The material is quoted using the acronyms ASI-BCI (Archivio storico Intesa San Paolo, Patrimonio Banca Commerciale Italiana, VCD (verbali del Comitato di direzione) and indicating the number of the volume where paperwork is kept and the date. Data for 1962 had been derived from ASI-BCI, VCD, Vol. 174, 23 January 1962.

27 The use of new techniques gave immediate results for some productions. Others were still at the experimental stage in the early 1970s; for example, the pasteurization of milk was quickly applied in the dairies that produced soft cheeses, provolone and Pecorino, whereas it was not actually used in plants manufacturing Grana Padano and Parmigiano-Reggiano; see Berti P., "Gli aggiornamenti tecnici nella produzione dei formaggi, del burro e del latte in polvere", in *Il mondo del latte,* 1970, 4, pp. 267–9.

milk, but now was provided with new technical tools. In particular, the implementation of the so-called "cold chain" now made milk products move directly and quickly from factories to the counters of shops. By using its own network of dealers and its own motor vehicles, in the mid-1950s the firm LIR, of the group Locatelli, was able to reach food product shops in 7,000 Italian municipalities twice a week.[28] The modernization of the distribution system encouraged the sales of soft and fresh products that had formerly been conditioned by the fast deterioration of this kind of cheese, especially at the times of year characterized by hot weather.

In the early 1960s, processes of integration among the six countries of the ECM started, these dynamics bring to a very serious problem (i.e. the price of milk).[29] The programs of the European Community, scheduling the setting of one price for the commodity in all the countries belonging to the common market, at a level Italian farmers thought was totally unsatisfactory. So far as the manufacturers of milk products were concerned, they denounced the increasing weight of European partners' competition. As a matter of fact, they had to compete with international producers who could take advantage of a milk price much lower than the one which (with much more difficulty) could be set in Italy at the beginning of every dairy year.[30]

With regard to production, it must be acknowledged that it underwent a significant fall-off during the war years; it returned to customary levels at the end of the 1940s, after public authorities had liberalized again the sale of milk.[31] After hitting a peak at 600,000 quintals in 1939, and having experienced an extreme fall-off during wartime with the lowest point in 1944 (353,000 quintals), the production of butter was recovering in 1950, although it was still below pre-war levels, hitting a peak at 579,000 quintals. In the same period, the production of cheeses moved from 2.7 million quintals in 1939 to 1.5 million in 1944 before rising again to 2.8 million in 1950.

---

28 ASI-BCI, VCD (Reports of the managing committee of Banca Commerciale Italiana), Vol. 143, 6 November 1957.

29 The EC instrument was set up in January 1962, but it was "a very gradual process, as the first regulation connected to dairy sector was the EC regulation of February 1964; note that, then, only since 1970 has the European market been completely liberalized (after remaining protections ensured to Germany and Italy had been removed) and that only since 1 April 1973 had the Italian market had the same organizational structure as that of other ECC's countries" (Tedeschi P., Stranieri S., "L'evoluzione del settore lattiero caseario lombardo dall'Ottocento al Duemila", in Archetti G., Baronio A. (eds.), *La civiltà del latte. Fonti, simboli e prodotti dal Tardo antico al Novecento*, Brescia, Fondazione Civiltà Bresciana, 2011, p. 742).

30 The 804/68 regulation concerned the prices by means of decisions that did not actually damage Italian producers of milk and manufacturers of Parmigiano-Reggiano and Grana Padano (*ibid.*, p. 746). As Andrea Maria Locatelli's chapter in this volume reminds us, the policy of the European Community contributed to a further increase in the imports of milk products in Italy in the 1970s and 1980s.

31 Until 1948 the trade of milk was subject to the system of stockpiles.

If we shift our attention to foreign trade, we may note that Italy of the early 1950s was a country that purchased butter on the international market. Around 43,300 quintals of this milk product were, at that date, imported from foreign countries.[32] The balance of cheeses trading was active again in 1950, with exports amounting to 185,400 quintals and imports amounting to 62,500 quintals. After the wartime slump, exports of Pecorino (75,100 quintals) and Gorgonzola (34,100 quintals) were significantly recovering, whereas foreign sales of Grana stabilized at 15,300 quintals. It was a much lower figure than the one that accounted for 46,800 quintals in 1939.

During the 1950s and 1960s, production of cow's milk increased. If we exclude the milk drunk by calves still feeding on their mother's milk, the average annual production during the five-year period between 1948 and 1952 amounted to 56 million quintals. At the end of the 1960s, the annual average was not far from 100 million quintals.[33] The increase was not linear: production significantly increased during the 1950s; between 1961 and 1964 there was a crisis in the sector,[34] with recovery beginning from 1965 and continuing in the following five-year period.

The production of cow's milk was concentrated, as formerly, in the Po Valley, in particular in the area included between the Rivers Sesia, Reno and Mincio; except for this territory, extended between Piedmont, Lombardy and Emilia, the only region that saw an increase was Latium, as a consequence of the land drainage started in the 1930s. Lombardy continued to be the region contributing the most to home production, by means of a share close to 30%.[35] The production of milk was increasing significantly in

---

32 Exports were reduced to few hundred quintals in that period.

33 The increase may be explained by the rise in the number of bovines being bred and the qualitative improvement of the subjects having to produce milk. In this connection, it is sufficient here to recall that the cows and the cow buffalos registered in Italy were 8,215,000 in 1940, 5,885,000 in 1945, 8,331,000 in 1950 and 9,845,000 in 1970; see Bonsembiante M., Gallo L., Schiavon S., *L'allevamento e le produzioni animali nel XX secolo*, in *L'Italia agricola nel XX secolo. Storia e scenari*, Rome, Italian Society of Farmers, 2000, p. 631, table 3. We must note, however, that, in spite of its significant increase, the production of milk in Italy continued to be very much lower than that estimated in the other two big EEC's countries; 277 million quintals of milk in France and 213.5 million quintals in Germany were produced in 1966.

34 Between 1961 and 1964 the high price of meat caused a significant increase in the slaughter of cows with a consequent fall in the productions of milk and a significant increase in the prices of this commodity. The fall in the production of milk after 1961 is also explained by the EC policies of inducement to kill milk cattle and to stop production activities in the zootechnical sector (Tedeschi P., Stranieri S., *L'evoluzione del settore lattiero caseario lombardo dall'Ottocento al Duemila*, *op. cit.*, p. 739).

35 In 1950 the global production of cow's milk was estimated to amount to 52.6 million hectolitres. At that date production in Lombardy amounted to 18.1 million hectolitres (34.4%), see Tentoni R., "Panorama sintetico della situazione lattiera italiana nel momento attuale con particolare riferimento al consumo del latte come tale", in *Il latte*, 30 (1950), 11, p. 6, table 2. Around 1964, production of milk amounting to around 96 million quintals was assessed, 73 million of which were available for human nutrition and transformation into milk products. 84% of these 73 million quintals were produced in the

Emilia Romagna, a region that contributed to the production of cow's milk by means of a share of 17% in 1966. On the other hand the production of sheep's milk was decreasing, in the first half of the 1960s, it stabilized at 5–5.5 million hectolitres a year.

With regard to processed milk, the amount increased between 1951 and 1962, moving from 37 million hectolitres to 50 million hectolitres. In 1963 this amount fell to 42 million hectolitres, but rose again to 52.2 million hectolitres in 1967.

According to data collected by Assolatte, at the end of the 1960s around 47,280,000 quintals of milk were used in its industrial processing, almost exclusively for home production (Tab. 7). Italian dairy, more than formerly, based itself on the transformation of cow's milk, as sheep farms contributed to the home production of cheeses by means of a share lower than 6% of the transformed commodity.

Table 7. Italy. Milk used for industrial transformations in 1969

Kind of product	Quantity q	%
Cow's milk for industrial use	44,493,000	94.1
of home production	43,897,000	(92.8)
imported	596,000	(1.3)
Sheep's milk for industrial use	2,369,000	5.0
Goat's milk for industrial use	419,000	0.9
Total milks for industrial use	47,281,000	100

Source: elaboration from Antonio Invernizzi's intervention, President of Assolatte and son of the president of Galbani's board of directors, on the topic "Production and exportation of dairy produce" at the round-table conference organized on the occasion of the First Show of Dairy Industries, Parma, 28 April-2 May 1971, in *Il mondo del latte*, 1971, 4, p. 250.

The increase of commodity availability as a result of the qualitative improvements in cattle-breeding in the north of our peninsula had no effect on butter processing; the average annual production for the three-year period 1963–5, amounting to 571,000 quintals, effectively turned out to be the same as that of the five-year period 1948–52, amounting to 560,000 quintals.[36] Better results, indeed, were achieved in the cheese sector, with a significant increase in production between the early 1950s and the middle of the following decade. During the five-year period 1948–52, every year 2.7 million quintals

Northern Italy and 34% only in Lombardy; see Ellena G., *Ministero dell'agricoltura e delle foreste. Ispettorato agrario compartimentale di Milano. Situazione attuale della produzione e del consumo di latte e derivati in Lombardia e prospettive per l'avvenire*, no pl., no pub., 1964, p. 1.

36 In the three-year period 1963–5, annually, 4.9 million quintals of butter were produced in Germany, and 4.3 million quintals in France; in these two countries, production had basically doubled relative to that of the early 1950s.

of cheeses were produced, in Italy, whereas an average annual production of 3.9 million quintals was estimated for the three-year period 1963–5. The production trend for cheese was quite similar to that for milk. There was a move from 3.1 million quintals in 1952 to 4.5 million quintals in 1962. In the following years, there was a fall in production of 17% and a subsequent recovery; in 1967 the production of cheeses had returned to the levels of 1962 (Tab. 9). In 1968 there was a new decrease, although less significant; at the beginning of the new decade production stabilized at 4,000,000 quintals (Tab. 8).[37]

Table 8. Italy. Produced cheese, processed milk and processing yields in 1971

Products	Cheeses obtained q	%	Milk used[1] hl	%	Average yield kg per hl of milk
Cheeses of cow's milk	3,095,027		39,048,968		7.9
Cheeses of "mixed milks"	709,378		5,961,163		11.9
Cheeses of sheep's milk	187,527		1,067,960		17.5
Cheeses of goat's milk	15,533		155,339		9.9
Other products derived from cow's milk	–		3,203,882		–
Total	4,007,465	100	49,437,312	100	–

1) With respect to the total, 47.1 million hectolitres of cow's milk entered processing, of which 1.96 million hectolitres of foreign origin; 1.94 million hectolitres of sheep's milk and 400 thousand hectolitres of goat's milk too were processed.

Source: *Produzione e consumi lattiero caseari italiani*, statistical appendix to Assolatte (ed.), *Assemblea generale dei soci, Milano, 16 giugno 1972. Relazione del presidente sull'attività svolta dall'Associazione italiana lattiero-casearia nel 1971*, Rome, Tip. Atena, 1972, p. 22.

Obviously, Italy was no longer at the peak of world cheese production as it had been at the end of the 1920s. The USA, which between the end of the 1940s and the early 1970s had basically doubled its production, now put 11.5 million quintals of cheese on the market. France had done even better. Starting from an average of 2.5 million quintals in the five-year period 1948–52, it had succeeded in manufacturing 8.3 million quintals of cheese at the beginning of the 1970s.[38] Only some kinds of cheese, in particular Grana in its different varieties, and soft cheeses, excepting Gorgonzola, contributed to this increase in production, which did not stand the pace of the USA and its European partners (Tab. 9).

---

[37] It should be noted that we are talking about estimates and we are comparing data gathered and elaborated according to criteria that did not completely coincide.

[38] Among EEC countries Germany had moved from 1.8 to 5 million quintals in the 1950–70 period and the Netherlands had moved from 1.3 to 2.9 million quintals. A similar trend to Italy's was registered in Switzerland, where production increased from 0.5 million quintals to 0.88; see *Produzione e consumi lattiero caseari italiani*, statistical appendix to Assolatte (ed.), *Assemblea generale dei soci, Milano, 16 giugno 1972. Relazione del presidente sull'attività svolta dall'Associazione italiana lattiero-casearia nel 1971*, Rome, Tip. Atena, 1972, p. 58.

Table 9. Italy. Cheeses produced in the years 1952, 1957, 1962, 1967 (000 of quintals)

Types	1952	1957	1962	1967
Hard cheeses	*1,165.0*	*1,196.0*	*1,650.6*	*1,858.0*
Grana	750.0	800.0	1,198.5	1,478.0
Pecorini	415.0	395.0	452.0	380.0
Semi-hard cheeses	*775.0*	*825.0*	*963.3*	*597.0*
Provoloni and similar cheeses	270.0	300.0	401.3	357.0
Asiago and similar cheeses	280.0	120.0	136.9	54.0
Montasio and similar cheeses		16.0	190.1	58.0
Caprini	35.0	35.0	31.5	14.0
Fontina	80.0	40.0	24.9	24.0
Emmental and Sbrinz	100.0	100.0	28.0	4.0
Other semi-cooked texture cheeses		70.0	123.6	86.0
Soft cheeses	1,160.0	1,340.0	1,930.1	2,069.0
Gorgonzola	380.0	320.0	330	250.0
Italico, Quartirolo		760.0	1,120.1	1,260.0
Other fresh cheeses		260.0	480.0	558.0
Total	3,100.0	3,360.0	4,517.0	4,524.0

Source: elaboration from Ferrari P., *L'industria del latte in Italia*, Piacenza, Piacenza's Chamber of commerce, no date [1971], pp. 253–63.

During the boom years, the processing of Grana exhibited a similar upward trend compared to the one of milk. The manufacture of this milk product continued to be carried out in "almost homemade level" dairies, which increased and decreased the production of this cheese depending on the price trend. In order to produce this kind of cheese, 41% of the milk transformed was used with yields of 6–7 kilograms/hectolitre. The production of Grana was split up, with a certain balance between the two most important kinds, Grana Padano and Parmigiano-Reggiano. For the first variety of Grana, the available estimates report a production of around 300,000 quintals in 1950, rising to 575,000 quintals in 1960 and 655,000 quintals at the beginning of the 1970s.[39]

There was a less significant increase in the production of Parmigiano-Reggiano, also because the starting levels were higher. Already in 1946 its production amounted to 400,000 quintals in the processing district for this

[39] It must be said that the average of the productions in the two-year period 1966–7 amounted to 751,000 quintals, as evidence of production that could encounter very big changes from one dairy year to another; see Iotti M., *Storia del formaggio di grana "Parmigiano reggiano"*, Reggio Emilia, Futurgraf, 1996, pp. 219 and 244, table A.

cheese; amounts of 730,000 quintals were exceeded in 1960, whereas in 1970 production amounted to a little more than 630,000 quintals.[40]

Production of Pecorino did not increase. It was also subject, in that period, to strong fluctuations. It must, be observed, however, that, within this group of cheeses, there was a significant increase in the production of Pecorino romano manufactured in Sardinia, risen from 55,000 quintals to 105,000, whereas production of that obtained from Latium countryside was cut by half. The different production trends in the various different areas is explained by the fact that traditional organization of production persisted in Central and Southern Italy, while in Sardinia in that period the industrial production of Pecorino also developed thanks to the financial support policy of this region.

A downward trend may be noted in the production of medium-hard cheeses in the 1960s. In this sector, the only growing production was that of provolone, also supported by an export flow towards North American markets. At the end of the 1960s, operators working for the manufacture of this cheese were engaged in promoting consumption and in the creation of a consortium of protection. The fall in the production of Montasio and Asiago was explained by the fact that a "home-made" production organization still persisted while foreign competition for similar products, which were cheaper, was increasing. Finally, the disappearance of the production of "Swiss-kind" cheese was linked to a change in consumer tastes. The original product, offered by the Swiss Confederation's producers at interesting prices, appealed more and more.

The production of soft cheeses, in particular of those whose taste was mild and refined, which were become more appreciated by the average consumer, was booming. It increased so much that, in 1971, nearly 635,000 quintals of Italico (Bel Paese), Quartirolo and Crescenza were produced, to which 709,000 quintals of "other fresh kneaded-curd cheeses and generally speaking fresh cheeses" may be added. This change in consumers' tastes also explains the fall in the production of Gorgonzola compared to the levels achieved in the early 1950s, which were not far from those registered in the 1930s.[41] The difficulty for farmers interested in this cheese also stemmed from the non-recovery of exports of this cheese, at least until the end of the 1960s, to European markets, where it could hardly withstand the competition from similar cheeses sold at lower prices.[42]

40 An oscillating trend in productions was also registered for Parmigiano-Reggiano, with an increase of 28% in 1960, a fall of 10% in 1963, an increase of 15% in 1966 and a following drop of 14% in 1968; *ibid.*, pp. 218, 244, table A.

41 The Gorgonzola produced until the early 1950s was different as regards to consistency, texture and taste from the current one. It was much stronger and much less soft. Concerning the production and the sales of this cheese, see Delforno G., "Produzione e commercio del formaggio Gorgonzola", in *Il mondo del latte*, 1970, 8, pp. 578–85.

42 As we have pointed out, before the effects of the 1929 crisis became completely manifest, more than 100,000 quintals of Gorgonzola were exported; between 1967 and 1969, every year, the amount of 36,000 quintals could not be exceeded, *ibid.*, p. 583, table 2.

According to the data provided by Assolatte, in the early 1970s nearly 48% of the milk transformed into cheese in Italian dairies became Grana cheese, whereas this product, in its varieties, contributed to sector production with a share slightly lower than 36%.[43]

Nearly 5% of cow's milk entering processing plants became Gorgonzola, whereas, by using the remaining 47% of the "processed" commodity, another forty kinds of cow's, sheep's and goat's milk cheeses were produced. In the cow's milk cheese sector, the importance of Grana and, to a small degree, Gorgonzola was so considerable that "the monthly calculation of the prices of milk [in the areas where the production of these cheeses was concentrated] referred to the prices of Grana, Gorgonzola and butter".[44]

The increase in cheese production was not the result of the greater presence of Italian milk products on international markets. Exports, which in 1952 were stagnant on a quota of 192,800 quintals – very far from the 400,000 quintals registered in 1931 – rose in 1962 to 266,700 quintals and then experienced a fall-off. In 1966, 224,700 quintals of cheese (accounting for 5–6% of global production) were exported for an exchange value of 22.5 billion lire.[45] Such a level is found again at the beginning of the 1970s (Tab. 10).

Table 10. Italy. Exports of cheeses in 1970

Designations of kinds	**Quantity** q	**%**
Grana in different varieties	60,400	26.2
Pecorino	59,900	25.9
Provolone and other ones	56,500	24.5
Gorgonzola	34,700	15.0
Melted cheeses	8,800	3.8
Fresh milk products	6,400	2.8
Other hard or medium-hard cheeses	4,100	1.8
Total	230,800	100

Source: compare Tab. 7.

43 In 1971, 649,000 quintals of Parmigiano-Reggiano, 655,000 quintals of Grana Padano and 131,000 quintals of "other kinds of Grana" were produced in Italy; see *Produzione e consumi lattiero caseari italiani*, statistical appendix to Assolatte (ed.), *Assemblea generale dei soci, Milano, 16 giugno 1972, op. cit.*, p. 22.

44 Antonio Invernizzi's intervention (President of Assolatte) on the topic "Production and exportation of dairy produce" at the round-table conference organized on the occasion of the First Show of Dairy Industries, Parma, 28 April-2 May 1971, in *Il mondo del latte*, 1971, 4, p. 249.

45 Exports were intended, most of all, for non-EEC countries; in 1966, out of 225,000 quintals of exported cheese, 75,000 quintals were sold in the USA and 31,400 quintals in Switzerland.

At the end of the period under consideration, Italian foreign cheese sales were still based on products of the national dairy tradition. In 1970, the different varieties of Grana, Pecorino, provolone and Gorgonzola covered nearly 92% of Italian cheese exports (Tab. 10).

Imports, instead, were soaring, because of the inability of home production to satisfy domestic demand. In 1952, purchases from foreign countries were still fewer than sales. At that date, imports of cheeses stopped, at 145,200 quintals. The following year Italy returned to being a net importer of cheese in terms of quantity.[46] In 1960, foreign purchases of cheese reached 387,000 quintals, and increased to 700,000 quintals in 1966, for an exchange value of 44.7 billion lire. In the same year, 235,000 quintals of butter were imported too with a disbursement amounting to 18 billion;[47] also 1,221,000 quintals of milk were imported, with an expense of 7.3 billion.[48] Imports continued increasing also in the following years as they succeeded in reaching a quota close to 930,000 quintals in 1970, and almost 1,132,000 quintals in 1971.[49]

The gradual deterioration of the balance of trade for cheese could be explained in different ways. First, it must be remembered that Italy used to export cheeses with average prices of 148 dollars/quintal, whereas the average prices of the cheeses exported by several European countries amounted to 80.75 dollars/quintal. In spite of production that in terms of value was lower in Europe only than French one, Italian dairy could not satisfy the domestic demand for cheese. The high cost of the most-produced cheeses in Italy was explained by the centrality of goods, such as Grana and Provolone, that required expensive processing and long ripening.

As a matter of fact, the 1961 production census identified a still-high number of industrial businesses: 3,417, with 23,750 workers (Tab. 11). Ten years later, a census was taken of 2,384 plants with 23,057 workers.[50] The fall in the number of businesses amounted to 30% in a decade, a far more significant decrease than that for the whole food sector, which, during the same period of time, saw production plants decrease by 13.7%. In spite of this, in the 1970s the dairy sector remained a sector with a low concentration. Fragmentation was always very strong in the processing of different kinds of Grana. In particular, if we consider the dairies producing

46 In terms of value, imports returned to outnumber exports in 1959.

47 In the period under consideration butter exports from Italy were always very limited.

48 Imports of milk increased from 5,900 quintals in 1950 to 76,500 quintals in 1964; the extremely significant increase in imports stemmed from 1965; in that year foreign purchases of milk amounted to 291,000 quintals.

49 *Produzione e consumi lattiero caseari italiani*, Statistical appendix to Assolatte (ed.). *Assemblea generale dei soci, Milano, 16 giugno 1972. Relazione del presidente sull'attività svolta dall'Associazione italiana lattiero-casearia nel 1971*, *op. cit.*, p. 44. Imports would definitively boom in the 1970s.

50 The 1971 census data are drawn from Gallo G., Covino R., Monicchia R., "Crescita, crisi, riorganizzazione. L'industria alimentare dal dopoguerra a oggi", in *Storia d'Italia. Annali*, 13, *L'alimentazione*, *op. cit.*, pp. 327–8, tables 1 and 2.

Parmigiano-Reggiano, we may note that they were 2,356 in 1955, one year after the establishment of the consortium, and 1,768 in 1968.[51] It must also be observed that, at the end of the 1960s, the production centers for Grana transformed milk produced in close proximity to dairies. At that date, only 4% of the production plants received milk from cowsheds situated further than 5 km from the small traditional dairy.

Table 11. Italy. Operating businesses and workers in the sector of dairy production in 1961

Businesses	**1961**				
Kind of business	**Businesses**	**%**	**Workers**	**%**	**Workers/Business**
Total businesses	3,417	100	23,750	100	7
Businesses with more than 10 workers	339	9.9	13,863	58.4	40.9
Businesses with more than 100 workers	22	0.6	6,135	25.8	278.9

Source: elaboration from Chiapparino F., "Tra polverizzazione e concentrazione. L'industria alimentare dall'Unità al periodo tra le due guerre", *op. cit.*, pp. 214–18.

Imports basically came from EEC countries, mainly from France and Germany; a significant import quota was the result of sales in Italy of cheese manufactured in Switzerland.[52] In 1971, for example, more than 400,000 quintals of cheese were imported from Germany, nearly 340,000 quintals from France, and a little under 158,000 quintals from Switzerland. The most-imported cheese was Emmental, formerly produced in large quantities in Italy. Many melted cheeses, nearly all produced in Germany, were bought abroad.

In order to try to revitalize exports and withstand increasing competition from foreign products in a domestic market that was no longer defended by high protectionist barriers, in the 1950s and in the 1960s initiatives for protecting traditional home productions multiplied with the signing of international agreements,[53] regulatory intervention by

51 The level below 1,000 was only achieved in the early 1980s; see Iotti M., *Storia del formaggio di grana "Parmigiano reggiano"*, *op. cit.*, p. 250.

52 The sale of Swiss products in Italy was completely arranged by the Union of the importers of Swiss cheeses (Swiss import) adhering to the Italian dairy association; the Union ensured "the general conditions of the rules and regulations governing the trading of Swiss cheese that were appropriate to get an even profitability for Italian operators [who used to sell these cheeses to consumers]" (Assemblea ordinaria della Swissimport. Milano, 22 ottobre 1970, in *Il mondo del latte*, 1970, 10, p. 819).

53 Initiatives to protect local productions and avoid trade fraud multiplied, starting from the years immediately after the end of the Second World War, while achieving important results such as in the case of Stresa's Conference in June 1951. For these initiatives see Burigana D., "L'Italie, la France et les fromages 'typiques' (1950–2009). Aux origines d'une politique étrangère européenne?", in Ceccarelli G., Grandi A., Magagnoli S. (eds.), *Typicality in History. Tradition, Innovation and Terroir*, Brussels, P.I.E. Peter Lang, 2013, pp. 248–61.

government,[54] and the establishment of consortia.[55] In the mid-1960s three cheeses benefited from consortia defending their own production: Fontina from Valle d'Aosta, Grana Padano, and Parmigiano-Reggiano. Via a presidential decree dating back to October 1955 (DPR, 30 October 1955, No. 1269), the "designation of origin" had been fixed, with the indication of area of production, qualitative standards, and manufacturing practices for six cheeses; we are referring here to Parmigiano-Reggiano, Grana Padano, Pecorino romano, Pecorino Siciliano, Gorgonzola and Fontina. For a further eight cheeses the same decree established the "designation of typicality" (i.e. product features), without suggesting the area for their production: we refer here to Taleggio, Provolone, Ragusano, Asiago, Montasio, Caciocavallo, Fiore sardo and Pressato. In 1965, standards and typical area were defined with regard to other cheeses, but only those for local consumption. The consortium for Parmigiano-Reggiano, whose establishment dates back to previous times, was involved in advertising campaigns to publicize the product to markets[56] such as those in Southern Italy, where sales continued to be limited. The Grana Padano consortium, which had been established in 1954 on the initiative of Ercole Locatelli and which, in the mid-1960s, counted on 250 associates, began operating in these years to promote sales of this cheese too. In the 1960s also the consortia for Pecorino and Gorgonzola were established. With regard to this latter cheese there was, during that period, a clear signal of the difficulty in aggregating Italian manufacturers of dairy produce. Until 1970, in fact, two Gorgonzola consortia operated at the same time. One of these was established in Novara in the local Chamber of Commerce and the other was created in Milan in the national association of the producers of milk products (Assolatte).[57] Related consortia were nearing establishment for other two products: we are referring here to the structure

54 In April 1954 the Law of protection of the designations of origin and typicality of cheeses was gazetted; see in this connection "Tutela delle denominazioni di origine e tipiche dei formaggi", in *Il mondo del latte*, 1954, 14, pp. 203–9. The law was the result of the implementation in Italy of what had been signed on the occasion of the Stresa Convention.

55 In the early 1920s a discussion began on consortia that were able to put together farmers and traders in order to offer for sale Italian milk products well identified and guaranteed as regards their origins. Among the first ones to bring action we find producers of Parmigiano-Reggiano, who were worried about the difficulty in catching up with the export levels of pre-war period; see Fascetti G., "Sull'organizzazione dei consorzi caseari a difesa dei latticini italiani", in *Resoconti dei lavori del III convegno nazionale dell'industria del latte organizzato dal Comitato della sezione italiana della Federazione internazionale di latteria*, 17 April 1924, Milano, tip. Agraria [1924], p. 32. The worry of the most important operators in the sector was about establishing consortia in order to guarantee "at most, the quality of product and not only its origin, as we would like somebody to do, both in guarantee of the consumer and also in guarantee of the existence of the constituent institutions" ("Riunione del Comitato tecnico per la produzione industriale del latte di vacca", in *Latte, latticini e conserve animali*, 1936, 1, p. 4).

56 See the chapter by Stefano Magagnoli in this volume.

57 See Vanessa Pollastro's chapter in this volume.

that put together the producers of Taleggio, whose offices were required to be in Bergamo, and to the consortium of protection for provolone cheese, in which producers from the provinces of Piacenza, Milan and Cremona were interested, although it was a product of the old dairy farming of Southern Italy.

Conclusions

This chapter has outlined a framework for the dairy sector in Italy in the late 1930s and early 1970s. The choice of chronological details is not incidental. In 1937, in the context of the *Production Census*, the first extensive survey of processing milk activities was carried out. The emerging picture is that of a sector dominated by small and very small enterprises. From the geographical point of view, the production of cow's milk cheese was concentrated in the plains of the Northern Italy, in particular in Eastern Piedmont, Lombardy and Emilia. On the other hand production of cheeses made using sheep's milk was concentrated in Latium and, primarily, in Sardinia. The variety of goods produced is very wide considering the wealth of regional traditions. Indeed, the cheeses manufactured in large quantities, especially those intended for export, were only Grana, in two varieties, Grana Padano and Parmigiano-Reggiano, Gorgonzola and Pecorino.

Also, the term "ad quem" is not incidental. Between the end of the 1960s and the beginning of the next decade the implementation of Community rules started a new phase, characterized by a strong growth of milk imports and by a gradual deterioration of the balance of trade for cheeses. Italy, which had been among the most important exporters of dairy products, was about to become a strong importer of cheeses, particularly of soft and fresh products that necessitated a fast ripening process and a low price.

From the point of view of production, the period from the end of the Second World War to the late 1960s is characterized by a growth in the cheeses just manufactured. The strong growth in production is accompanied by a significant change in the varieties produced. Grana holds up well while, in the recent years, mainly increases the fresh cheeses production, increasingly favored by consumers. The dynamics of Italian dairy farming are characterized by the huge importance of small and large sized firms.

The Influence of the European Institutions and Policies on the Italian Dairy Sector

Stefanella Stranieri

University of Milan

Paolo Tedeschi

University of Milan-Bicocca

1. Introduction: topics and aims

The aim of this chapter is to show how European institutions and policies influenced the Italian dairy sector from the 1960s up to the new millennium.[1] More precisely, this chapter illustrates the effects of the Common Agricultural Policy (CAP) on the reorganization and competitiveness of European agricultural activities. A special focus on the diary sector is provided. The CAP progressively reduced tariff barriers on trade of agri-food products among European countries, and at the same time strictly regulated the competition between European and extra-EU firms. The enlargement of agri-food trade enhanced foreign competition for the Italian dairy products: besides, it strongly influenced the production and the price of the milk. At the same time, the European Parliament and Commission progressively modified all the phases of the productive systems of dairy products in order to guarantee the health of consumers and correct competition between producers. To achieve this goal, a normative framework oriented towards increasing information among the operators of the dairy chain and among consumers was progressively established: so these latter were able to know all the product ingredients and to have information about the supply chain. All the decisions taken by the European institutions modified the existing conditions of European dairy markets and obliged the dairy firms to adapt in order to maintain full competitiveness. All these different policies influenced competition with foreign producers who abusively used Italian brands to attract consumer preferences.

1 Even though this article is the result of common research, analysis and reflection by its authors, Stefanella Stranieri (University of Milan) drew up paragraphs 3 and 4, while Paolo Tedeschi (University of Milan-Bicocca Dems) wrote the remaining text.

The present chapter also shows how a relevant European institution, the European Investment Bank (EIB), indirectly helped the Italian dairy sector through the modernization of existing infrastructure (such as new electric railways and new freeways), which allowed fast transportation of dairy products. Such interventions favored the access of Italian dairy factories to European dairy markets. Moreover, the EIB partially financed the renovation of some dairy firms in Southern Italy: its loans in particular allowed the development of brands such as "Sole" and "Vallelata", which progressively became leaders in the Italian dairy sector.

2. The birth of CAP and its effects on the Italian dairy sector

The CAP officially started in 1962, and it began to regulate all the agricultural products which were included in the European Common External Tariff F List since 1959. Agricultural production was considered a strategic issue for the European economy. Due to its specificity, it was regulated by special rules which were very different from those for other productive sectors. The tariffs concerning the agricultural products present in the F list were directly established by the European institutions, and were not the arithmetic mean of the previously tariffs existing in the European countries as happened with other products. Since 1962, the CAP increased protection for EEC agricultural products, and it concentrated attention both on the support of producers and on the valorization of products. Policy interventions based on "coupled support" offered the safeguard of European agricultural productions through price protection, that is, the price normally adequate and profitable for all producers. Such measures led to both positive and negative economic consequences for the European agricultural sector. First, they allowed European agriculture to realize a safe economic income, to invest in increasing productivity, to align agricultural incomes with those of other sectors, to stabilize prices, and to augment agricultural exports. However, such rules also brought some market distortions, which were mainly linked to the creation of agricultural product surplus and an increase in costs for their disposal.[2]

[2] Regarding the history and effects of the CAP there exists a wide bibliography which it is not possible to quote here. See, among others, Ledent A., Burny P., *La politique agricole commune des origines au troisième millenaire*, Gembloux, Presses agronomiques de Gembloux, 2002; Andreosso-O'Callaghan B., *The Economics of European Agriculture*, Basingstoke, Palgrave, 2003; Patel K. (ed.), *Fertile Ground for Europe? The History of European Integration and the Common Agricultural Policy since 1945*, Baden Baden, Nomos Verlag, 2009; Knudsen A.C., *Farmers on Welfare: The Making of Europe's Common Agricultural Policy*, Ithaca, Cornell University Press, 2009; Kay A., Ackrill R., "Problems of Composition, Temporality and Change in Tracing the Common Agricultural Policy through Time", in *Journal of European Integration History*, 2010, No. 2, pp. 123–41,

The European policy on dairy products gradually entered into force, and its influence on Italian dairy producers became effective when the implementation of the principles of the CAP actually modified the rules existing in Italy.[3] The first rule concerning the dairy sector was the European Regulation 13/1964 on the establishment of new rules for the common reorganization of the milk market and the dairy sector. Only in the 1970 the European dairy market was fully free from the abrogation of the last protections guaranteed for German and Italian products. Since 1 April 1973 the Italian agricultural market has actually had the same real structure and organization as markets existing in other EEC countries. Up until the early 1980s the CAP promoted, especially, the protection of European dairy products and followed three main principles: the establishment of the European Common market, the preference for EEC products, and financial solidarity.

Several normative interventions were adopted by the European institutions to increase the economic efficiency of the farms producing milk and to guarantee their competitiveness.[4] The most relevant support for the dairy sector concerned milk price. Due to relevant differences in the EEC countries, Regulation 13/64 established different prices for milk, ranging from a minimum to a maximum price. Not all European countries had the same price, even if the objective of the European intervention was to

---

Spoerer M., "*Fortress Europe in long-term perspective: agricultural protection in the European Community, 1957–2003*", *ibid.*, pp. 143–55; Tedeschi P., "La Politique Agraire Commune ou le grand cartel agricole communautaire", in Müller M., Schmidt H.R., Tissot L. (eds.), *Regulierte Märkte: Zünfte und Kartelle/Marchés régulés: corporations et cartels*, Zürich, Cronos Verlag, 2011, pp. 243–59. See also the references in Notes 3, 10, 12 and <http://ec.europa.eu/agriculture/cap-history/index_en.htm>.

3 Concerning the effects of the CAP on the Italian diary sector, besides the chapters in this volume (in particular Claudio Besana's) see: Assocaseari, *Comunità Economica Europea. Regolamenti del settore del latte e dei prodotti lattiero-caseari* [Milan] 1964; *Atti del 4° Convegno nazionale sul latte alimentare. Problemi delle centrali del latte italiane in rapporto alla regolamentazione comunitaria del latte alimentare. Roma 19 novembre 1970*, Rome, Federazione italiana aziende municipalizzate centrali del latte annonarie e farmaceutiche, 1970; Ferrari P., *L'industria del latte*, Piacenza, Camera di commercio, industria e agricoltura di Piacenza, 1970, pp. 229–38, 360–96; Fanfani R., *Lo sviluppo della Politica agricola comunitaria*, Roma, Carocci, 1998, p. 75–351; Felice E., "Il settore lattiero-caseario in Italia dal dopoguerra al Duemila", in Bertagnoni G. (ed.), *Una storia di qualità. Il Gruppo Granarolo fra valori etici e logiche di mercato*, Bologna, Il Mulino, 2004, pp. 35–91; Stranieri, S., Tedeschi, P., "L'evoluzione del settore lattiero-caseario lombardo dall'Ottocento al Duemila", in Archetti G., Baronio A. (eds.), *La civiltà del latte. Fonti, simboli e prodotti dal Tardoantico al Novecento*, Brescia, Fondazione Civiltà Bresciana, 2011, pp. 742–57; Canali M., "L'Italia e l'integrazione delle agricolture europee (1960–1990)", in Canali M., Di Sandro G., Farolfi B., Fornasari M., *L'agricoltura e gli economisti agrari in Italia dall'Ottocento al Novecento*, Milan, Angeli, 2011, pp. 71–170; Burigana D., "L'Italie, la France, et le fromages 'typiques' (1950–2009). Aux origines d'une politique étrangère européenne?", in Ceccarelli G., Grandi A., Magagnoli S. (eds.), *Typicality in History. Tradition, Innovation and Terroir*, Brussels, P.I.E. Peter Lang, 2013, pp. 248–61.

4 For all EEC/EU regulations, recommendations, directives and laws which are indicated in this chapter see: <http://eur-lex.europa.eu/browse/summaries.html?locale=en> and <http://eur-lex.europa.eu/homepage.html>.

progressively align different product prices within all European countries. Regulation 804/68 introduced three different prices: indicative price, entry price and intervention price. The first was considered the "equilibrium price" (the price at which the market should aim). The second was the minimum price for milk imported from non-European countries. It was a real "threshold price" and it was always higher than the market price of European milk. The third price concerned the price that the CAP guaranteed for all EEC producers. More precisely, it referred to milk powder, butter, and two of the most important Italian cheeses, that is, Parmigiano-Reggiano and Grana Padano. Thus the CAP aimed at protecting EEC milk producers, and also the main and profitable products of the dairy sector.

During the phase of "coupled support", milk farms had relevant profit advantages through the European price support. Consequently, the production of milk increased and the final price decreased. Such a trend concerned dairy factories too. The reduction of milk prices led to the higher competitiveness of dairy products and a growth in their consumption. Moreover, the CAP support made the stipulation of contracts between the dairy factories easier. Thanks to European subsides (such as price intervention, or the minimum price granted by the CAP), milk producers' earnings were in fact related to production and not to the real price existing on the market or paid by dairy factories.

During the 1960s, European incentives and subsidies and the new rules about milk safety favored the exit of less efficient breeding farms from the market. Such structural reorganization brought a temporary reduction in the numbers of Italian cows and, at the same time, implied a renovation of farms and an improvement in milk quality. Such phenomena led also to a new generation of breeders, who invested in new, modern plants and equipment: besides, breeders bought cows producing more milk. By contrast, in other regions old breeders used incentives for closing their activities. New rules modified the Italian livestock sector and also assured consumers (in particular foreign ones) of Italian milk quality: in 1968 breeding was in fact prohibited in the case of cows affected by tuberculosis.

The European institutions also influenced the institutional context in which the dairy firms operated. They in fact obliged the Italian government to remove the monopoly of the "municipalizzate", that is, those dairy factories belonging to municipalities. Such firms, which were also called "Centrali del latte", collected milk for healing, packaging and supplying to consumers. So the new rules increased market competition, and they partially reduced the price for cheese producers. Moreover, European institutions removed all obligations for milk producers to give their milk to the provincial public warehouse (the "ammasso"). At the same time, producers received financial aid for building the private warehouse of the main Italian cheeses. This incentive further increased firms' competition, because producers had permission to buy milk in other Italian provinces and European countries. Such an intervention led to the definitive end of the monopoly of the "Centrali del Latte", which normally produced a wide product range (yoghurt and cream for cooking included). Even if for many years the "Centrali del Latte" had continued to receive most of the milk (and this obviously limited the actual availability of milk and increased its price, in particular for the small and medium sized dairy factories), the general growth of competition favored the diversification of Italian dairy products. The improvement of the quality of cheese and the actual costs for pasteurization and packaging (which were not higher than those paid by foreign competitors) allowed Italian diary firms to remain competitive: so they were able to defend their market share and to increase sales within other European countries.

The CAP supported also the growth of butter and milk powder productions. It established new rules for their storage and sale at a guaranteed and profitable price. New rules were introduced also for skimmed milk and new subsides were granted for the casein used by the factory making cakes and confectionery. The FEOGA (a French acronym for Fonds Européen d'Orientation et Garantie pour l'Agriculture) introduced subsides for economic subjects, who intended to invest to create or to renew dairy companies, in particular the cooperative one. The FEOGA paid half of the investment and the remaining half was covered by a low interest rate loan.

All these European rules facilitated the increase in Italian dairy production and export of high quality Italian cheeses to EEC countries.[5] The production and quality of Italian dairy products improved, and the most successful dairy firms continued to increase their sales within the European Common Market (ECM). This last in fact enlarged the market for high quality Italian cheeses, which could benefit from increasing demand from Italian migrants living in other European countries. As had happened in the past, the migrants helped the Italian producers to promote their products, even if the effect was amplified by the progressive reduction of tariffs. The growth of competition in the Italian market led to some cooperatives defaulting due to their lack of capacity in maintaining their profitable earnings: however, the main cooperatives continued to increase their profits and sometimes incorporated the cooperatives in crisis.

The competitiveness of Italian dairy factories was also favored by the reduction of transport fees, thanks to the improvement of infrastructures like railways and highways. During the 1960s and 1970s the EIB co-financed the building of these latter, allowing the creation of a new rapid link between Italy and its European partners.[6]

5 Concerning the increase in Italian dairy production and the export of Italian cheeses in the EEC countries and the importation of communitarian dairy products see in particular A.M. Locatelli's chapter in this volume.

6 Concerning the new infrastructures crossing the Alps which were co-financed by the EIB's loans see: Tedeschi P., "The EIB and the Economic and Social Development of Italy from 1958 to the Beginning of the 1970s", in Dumoulin M., Bussière E., Willaert E. (eds.),

The EIB's aims also included financial aid to renovate the equipment and plants of enterprises located in Southern Italy. This strategy implied the creation of new factories in the chemical, petrochemical, iron and steel sectors. Also, the food industry received EIB funding for the creation of new plants aimed at the refining of olive oil and other vegetal oils, the transformation and storage of fruit and vegetables, and the production of beer, wine, pasta, confectionery, biscuits, monosodium glutamate, etc. The EIB financed only a part of the investment (no more than 50%), and dairy producers had to reimburse the loan (interests included). The EIB's loans had different rules from those of FEOGA subsides: the EIB, in fact, financed a very expensive renovation of the equipment and plants of the firms and, moreover, some major EEC commercial banks participated in the financing project.[7]

The EIB financed some Italian dairy firms too: the most relevant case was represented by the "Latte Gala". In 1963 this company, which was based in Catania, started to produce new kinds of milk, such as micronized milk (it was crumbled into one micrometer particles and so became the easiest to digest) and milk enriched with the D3 vitamin. The growth of sales obliged managers to renovate the livestock area and the dairy fix investments. A part of the related investment was financed by the EIB in the second half of the 1960s. The EIB lent one million of the UCE (European unit of account), that is, 625 million Italian lire, which had to be reimbursed within ten years (including two of "grace", that is, without reimbursement). The new Gala factory was able to produce both sterilized milk in glass bottles and UHT milk (long life milk) in Tetra Brick package (and, in the 1970s, in Tetra Brick Aseptic package). This last packaging was considered the best solution for maintaining the safety characteristics of milk: it was also the most convenient

The Bank of the European Union. The EIB 1958–2008, Luxembourg, EIB, 2008, pp. 73–90; *Id.*, "Attraversare le Alpi per formare l'Europa: la BEI e il finanziamento dei progetti relativi alle vie di comunicazione alpine (anni 1960 e 1970)", in *Geschichte und Region/Storia e Regione*, 2016, No. 2, pp. 82–101.

7 The EIB had to finance only the 50% because European Institutions wanted to promote the renewal of the EEC infrastructures and firms and at the same time to guarantee relevant contracts for the main EEC commercial banks. Concerning the EIB's loans for improving the food industry in the south of Italy see the following cases in the EIB's Archives: Tyndaris (CM 006, No. 10018, signed on 14 February 1967), Also (CM 151, No. 10422, 25 June 1974), confectionary; Birra Dreher (CM 014, No. 10029, 18 February 1963), Malteria Tirrena (CM 028, No. 10045, 24 September 1963), Hemned (CM 039, No. 10063, No. 4 May 1964), Birra Peroni Adriatica (CM 047, No. 1074, 7 October 1964), breweries; IDAC (CM 016, No. 10031, 18 February 1963), Surgela (CM 018, No. 10033, 14 March 1963), Findus (CM 022, No. 10039, 24 September 1963), transformation and storage of fruits and legumes; SVOA (CM 040, No. 10064, 4 May 1964), Costa (CM 053, No. 10083, 30 April 1965), refining olive or vegetal oil; Plasmon Sud (CM 079, No. 10144, 23 October 1967), cookies; Sabea Etnea (CM 083, No. 10151, 21 December 1967), wine; Buitoni (CM 094, No. 10169, 24 July 1968), De Cecco (CM 122, No. 10246, 2 October 1970), pasta; Ajinomoto-Insud (CM 043, No. 10070, 7 October 1964: monosodium glutamate (chemical industry).

packaging and allowed the final price of milk to be reduced. Moreover, the presence of new equipment and plants allowed a widening of product range. Milk was not the only product of the Gala factory. Yoghurt, cream, butter and new cheeses were some of the product lines of the firms. The "Latte Gala" became the "Gala Italia" and created the brand "Latte Sole" (literally, "milk sun" to indicate that its dairy products were produced in Sicily, the land of the bright sun). Sales strongly increased and allowed the Sicilian factory to buy other Italian dairy firms. This represented a relevant exception in an economic context where the factories of the Southern Italy survived with great difficulty. This also meant that the EIB's loan for "Latte Gala" achieved its goal (the development of industrial activity in the economic and social backward regions of the EEC) and had a positive influence on the evolution of the Italian dairy market, in particular in the Italian Mezzogiorno.[8]

In this regard, in 1975 the EIB also financed new plant and equipment for the Vallelata dairy firm in Aprilia (Latium). In this case, the EIB lent 800,000 UCE (650 million Italian lire) and this allowed the renovation and success of the Vallelata company.[9] The EIB's loan in fact arrived during the negative economic trend related to the oil shock, when it was necessary to reduce production costs and to diversify products: only a modern firm could realize these two conditions and take advantage of competitors' weaknesses. Up until the early 1970s Vallelata was simply the name of the resort where the dairy factory was situated: in the following years, the name progressively became one of the most relevant brands of Italian dairy product (like ricotta, buffalo milk mozzarella, Scamorza, etc.).

For other Italian dairy firms, the negative economic trend, which started in the autumn of 1973, was not an opportunity for growth. The oil shock provoked an increase in the price of dairy products and in the numbers of jobless: the consequent reduction in sales undermined many dairy producers, in particular small and medium sized ones. Big producers were able to make profits through economies of scale and the devaluation of the Italian currency in the international markets. They reduced unit costs and diversified their production: they expanded their range with new and less expensive dairy products, and they were able to meet new consumer demands. They also had strong contractual power and so they paid less for milk. By contrast, at the same time, small and medium sized dairy factories could remain competitive only if they were able to greatly increase the quality of their products and to create new market niches where price was less important than quality. However, only a few small and medium sized dairy producers managed to reorganize their activity and to improve their dairy products:

8 Concerning the EIB's loan to the "Latte Gala" see in the EIB's Archives: Gala, CM 082, No. 10150, signed on 21 December 1967; Comité de Direction, CD 24/67, Procès-verbal de la réunion du 20 septembre 1967.

9 Concerning the EIB's loan to the "Vallelata" see in the EIB's Archives: Comité de Direction, CD 41, DS 10/75, Procès-verbal de la réunion du 17 au 21 mars 1975.

thus, most of them did not overcome the crisis and were closed or sold to the great producers. So, the negative trend provoked a reduction in the number of Italian dairy producers and an increase in the average dimension of the companies. Besides, consumer choice widened, thanks to new high quality (and more expensive) dairy products and new cheap ones.

The CAP rules about price intervention (and some subsides too) were fundamental, because they enabled dairy companies to limit increases in the prices of their products and, at the same time, avoided a serious crisis for Italian milk producers and granted their survival. However, as the same situation arose throughout all EEC member states, the cost of CAP support became unbearable and a reform of the rules in favor of milk producers became necessary. Thus, the new European rules promulgated in 1984 introduced productive quotas and penalties for excess production levels were established. This new intervention enlarged the gap between large and small producers. Regulations 856/84 and 857/84 introduced a "decoupled policy" and allowed European institutions to control the milk offer for the monitoring of the quantity of milk produced. Every European country could produce a maximum quantity, and all producers had to respect their maximum production limit. A responsibility levy (called "milk quotas", whose value was amended several times) was established for exceeding quantities. Such a policy allowed the limit established to be overcome. Initially, the penalty was for 75–100% of the indicative milk price. The actual percentage depended on the system for the assignation of quotas. These latter could be assigned to the milk producers (the first case) or to dairy factories (the second case). The "milk quotas" strongly influenced the number of Italian milk producers: from 1983 to 1988 these latter decreased from 330,000 to 200,000. Most of them were old, and their sons did not want to continue the activity if they saw a progressive reduction in production. However, the remaining producers did not respect the "milk quotas": they in fact produced almost 20% more than the total production assigned to Italy. Besides, even if they were respected, the quantitative limits were higher than the real need for milk: thus the new policy did not solve the problem of the EEC milk production surplus and the related costs paid by the FEOGA. The European institution wanted to avoid a shock for milk producers, but this implied European citizens continuing to pay more for the price intervention.[10]

This situation allowed a reduction in the price of milk and favored the dairy factories, in particular the big ones. Even if in 1987 the associations of producers obtained the same rights as the single producers, the big dairy producers continued to have an obvious advantage. They in fact purchased a

---

10 Concerning the introduction of milk quotas in the CAP see Petit M. *et al.*, *Agricultural Policy Formation in the European Community: The birth of milk quotas and CAP reform*, Amsterdam, Elsevier, 1987; Borroni R., Scoppola M., Sorrentino A., *Le quote latte in Italia. Una disavventura nel cammino verso l'Europa*, Milan, Angeli, 2001.

lot of milk, so they were able to obtain a price reduction from cattle breeders. By contrast, the small and medium sized dairy factories suffered a reduction in the price of dairy products in the wholesale market. If managers were not able to create their own market niche, they had to close or to sell the factory, as had happened during the negative economic trend of the 1970s. These factories were sometimes incorporated into big companies and so the number of Italian dairy firms reduced again. Besides, the new property was not Italian: thus, the brand and quality were Italian, but the factory belonged to a foreign company. As multinational companies also bought some big Italian dairy firms, this situation was no more the exception, but became the "rule".[11]

The new "milk quotas" were not the only relevant European intervention in the dairy sector. European institutions promoted and developed new rules concerning European dairy production, like, for example, the CDO (Controlled Designation of Origin), the PDO (Protected Designation of Origin), the PGI (Protected Geographical Indication), the preservatives that were allowed to be added to dairy products, and all the prohibited products. All these rules favored big firms, which were able to invest and adapt their productions on the basis of the new normative framework in order to differentiate the quality of production through the use of specific labels. By contrast, small and medium sized producers had to join new producers' associations which were created to defend some dairy products against fraud and to assure consumers about the quality of their products. These associations (normally called "consorzi di tutela") established specific production rules (called "disciplinari") concerning the guidelines regarding production procedures and the geographical production area for the use of CDO/PDO certifications. These associations also engaged in lobbying activities with members of the European Parliament in order to protect the CDO/PDO products. The new rules about CDO/PDO also allowed enlargement of the product range. Besides, new European rules allowed producers' organizations to buy "foreign" milk. In particular, dairy factories could decide to produce a new cheese using local milk and/or foreign milk. In the first case, products could become CDO/PDO or simply PGI.

Such new rules enriched discussion among Italian producers associated in the "consorzi di tutela". Some thought that the best strategy was just to buy a higher quantity of milk (and so to pay less) and, moreover, to obtain higher prices by exploiting their brands and, obviously, their high quality dairy products (in particular cheese). By contrast, for other producers the CDO/PDO certifications were not considered useful strategies for product differentiation and they did not improve sales. Besides, these producers affirmed that producers' organizations helped the worst firms only and limited, for the best ones, the possibility of diversifying. Thus, they were

---

11 About the changes concerning the Italian dairy sector during the 1980s and 1990s, see in particular A.M. Locatelli' s chapter in this volume.

opposed to restrictive guidelines and stated that quality could be improved by the use of alternative natural ingredients. They considered that their success in the new European dairy market depended mostly on the intrinsic quality of products and only partly on the brand of the producers' associations and their guidelines.

"Coupled support" ended definitively in 1992 with the McSharry Reform, which established "partially decoupled support". The European Union (EU) was established and new rules (in particular Regulation 100/91) concerning agricultural products were set up during these years. The main aims of European institutions were the reduction of the cost of the CAP (by the limitation of financial support to the small producers only, that is, the creation of the small scale aid for farming) and the sustainable development of agriculture (through real protection of the countryside and of biodiversity). The new regulatory framework was not oriented towards price support but towards the safeguarding of farms' income. Indeed, after 1992 the new CAP helped only small producers and all other subsides disappeared. Moreover, the "threshold price" for non-EU products was cancelled, and this obviously influenced also the new EU common dairy market. Competition increased and prices decreased: dairy producers had to improve the quality of their products, and it was also necessary to develop an efficient supply chain. This obviously favored the big multinational companies.[12]

The "milk quotas" remained into force, but they were assigned only to producers. At the same time, the responsibility levy was increased. Even if the production of milk decreased, it remained higher than the demand for milk. As in previous years, European institutions did not want to provoke a serious crisis for milk producers. The intention of institutions was, in fact, to reduce milk protection policies gradually and to minimize negative consequences for diary farms' income. The maintenance level of "milk quotas" allowed less competitive farms to survive in the EU market and, at the same time, it guaranteed an adequate price for the milk purchased by EU dairy factories. So, all EU consumers paid for the social and economic equilibrium of the dairy sector. In particular, they had to accept higher taxation in order to finance the CAP budget.

---

12 Concerning the McSharry Reform and its effects see: Cunha A., Swinbank A., *An Inside View of the CAP Reform Process: Explaining the MacSharry, Agenda 2000, and Fischler Reforms*, Oxford, OUP, 2011, Ch. 5; Daugbjerg C., "Policy Feedback and Paradigm Shift in EU Agricultural Policy: The effects of the MacSharry reform on the future reform", in *Journal of European Public Policy*, 2003, No. 3, pp. 421–37; Coleman W.D., Tangermann S., "The 1992 CAP Reform, the Uruguay Round and the Commission: Conceptualizing Linked Policy Games", in *Journal of Common Market Studies*, 1999, No. 3, pp. 385–405; Daugbjerg C., "Reforming the CAP: Policy Networks and Broader Institutional Structures", *ibid.*, pp. 407–28; Ingersent K., Rayner A., Hine R. (eds.), *The Reform of the Common Agricultural Policy*, Basingstoke, Macmillan, 1998; Fennell R., *The Common Agricultural Policy: Continuity and Change*, Oxford, Clarendon Press, 1997; Josling T., "The Reformed CAP and the Industrial World", in *European Review of Agricultural Economics*, 1994, Nos. 3–4, pp. 513–27.

In this context, the revision of the McSharry reform in the second half of the 1990s was not unexpected. The aims were the further reduction of the cost of the CAP (also done from the perspective of the entry into the European Union of the countries of Eastern Europe, where a high percentage of the population worked in agriculture) and the new growth in competition in the EU single market (which had to gradually rationalize the EU dairy sector and eliminate less efficient producers). The perspectives for the Italian dairy factories were not negative if they followed the strategy of renewing plant and equipment, the diversifying their product range, and reducing internal costs. This also meant that the new European institutions' policies obliged the Italian dairy sector to renew itself and, at the same time, favored the progressive closure of the less efficient factories (or their incorporation in big companies).

3. The CAP in the new Millennium and the Italian dairy sector

With the Mid Term Review, the Agenda 2000 reform and the Fischler reform, the decoupling of agricultural payments was confirmed and reinforced within a consolidated normative framework.[13] Moreover, after such reforms, rural development of agricultural products has begun to play a significant role. The impact of the reform was strong on all rural activities and the effects were amplified by EU enlargement towards Eastern Europe. Since 2003, the "policy of decoupling" aimed at providing measures to sustain farm incomes rather than agricultural production. These policies considered agricultural and food differentiation to be one of the main elements for the competitiveness of EU firms.[14]

During the last decades, different normative interventions have been introduced into the EU in order to allow agri-food firms to differentiate the quality attributes of their products. Some rules are mandatory and other voluntary. While the "milk quotas" remain into force, the main normative

---

13 Concerning the Agenda 2000 Reform and the Fischler Reform and the new perspectives for the CAP for the early 21st century see: Loyat J., Petit Y., *La politique agricole commune (PAC): une politique en mutation*, Paris, La Documentation Française, 2013; Cunha A., Swinbank A., *An Inside View of the CAP Reform Process*, *op. cit.*, chapters 6–11; Sorrentino A., Henke R., Severini S. (eds.), *The Common Agricultural Policy after the Fischler Reform. National Implementations, Impact Assessment and the Agenda for Future Reforms*, Farnham, Ashgate, 2011; Swinnen J., "The Political Economy of the Most Radical Reform of the Common Agricultural Policy", in *German Journal of Agricultural Economics*, 2010, No. 59, Supplement 2, "Perspectives on International Agricultural Policy – In Honor of the Retirement of Professor Stefan Tangermann", pp. 37–48.

14 See ISMEA, *L'impatto della riforma PAC sulle imprese agricole e sull'economia italiana*, Milan, Angeli, 2004; De Castro P., *Verso una nuova agricoltura europea. Quale politica agricola nella UE allargata?*, Rome, Agra, 2004.

interventions concerning product diversification (which also apply to the dairy sector) relate to: (a) recognition and protection of product origin; (b) certification of the environmental attributes of products; (c) valorization of the nutritional characteristics of food products; (d) the traceability of products. These rules involved the Italian dairy products specifically, which had to face increasing competition from foreign producers and unfair practices relating to imitations of Italian dairy products.

With regard to product origin, Regulations (EU) 2081/92 and 510/2006 introduced three voluntary quality schemes for geographical indications and traditional specialties. Such EU certifications are: PDO, PGI and TSG (Traditional Specialty Guaranteed). Each of these aims at safeguarding and promoting the names of agricultural products with specific quality attributes. The PDO covers those products which are produced, processed and prepared in a given geographical area. Moreover, this quality scheme guarantees that the certified food products use well-recognized know-how. The PGI scheme refers to less stringent rules for product certification. To obtain this quality standard it is sufficient that at least one of the stages of production, processing or preparation refers to the geographical area involved within the technical specification. The TSG does not refer to the valorization of product origin. Such a label aims at promoting the traditional characteristics of food products, that is, their composition and/or means of production. The majority of PDO/PGI are concentrated in Italy and France. Thus, such schemes constitute a strategic source of differentiation for the agri-food products of such countries. Moreover, the dairy sector is one of the most important sectors in terms of PDO/PGI certifications. In 2014, in EU countries there existed almost 1,250 PDO/PGI products: one sixth of them were represented by dairy products, cheese in particular.[15]

Another EU normative intervention relating to product origin is Regulation (EU) 1151/2012, which refers to schemes for agricultural products and foodstuffs. Among the rules presented by this Regulation to promote the quality characteristics of food products, the "mountain product" quality scheme has been also introduced. Such a label can be used when both product raw materials and the feedstuffs for farm animals come mainly from mountain areas. In the case of processed products, like dairy products, the processing also takes place in mountain areas. As many dairy products are partially or totally produced in mountain areas, this quality label can represent a good instrument with which to differentiate the quality characteristics of these products.

In Italy, it is possible also to distinguish between conventional and traditional agri-food products. Traditional products have been officially recognized by the Italian Parliament with Law 173/98. This national rule recognized a national list at the Ministry for Agriculture, Forestry and Fishing,

15 See De Castro P., "Prodotti DOP/IGP in Europa: presente, futuro, fattore di competitività", in Nardone A., Piva G. (eds.), *Prodotti a denominazione di origine. Fattore di competitività e qualità: i formaggi*, Florence, Accademia dei Georgofili, 2015, pp. 11–12.

which is updated annually by the Regions. The traditional agri-food products relate to those products which have not been recognized at EU level as PDO, PGI or TGS. They have limited production and a high cultural and culinary value, and their processing methods are characterized as "traditional" (i.e. the processing phase must be stable for a period of at least 25 years). In 2014, 4,813 traditional food specialties were present within Italian territory. A great number of dairy products (in particular cheese) are reported in this list.[16]

Concerning the certification of environmental product attributes, the main EU normative contribution relates to Regulation (EU) 834/2007 on organic production and the labeling of organic products, which repeals Regulation (EU) 2092/91. The organic labeling scheme refers to a way of producing food that minimizes the human impact on the environment and operates as naturally as possible, in accordance with the following principles. First, crops are rotated in such a way that on-site resources are used efficiently. Second, chemical substances are severely restricted. Third, disease-resistant plant and animal species adapted to the local environment are used to protect agricultural production. Fourth, genetically modified organisms (GMOs) are banned.

Supplementary conditions concern the use of the EU organic logo. First, products must be produced mainly from ingredients of agricultural origin. Moreover, during the processing phase firms can use only those ingredients which have been authorized by Regulation (EU) 834/2008.

The norm on organic production is under reconsideration, as the quantity of certified food products has increased significantly over the last ten years. Also, the dairy sector is involved in this trend towards environmentally sustainable production. In 2011, there were 0.7 million certified organic dairy cows in the EU, which account for the 3% of all EU dairy cows. In the EU-15, the organic sector represents 3.7% of the whole dairy sector. The Member States where the organic sector holds the largest share are: Austria (18%), Sweden (12.7%), Denmark (10.9%) and the UK (8.1%). Moreover in France, the second largest EU dairy producer, the number of organic dairy cows toward the total dairy cow herd stands at 2.1%.[17]

With regard to the nutritional attributes of dairy products, the most important normative intervention refers to Regulation 1924/2006 on nutrition and health claims. This Regulation aims at improving consumer awareness on food choices through the introduction of standardized information relating to nutrition and health attributes to be put on food packaging. More precisely, it has introduced a mandatory scheme for using nutrition and health claims for food products, requiring specific homogeneous claims in line with the

16 See Istituto Nazionale di Economia Agraria, *Italian agriculture in figures 2014*, Rome, Inea, 2014.

17 See European Commission, *Inventory of certification schemes for agricultural products and foodstuffs marketed in the EU Member States – Data aggregations*, Brussels, European Commission, 2010.

nutrition properties. Regulation (EU) 1924/2006 provides both a list of permitted nutrition claims and the relative conditions to be followed for using these cues. There exist different claims identified by the Regulation, which can be grouped into several categories. Any nutritional claim which is not provided for by the Regulation may not be put on labels.[18] On the basis of Regulation (EU) 1924/2006, dairy products can also take advantage of many intrinsic function characteristics which are communicated to consumers through health claims. Such attributes relate to the presence of calcium, probiotics, prebiotics, and low saturated fat and sodium content.

With regard to product traceability, EU legislation introduced mandatory traceability for all food products, including feed, food-producing animals, and any other substance incorporated into a food at any stage of production, processing and distribution. On the basis of the general Food Law, Regulation (EU) 178/2002 provides one of the most complete normative frameworks for food traceability.[19] On the basis of this Regulation, food firms must identify from whom and to whom an agri-food product was supplied. No rules regarding the kind of system to adopt are mentioned by Regulation (EU) 178/2002. This indicates only the information which must be recorded by food operators within the food supply chain. More precisely, the traceability standard introduced by this Regulation implies the implementation of a traceability system established at all stages of production, processing and distribution. Moreover, it obliges business operators to be able to identify any person from whom they have been supplied with any substance that has been incorporated into a food product. Finally, business operators must have in place systems and procedures to identify businesses to which their products have been supplied. The new traceability requirements specify the results that have to be achieved, but do not prescribe how such results can be reached. This allows a certain flexibility to business operators in the implementation of these requirements.

In addition to mandatory traceability, also voluntary traceability standards can be applied within the EU territory. Such standards offer firms the ability to deliver more information than that provided by the EU normative framework. There are several voluntary traceability standards. A

---

18 The claims identified by the regulation which can be grouped into the following categories: energy (low energy, energy-reduced, energy-free), fat (low-fat, fat-free, low-saturated fat, saturated-fat-free), sugar (low sugar, sugar-free, with no added sugar), sodium (low sodium/salt, very low sodium/salt, sodium-free or salt-free), fibre (source of fibre, high fibre), protein (source of protein, high protein), vitamin (source of vitamins and/or minerals, high vitamins and/or minerals), nutrient (contains nutrients or other substances, increased nutrients, reduced nutrients), light and naturally/natural. See Stranieri S., Baldi L., Banterle A., "Do Nutrition Claims Matter to Consumers? An Empirical Analysis Considering European Requirements", in *Journal of Agricultural Economics*, 2010, No. 61, pp. 15–33.

19 See Charlebois S., Sterling B., Haratifar S., Naing S.K., "Comparison of global food traceability regulations and requirements", in *Comprehensive Reviews in Food Science and Food Safety*, 2014, No. 13, pp. 1,104–123.

study conducted by European Commission in 2010 has revealed the presence of 145 food traceability standards. The main countries represented are Spain, Italy and Germany. In Italy, such rules are mostly implemented by retailers for the management of private label supply chains and by institutions for the development of local economies through the implementation of systems aimed at differentiating traditional food products.

Voluntary traceability standards can differ in terms of the target actors involved (B2B or B2C certifications), the types of product involved, the schemes' owners (private, sector organization, NGOs, etc.), the standard legal validity (national or international), and the level of the system complexity adopted (supply chain or supply chain and product traceability). With regard to the level of system complexity, it is possible to distinguish among traceability standards in terms of the degree of system batch dimension (degree of accuracy of product identification), the amount of information to be recorded for the standard compliance, and the sectors traced.[20] The dairy sector is one of the most important sectors in terms of number of voluntary traceability standards implemented, as this is evidenced by third-party certifications owned by NGOs. Now, ISO 22005: 2007 is one of the most popular traceability standards present in the dairy sector.[21]

Such voluntary traceability can be considered not only as a tool with which to provide specified safety and quality levels during the different phases of food production, but also as a way of ensuring better management of information flows along the supply chain and, therefore, as a useful means of achieving a higher level of organizational efficiency between the economic subjects involved in dyadic vertical relationships within the food supply chain. Via the empirical analysis they conducted, a reduction in information asymmetry among partners, a better allocation of specific liabilities to the main actors within the chain and an increase in the degree of confidence were observed.[22]

These results are due to a variation of the characteristics of the transactions, that is, a reduction in the level of uncertainty and an increase in the degree of idiosyncrasy of vertical exchanges, leading to a more responsible and efficient coordination within the sector. Such changes lead also to a reduction of transaction costs, in terms of information costs, negotiating and monitoring after the implementation of a voluntary traceability standard.

---

20 See Golan E., Krissoff B., Kuchler F., Calvin L., Nelson K., Price G., *Traceability in the US food supply: Economic theory and industries studies. Agricultural Economic Report Number 830,* Washington DC, United States Department of Agriculture, 2004.

21 See Banterle A., Stranieri S., Baldi L., "Traceability and vertical co-ordination in the Italian dairy chain: A transaction cost approach", in *Journal on Chain and Network Science,* 2006, No. 6, p. 69–78.

22 See Banterle, A., Stranieri, S., "Sustainability standards and the reorganization of private label supply chains: A transaction cost", in *Sustainability,* 2013, No. 12, pp. 5,272–288.

4. Conclusions: European institutions' choices and policies strongly influenced (and will influence) the evolution of the Italian dairy sector

European institutions' attitudes towards the dairy products influenced the evolution of the Italian dairy sector. Since the early 1960s, European institutions, in fact, have supported and regulated the activity of all European farms and dairy factories. The relevance of the European institutions' action was clearly confirmed on 15 June 2015. A few weeks after the end of the "milk quotas era", the Committee on Agriculture and Rural Development of the EU Parliament called on the European Commission to create new rules for the European dairy market. In its report on prospects for the EU dairy sector, the review of the implementation of the Dairy Package (2014/2146(INI)), the Committee underlined the fact that the EU dairy sector (obviously including the Italian one) suffered the effects of the negative trend following the financial crisis of 2008, which was made worse by the Russian embargo. In order to help the dairy sector face up to the reduction in price of its products, the Committee asked for new measures for the maintenance of a sustainable dairy sector in disadvantaged, mountainous, insular and most remote regions. Moreover, it invited the European Commission to establish new aid to facilitate on-farm investment and modernization. In terms of the financing of young farmers, the EIB was called upon to use new specific funds to renew the dairy sector.[23] The Committee also added that global dairy demand was predicted to grow and offer opportunities for European products, but, at the same time, it stressed that all export opportunities had to be balanced by a stable domestic market which was increasingly dominated by dried dairy products. Besides, the Committee underlined that small-scale producers were too vulnerable in the supply chain and that all UE member states had to protect farmers from

[23] See in particular point 14 of the report inviting the European Commission "to establish new financing opportunities for Member States, including with the aid of the European Investment Bank (EIB), by means of which the dairy industry will be reformed; considers financial support, such as guarantee funds, revolving funds and investment capital, to be essential, along with resources provided by the EIB, in order to intervene at the level of structural and European investment funds, in particular in harmony with rural development; this would enable a multiplier effect to be achieved in terms of growth and income, as well as facilitating access to credit for dairy farmers; welcomes in this regard the financing opportunities presented to farmers in the dairy sector by the EIB's new fund, which offers lower interest rates to facilitate on-farm investment and modernization, while offering financing opportunities to young farmers to grow their businesses; further highlights the complementary nature of the European Fund for Strategic Investments financing, which would help to develop the milk sector by attracting private capital with a view to expense accountability and increased investment effectiveness". For the whole report of the Committee on Agriculture and Rural Development of the European Parliament see: <http://www.europarl.europa.eu/sides/getDoc.do?pubRef=-//EP//TEXT+REPORT+A8-2015-0187+0+DOC+XML+V0//EN>.

the abuse of industry and distributors and their dominant position in the retail market.[24]

This chapter has shown that the EU Parliament was conscious of the relevant influence on the European dairy sector of the European institutions' choices and policies. Through the publication of the report of the Committee on Agriculture and Rural Development, the EU Parliament invited the European Commission to modify the rules governing the European dairy sector and the process of creating the prices for its market. Moreover, it stressed that it was necessary to change the European Commission's strategies for the development and the future of the European dairy sector. All these recommendations obviously concerned the perspectives of Italian dairy producers too.

24 See in particular points 54 and 74 of the report. In the first, the Committee "points out that global dairy demand is predicted to grow by 2% per annum, offering opportunities for products of EU origin, but stresses that these export opportunities must be balanced by a stable domestic market, the latter representing more than 90% of the dairy product market in Europe; notes, however, that the market is increasingly dominated by dried dairy products". In the second the Committee "stresses that dairy producers, especially small-scale dairy producers, are particularly vulnerable to imbalances in the supply chain, in particular owing to fluctuating demand, rising production costs and decreasing farm-gate prices, but also to the economic priorities in each Member State". The Committee also "considers that the downward pressure on prices by retailers from own-brand labeling and the persistent use of liquid milk as a 'loss leader' by retailers undermines the work and investment of producers in the dairy sector and devalues the end product for the consumer". Furthermore the Committee defends "the need to introduce codes of good practice among the various participants in the food supply chain". It finally stresses "the need to find mechanisms that effectively protect farmers from abuse by industry and distributors and their dominant position in the retail market, and asks the Commission to present its proposal on the containment of unfair trading practices as soon as possible and to consider a sector-specific approach to competition law and unfair trading practices". In this regard see *ibid.*

The Canadian Dairy Sector in the Twentieth Century

From Openness to Closure and Self-Sufficiency

Benoit Mario PAPILLON

Université du Québec à Trois-Rivières

1. Introduction

For a long time, dairy products have been an important part of human alimentation in Europe and elsewhere. During the Neolithic period, sometimes represented as the first economic revolution, a wide proportion of humankind switched from hunting to agriculture and "the domestication of animals [...] added dairy products to the human diet",[1] making dairy products an accessible source of protein and, indirectly, of filtered water. Unprocessed milk has a short life and a low ratio of *value over weight*, making transportation costly. Up until the nineteenth century, this was the case with other dairy products, and, historically, consumption was satisfied from local production.

With the onset of industrialization and innovations in land transportation and communications, a long-term process of reductions in natural barriers to trade was initiated; two centuries later, it is still going on. Dairy products were not an exception, especially with advances in the cold cycle process from the late nineteenth century. Although the variety of dairy products was limited locally, globally there was some diversity; for instance, yoghurt was commonly consumed in some areas and much less known in others. Besides the reduction in transportation costs, mass migrations, particularly to the Americas, have increased dairy product diversity locally and stimulated demand for and trade in other dairy products. Canada has been part of these transformations.

Consumption of milk and butter directly from the farm had practically vanished in Canada by the early twentieth century because of urbanization, factory methods in butter and cheese-making, and disease controls. This

---

1 Graaf M., Kenwood A., Lougheed A. (5th edn), *Growth of the International Economy, 1820–2015*, London and New York, Routledge, 1994, p. 7.

made dairy farm income dependent upon decision makers further down the value chain, in the carrying, processing and distribution of products. Processing ranges from the pasteurization and conservation, with methods like UHT, of table milk to the production of yoghurt, cheese, ice cream, cheese cakes and other foods with a substantial dairy content. Decisions at one level have cost and income implications at other levels, upstream or downstream of the value chain. This dependency can be conflictual; sector performance requires stable relations between actors at various levels of production and distribution.

Traditionally, farmers have received greater public attention than people at other levels of the value chain. For a long time, the majority of the population was rural. Furthermore, productivity gains since the second agricultural revolution and industrialization have created a lasting state of excess supply of rural land and rural labor inputs in many regions of the world, placing downwards pressures on rural and farm income. Among factors differentiating the organization of agricultural sectors, there are various institutional factors along with precedents regarding adjustment policies for farms and rural areas.

The first part of the chapter presents the evolution of the Canadian dairy sector and its participation in international trade. Transactions and trade are based on rules, from the simple social norm of exchanging from free will rather than grabbing by force to more complex rules on origin, product characteristics, prices, etc. The second part summarizes the rules governing the dairy sector in Canada, known as the *supply management* system. These rules include stiff import restrictions. The third part analyses dairy product accessibility in Canada from a general perspective, as well as in terms of prices; it singles out the Canadian dairy sector among other food sectors in Canada, as well as among other comparable countries. The last part concludes.

2. The Canadian Dairy Sector: Production and Trade

Overview: Farms, Processing Plants and Product Composition

In 2013, the annual report of the Canadian Dairy Commission reported 12,231 dairy farms, down to 11,683 in 2016, for a country herd slightly above 959,000 cows, which put the average farm size at around eighty cows. Butterfat is the main unit used in Canada to measure milk production, and for the year 2013–14 the total production was 313 million kilograms of butterfat; measured in milk units, this represents 78.2 million hectolitres.

Figure 1. Dairy Farming in Canada (2016)

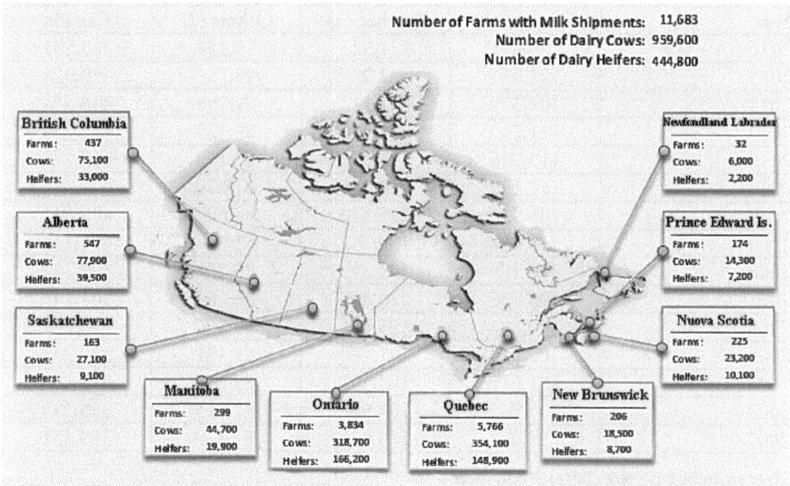

Source: Government of Canada at <http://www.dairyinfo.gc.ca/images/farm_mapofcanada_eng.jpg>.

Milk is either consumed after pasteurization as liquid (fluid milk) or classified as industrial milk and processed into butter, cheese, yogurt or other dairy products. Fluid milk accounts for a little more than one third of dairy farms' production, more precisely 37% in 2013–14; the other 63% share of production goes into industrial milk. This has been fairly stable over time with a slight move towards industrial milk; its share in 1990, for instance, was 61%.[2] The use of industrial milk has also changed over time, with the share of butter going down significantly and the share of cheese, more specifically variety cheese (Brie, Gouda, etc.), and yoghurt going up. For instance, in the 1950s, the butter share was 75%;[3] in 1990, it was down to less than 50%.[4] The growing popularity of variety cheese can be traced back to the 1980s; from 1980 to 1990, the production of variety cheese had increased by 80% while the production of other cheese was almost stable.[5] Roughly speaking, one kilogram of cheese will require, on average, a little more than 10 liters of milk. Canadian cheese exports have become marginal since the middle of the twentieth century and this rise in cheese production is a rise in domestic consumption. The current geographical distribution of cheese production in Canada reflects the historical concentration of dairy processing in the Province of Quebec.

---

2 Canadian International Trade Tribunal (CITT), *Rapport descriptif du personnel Une enquête concernant la répartition des contingents d'importation*, Ottawa, 1991.

3 Ministère de l'Agriculture, des Pêches et de l'Alimentation du Québec (MAPAQ), *Chronologie de l'industrie laitière du Québec (1608–1992)*, Québec, 1992.

4 CITT, *Rapport descriptif du personnel, op. cit.*

5 *Ibid.*

Figure 2. Production Cheese by Province: 2002–2016 (000 kg)

Year	Ontario (1)	Quebec (1)	Others (2)	Canada
2016 (JA – JL)	79,401	121,923	59,257	260,581
2015	122,061	222,070	95,797	439,928
2014	107,329	241,327	89,798	438,454
2013	111,792	224,333	89,425	425,550
2012	115,246	213,435	90,494	419,175
2011	112,898	210,276	81,223	404,397
2010	108,936	215,110	84,601	408,647
2009	110,093	208,753	85,327	404,173
2008	104,109	209,399	86,254	399,762
2007	103,738	210,807	86,604	401,149
2006	100,881	195,861	90,195	386,937
2005	102,014	187,780	89,451	379,245
2004	101,612	180,551	89,642	371,805
2003	95,781	182,933	87,515	366,229
2002	99,017	178,855	81,479	359,351

1 Not including cottage cheese production.

2 Includes the production of cottage cheese for all of Canada.

Source: Statistics Canada at <http://www.dairyinfo.gc.ca/pdf/prod_cprov_e.pdf>.

Calculations done by AAFC-AID, Market Information Section.

As with other agricultural productions, the number of dairy farms in Canada has been decreasing significantly over the long term: in the early 1990s there were 33,000, almost three times the number for 2013, in 1981, 55,000, in 1971, 123,000, and in the mid-1950s the number of dairy farms was estimated at 400,000.[6] Further back in time, the typical dairy farm specialized less wholly in milk, combining other market productions, as well as non-market productions directly addressing some household needs.

The Canadian Dairy Commission annual report estimates there were, in 2013, 478 dairy processing plants, employing 22,700 people. There is a significant degree of vertical integration in farming and processing; it goes from the farm level to the processing level, with cooperatives of dairy farmers owning processing plants. In the early 1990s, it has been estimated, cooperatives of dairy farmers owned approximately 50% of the processing facilities, another 35% was owned by large industrial businesses, and the other 15% was owned by small businesses. It was also found that cooperative ownership was particularly present in the industrial milk segment of the processing industry.[7]

6 *Ibid.*

7 *Ibid.*

Determinants of Production: Trade, Innovation and Demography

Dairy farming in Canada can be traced back to the arrival of French colonialists at the beginning of the seventeenth century; from the middle of the nineteenth century, under the influence of colonialists from the British Isles, dairy farming starts to develop into a commercial activity.[8] Calories of food energy produced gives a measure of the importance of milk in farm production, and it has been estimated that it has more than doubled over the period 1850–1910, going from 10% in 1850 to 25% in 1910, and to around 35% in 1970.[9]

Canada is a wide country with a low population density, and, given the low *value on weight* ratio of fluid milk, local needs have generally been satisfied from local production. On the other hand, and partly for historical reasons related to international trade, dairy farming for industrial milk has been concentrated in the center of the country, especially the Province of Québec. For instance, in the last decade of the nineteenth century, out of the 2,279 factories producing butter and/or cheese reported for all provinces of Canada, more than 2,055 were located in Québec or in Ontario, with the larger proportion, 1,485, located in Québec.[10] Although Canada is a federal entity, widely decentralized with provincial states enjoying substantial autonomy in numerous domains, including inter-provincial trade, the geographical specialization of central Canada in industrial milk is still observable today, the Province of Québec accounting currently for more than 45% of Canadian industrial milk production, twice its share of the Canadian population.

A notable event in the history of the Canadian dairy sector is the Reciprocity Treaty between, on the one hand, the USA and, on the other, Great Britain and the British North American colonies, made up at the time of Ontario (Upper Canada), Québec (Lower Canada), Nova Scotia, New Brunswick and Newfoundland, signed in 1854 and effective until the end of the American civil war in 1866. The political motivations for it ranged from conflicts regarding fisheries on the Atlantic seaboard to Southern States' perceptions that such a treaty would make less likely a political union between the Northern States and other non-slave continental regions.[11] Although somewhat remote from agriculture in its motivations, the Treaty contributed to the early transformation of the Canadian dairy

---

8 MAPAQ, *Chronologie de l'industrie laitière*, *op. cit.*

9 Isbister J., "Agriculture, Balanced Growth and Social Change Central Canada since 1850: An Interpretation", *Economic Development and Cultural Change*, 25 (4), 1977, pp. 673–97, cited in Marr W., Paterson D., *Canada: An Economic History*, Toronto, Gage, 1980, p. 110.

10 *Census of Canada 1901*, Vol. II, cited in Armstrong R., *Structure and Change: An Economic History of Quebec*, Toronto, Gage, 1984, p. 198.

11 "The Reciprocity Treaty", in Easterbrook W., Aitken H., *Canadian Economic History*, Toronto, Macmillan, 1956, pp. 361–5.

sector. Annual butter exports from the Province of Québec to the USA from 1861 to 1865 oscillated at around one million kilograms, while, in quantity, cheese exports never went beyond 15% of the amount of butter, barley, oat and wool also exported, making the 1860s a period of transition from self-sufficiency agriculture to market specialization for the province.[12] It is too early, however, to talk about factory production in dairy; this would come with the innovations of the ensuing decades, including for instance the cream separator, usable on the farm and which reduced transportation costs for creameries. Another example is steam power in shipping; this made marine transport faster and, in combination with refrigeration, facilitated the shipping of dairy products over long distances.

Commonwealth tariff preferences in the late nineteenth century were a big stimulus for dairy farming and processing in the Province of Québec. As a result, dairy products became, between 1870s and 1914, one of the two main agricultural exports from Canada, parallel with grain and grain products and before meat and live cattle; in some years during this period, it was even the most important export category, before grain and grain products.[13] Half a century later, this export performance by the Canadian dairy sector was a memory, and by the mid-1970s the dairy sector had withdrawn to a large extent from international trade. In the previous years, surpluses, which accumulated through government support programs and were sold at a world price substantially below the Canadian domestic price, came to represent a growing share of Canadian exports.

An assessment in 1988, little more than a decade after the introduction of stiff restrictions on imports of cheese to Canada, gives a good sense of the radical change in the trade orientation of the Canadian dairy sector from the beginning to the later decades of the twentieth century. For instance, in 1986, farm dairy production was valued at $3 billion, and dairy product exports were $193 million, a long way behind pork exports to the USA alone of $623 million, representing 30% of farm cash receipts for hog production.[14]

Milk production in Canada had increased almost steadily during the twentieth century up to the 1960s. Aside from fluctuations related to the war years, 1939–45, it was almost a continuous growth, year after year, bringing annual production from 48 million hectolitres in 1920 to 80.1 million in 1961, which means a 70% increase over 40 years. Over the same period, the population went from 8.8 million to 18.2 million, a 110% increase. During the 1960s, annual production fluctuated at around the 80 million hectolitre mark; after 1970, it fell under this mark. The dairy production level of 78.2 million hectolitres reported above for 2013 represents a 2.4%

[12] Hamelin J., Roby Y., *Histoire économique du Québec 1851–1896*, Montréal, FIDES, 1971, p. 194.

[13] Marr and Paterson, *ibid.*, p. 112.

[14] Agriculture Canada, *The Canada-U.S. Free Trade Agreements and Agriculture, An Assessment – TRADE: Securing Canada's Future*, 01–08–88, Ottawa, 1988, 65 p.

decrease from the 1961 level, even though the Canadian population had increased by 93%, going from 18.2 million to 35.2 million. Not only did milk production not keep up with population growth, but both went in the opposite direction, dairy production decreasing slightly while the population was almost doubling in size. While the weight of the 0- to 19-year-old group in the total population has been decreasing in the last few decades, it has kept up in absolute terms; furthermore, the dairy processing industry has substantially increased the diversity of dairy products offered to consumers. Therefore, the reduction in milk production from 1961 to 2013 cannot be accounted for by demand factors. It does not seem that, on the supply side, it can be accounted for by considerations related to productivity either.

Productivity gains translate into an increase in production, since productivity increases reduce unit costs and unit prices, increasing the quantity demanded. Productivity indicators for milk production reveal substantial productivity gains. In comparison to the early 1970s, the production of milk per cow has nearly tripled.[15] Other examples of sources of productivity gains could be given, for instance with dairy farm equipment. In order to account for the 1961–2013 reduction in milk production, other factors have to be taken into consideration. Prices have a central role in the process allocating resources in production, as well as in the determination of consumers' real income and, consequently, in their budget decisions, including choice of products.

3. Process Allocating Resources for the Canadian Dairy Sector

Since the 1970s, milk production in Canada has been organized through quotas defined in terms of milk fat kilograms; quotas determine the amount of milk a dairy farmer can sell. More precisely, one quota is for one kilogram of butterfat per day. Quantity restrictions on the transaction of milk and dairy products are part of a system which has come to be known as the *supply management* system. Although there are notable differences in the way restrictions are defined or adjusted over time, this system was also introduced at various dates in the 1970s and 1980s for the Canadian egg and poultry sectors.

15 For instance, for the 1976–93 sub-period alone, it went from 3,475 kilograms of milk per annum to 5,500 kilograms (GREPA – Laval University, *94-10 Annuaire statistique laitier du Québec 1994*, Ste-Foy, 1994, p. 118).

Components and Operation of "Supply Management" of the Dairy Sector

The Canadian Dairy Commission (CDC) was created in 1966 and became in the 1970s the main administrative body of supply management in dairy, coordinating dairy marketing boards' decisions in various provinces. Among other things, it assesses "changes in demand for milk and dairy products and production of milk", in order to recommend to the Canadian Milk Supply Management Committee (CMSMC), which it chairs, the amount of milk which should be produced in Canada in the coming year.

The CDC 2013–14 annual report summarizes its role in the following terms:

> Since supply management was first applied to the dairy sector, the CDC has been in charge of [...] support prices [...] prices at which the CDC buys and sells Canadian butter and skim milk powder under its Domestic Seasonality Programs and industrial milk quota. Once a year, the CDC sets the support price of butter and skim milk powder based on consultations with industry stakeholders and a study on the cost of production. These prices are used as references by the provincial milk marketing boards to establish the price of industrial milk in each province. Through its many roles and central facilitating function in the dairy industry, the CDC ensures that Canadian dairy farmers receive sufficient revenues from the marketplace to achieve decent living standards and invest in their businesses.
>
> The CDC also monitors national production and demand and recommends the necessary adjustments to the national production target for industrial milk. It ensures that milk production in Canada matches demand from Canadian consumers. It is important to avoid any shortage of dairy products and just as important to avoid production surpluses that are costly to handle, store and market (p. 9).

The computation of the support price, or "Target Price for Industrial Milk", has varied over time. The CDC 1990–1 annual report presents this computation in the following terms:

> The target price is set at a level intended to provide efficient milk producers with a return which will cover their labour, investment, and cash costs related to the production of industrial milk. [...] The cost of producing milk is calculated from data collected by surveys on approximately 350 farms in Ontario, Quebec, New Brunswick, and Manitoba. Thirty percent of the sample, made up of producers with the highest per-hectolitre costs, is eliminated for the purposes of calculating the target price. [...] Since the mechanism measures actual costs on the farm on an annual basis, changes in the cost of producing milk can be reflected promptly in the pricing structure. This is intended to pass productivity gains directly through to consumers.

The 2013–14 report indicates that producer sampling is now made in the nine provinces producing industrial milk: a sample of 216 dairy farms, representative at the Canadian level, was taken. While a small number has been dropped, nine in this sample, there is no indication that the 30% highest cost producers are still dropped from the sample.

On the CDC website[16] the following observations are made, under the heading "Supply Management":

> All industries, from car manufacturers to shoe factories, attempt to strike the most accurate balance between supply and demand. Under supply management, farmers attempt to do the same thing. They manage their production for a certain period so that it coincides with forecasts of demand for their products in the same period. By effectively controlling production, costly surpluses are avoided. Surpluses result when the total market demand turns out to be less than what has been produced. Any excess production generates storage and disposal costs. [...] Managing supply is more complex in agriculture than in many other industries, primarily because of the large number of production units involved. Weather and other uncontrollable factors also affect production cycles. Given the high degree of planning and investment required for efficient production, stability of markets and returns from sales are of paramount importance. [...] In dairy, balancing supply with demand largely implies balancing milk production from all farms with domestic consumption of dairy products. Controlling national production at the farm level is achieved through the establishment of marketing quotas through the framework provided in the National Milk Marketing Plan. Supply management also takes into account certain imports which enter Canada, as well as some production which is shipped to export markets.

Trade Restrictions: an Essential Complement to Supply Management

Shortly after the introduction of the milk production quotas under *supply management*, other restrictions were requested by dairy farmers to limit dairy product imports to Canada. A regime of import quotas was put in place. In the case of cheese, which is a major use of milk available for transformation, a global import quota of 22.7 million kilograms was put in place in 1972, on the basis of the previous years' levels of imports; this quota was revised down in 1975 to 20.4 million kilograms and has not changed since then. The weight of this fixed quota in national consumption has been going down over the years, as total cheese consumption in Canada has kept increasing; it is now under 5%. There are similar import quotas for other dairy products: fluid milk, butter, yoghurt, ice cream, condensed milk, powdered buttermilk.

In the case of ice cream, as well as yogurt, it was only in January 1988 that the product was added to the Import Control List, an instrument used by the Canadian Government to regulate, for various purposes, the entry of products in the country. The import quota originally defined for ice cream

[16] <http://www.cdc-ccl.gc.ca/CDC/index-eng.php?id=3806/>.

was 345,000 kilograms, based again on previous levels of imports, and it has been slightly raised since then to 484,000 kilograms, but it still represents a negligible share of less than one third of 1% of the domestic market. In the case of fluid milk, there is an import quota which the Government of Canada considers "'filled' through cross-border shopping, e.g. Canadian travellers returning across the border to Canada with milk for personal consumption".[17] Milk, mostly from the USA, is commercially imported under the Import Re-export Program.

Dairy products could be imported to Canada in a quantity exceeding the import quotas referred to above, but this is unlikely to happen. As a member of World Trade Organization (WTO), Canada has been asked to convert its import quotas, allowing access to given maximum quantities, into a system of tariff rate quotas (TRQs). Under this system,

> within access commitment and over access commitment tariff items were created and included in the List of Tariff Provisions set out in the schedule to the Customs Tariff to classify these agricultural products. Imports within the quota amount are subject to the low rates of duty of the within access commitment tariff items and imports over the quota amount are classified under the higher rates of the over access commitment tariff items.[18]

For dairy products, the higher rates are between 200% and 300%. For butter, for instance, it is 289.5%, and for yoghurt it is 237.5%. For cheese, it is around 240%. It is calculated on a f.o.b. basis and can represent many dollars per kilogram. The same cheese, falling into the "within access commitment tariff item" category, will have a tariff rate of a few cents per kilogram. Following various trade negotiations, some proportion of global import quotas have been allocated to some specific countries or group of countries. At the GATT Tokyo round, 60% of the Canadian global import quota for cheese was allocated to members of the European Community.[19]

Product Accessibility and the Welfare of Market Participants

While poverty and queues were frequent in centrally planned socialist countries, private sector industries in less centralized countries, for instance North America and Western Europe, are in a nearly permanent state of

---

17 <http://www.thedairysite.com/articles/2750/dairy-trade-flows-between-the-us-and-canada/#sthash.3tu7cf6W.dpuf>.

18 <http://www.cbsa-asfc.gc.ca/publications/dm-md/d10/d10-18-1-eng.html>.

19 If the trade agreement negotiated recently between Canada and the European Union is adopted by the European political authorities, the European percentage share of the Canadian cheese global import quota would increase because of a transfer of 0.8 million kilograms from non-members to members of the European Community and because of a gradual increase over a six-year period of the European share in kilograms. The combined effect could more than double in volume European Union cheese exports to Canada.

excess supply. Monopolistic competition is a general characterization of these industries, in the sense that consumers can choose among a number of sellers eager and ready to sell. This degree of excess supply, a corollary of competition, is an important consideration in policy analysis, especially for the agriculture sector, constrained by short-term inelastic demand.[20] Access to products is not given by the position one has in a queue, but by product price; this price is a main determinant of real income and living standards.

Occasionally, the high price of milk and dairy products in Canada is a public issue. Comparisons with the consumer price index (CPI) are made as a means of justification. CPI is not a valid comparator as it includes services, for which productivity gains have been much less important than for goods, including food items; it also includes sales taxes. Most dairy products, as food items, are not taxed, while other product categories (gas, tobacco, alcoholic beverages, clothing, etc.), accounting for significant shares of CPI, are; besides general tax increases, some categories have also known very substantial tax increases over the last decades. Comparable categories give a different picture.

Within the twenty year period of the 1960s and 1970s, the level of consumer prices in real terms for dairy products in Canada has moved upwards significantly. The time series for these prices indicate large increases, between 50% and 90%, concentrated in the early or middle years of the 1970s. The first panel of Fig. 3 represents fluid milk, and a similar price trend is observable for cheese, butter and "other dairy products". The evolution of consumer prices, in real terms, for the other three main animal product food categories in Canada, beef, pork and poultry, has been quite different, as illustrated by the figure for pork in the second panel (Fig. 3).

20 The farm product will usually account, in value, for a small portion of consumer product from which farm product demand is derived; big changes in farm price will have little effect on consumer prices.

Figure 3. Evolution of the real consumer price (in $1986) and of the consumed quantity (per-capita real expenditures in $1986) of milk and of pork over the period 1961–1988

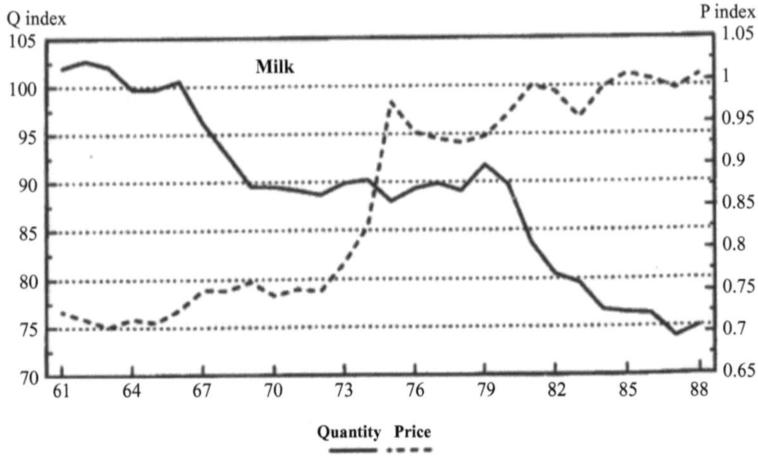

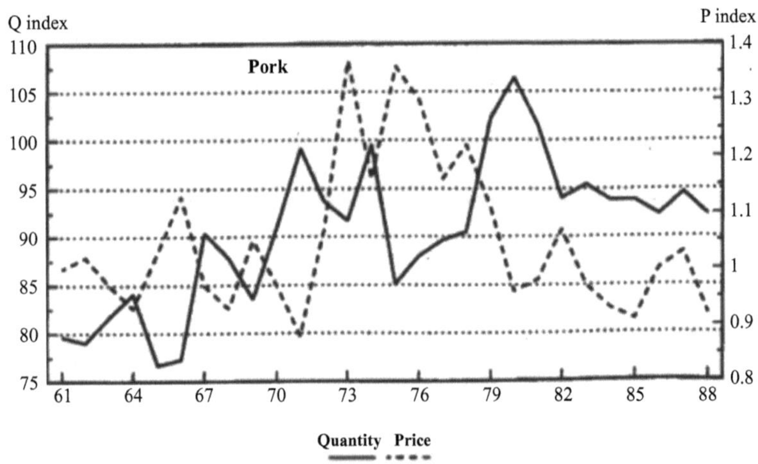

Source: Moschini G., Moro D., *A Food Demand System for Canada*, Ottawa, Agriculture Canada, Technical Report 1/93, 1993, p. 121.

While there were large fluctuations in the 1960s and 1970s, the consumer price in real terms from the beginning to the end of the period increased slightly for beef by around 5%, while it decreased for pork by around 10%, and for poultry by nearly 15%. The quantities consumed evolved in the opposite direction, as expected, given consumers' budget constraints. Therefore, milk and dairy product prices, in real terms, not only increased by a very large magnitude during the 1960s and 1970s, but these prices, relative to the prices of other important sources of proteins in human alimentation, have also increased by a large amount.

The timing of price increases reported above for dairy products coincides with the introduction of supply management in dairy in the early 1970s, eliminating a direct and indirect source of competition for Canadian dairy farmers. As is indicated in the first panel of the figure, and as predicted by economic theory, the quantity of dairy products consumed has evolved in the opposite direction.[21] Chronologically, supply management came on the top of a long series of subsidies and of price support measures stimulating production while increasing consumer prices; the related evolution of quantity produced illustrates an early 1980s conjecture of Paul Josling who compared Canadian dairy policy with the Canadian wheat program:

> In contrast to the wheat program, where both producers and consumers tended to benefit from government programs, in the case of milk the burden rests squarely on consumers. […] the present market imbalance is largely a result of policy. […] over time the level of price support has undoubtedly attracted both labor and investment […] into dairying. Similarly, consumption is clearly restrained by the support price of dairy products.[22]

Price increases imply income transfers from consumers to dairy farmers. These transfers, on an annual basis, have been estimated using data for 1980, less than a decade after the introduction of supply management, at nearly $700 million;[23] more recent figures for the "rental rate of milk production quota per hectolitre", based on data for the period 1994–2009, imply annual transfers above $2 billion.[24]

Milk production quotas can be traded, and their market value gives a measure of the enrichment of dairy farmers, following the elimination of competition and price increases under supply management. Before restrictions started to be imposed in 2006, quota prices could easily be traded above $30,000, which makes the total market value of milk production quotas over $25 billion.[25] Such enrichment is questionable on an ethical basis. As Cairns and Meilke recall, this has been "a controversial aspect of supply management since its inception […] (in the beginning), attempts

21 Among other categories of dairy products, the case of cheese is different, with numerous new varieties of cheese being introduced on the Canadian market during the period.

22 Josling P., *Intervention and Regulation in Canadian Agriculture: A Comparison of Costs and Benefits among Sectors*. Ottawa, Economic Council of Canada – Regulation Reference, Technical Report No. E/I 2, 1981.

23 From table 7, p. 45 in Barichello R., *The Economics of Canadian Dairy Industry Regulation*. Ottawa, Economic Council of Canada – Regulation Reference, Technical Report No. E/I 4, 1981.

24 Cairns A., Meilke K., "Price Ceilings on Milk Production Quota Values: Future or Folly?", *Canadian Journal of Agricultural Economics*, 60, 2012, pp. 93–112.

25 As a straight approximation, total annual production for 2013–14 is divided by 365, since the production quota is for one kilogram per day, and the result is multiplied by $30,000.

were made to keep the value of quota at zero – that is, what the original recipients paid for it".[26] Among other related issues:

> new entrants are essentially blocked from starting dairy operations as a result of the high quota prices, effectively creating a gentry of dairy producers and their families [...] the nonsupply managed sectors of agriculture have actively criticized supply management claiming that the system permits dairy farmers to leverage their quota holdings in order to purchase land at prices cash crop farmers cannot afford.[27]

Some provinces, for instance Ontario and Quebec, have capped milk quota price at $25,000. With this cap value, the quotas themselves will still represent around 50% of the total value of a dairy farm.[28] Quota values and related income transfers, while questionable, are not a loss to society as a whole. There is, however, a welfare loss associated with food items of lesser quality that consumers substitute for dairy products. This welfare loss would also include some of the administrative costs of agencies, CDC and Provincial Marketing Boards, involved in the operation of supply management, and some of the $80 to $100 million spent annually by Dairy Farmers of Canada for advertising and promotion.

More recent data on prices do not indicate a reversal of the significant upwards adjustment of the 1970s and 1980s in the real and the relative prices of dairy products. Saputo, a firm of Canadian origin and a large milk processor, pointed out recently that the farm price of milk in Canada is double that in the USA.[29] Beyond exchange rate fluctuations, the consumer price of fluid milk has also been, for many years, twice as high in Canada as in the USA, and if comparisons are made with France the gap is even higher. For instance, a liter of UHT milk bought in a small grocery in downtown Paris for 0.7 euros in October 2015 was selling in a large discount grocery in Trois-Rivieres, a medium sized city in eastern Canada, at the same period, at a price around $2.6, or 1.8 euros, which is 2.6 times the price in Paris; for dairy products like cheese, the gap could be even larger. A likely consequence of the elimination of competition at the farm level is a reduced level of competition further down the value chain, in milk processing and dairy product distribution, which can contribute to the big gap in consumer prices between Canada and other comparable countries.

---

26 *Ibid.*, p. 94.

27 *Ibid.*, p. 101.

28 For instance, in the price of $7 million for a dairy farm of 150 cows recently put on sale in the Province of Quebec, the value of milk production quotas, at a capped unit value of $25,000, was $3.5 million. One way to circumvent the cap value is to condition the transaction on the quotas to the purchase of other assets, for instance the herd and the equipment.

29 Jean-Philippe Décarie, "Saputo pourrait se passer de la gestion de l'offre", in *La Presse* (Business Section), 5 August 2015. The high price of milk in Canada explains partly the expansion of Saputo abroad. Half of his production is now in the USA; it also has production facilities in Australia and Argentina.

4. Conclusion

There are numerous regulations in agriculture as well as in other sectors of activities, from primary sectors to services, including transportation, for instance taxi services; a number of professions are also highly regulated. Suppliers' competence, product quality, health and safety are common justifications for regulation, as a complement to competition, considered sometimes insufficient to bring the best choice of products because of high information costs. Within the regulation universe, there is a subset which aims directly and explicitly at restricting the quantity put out for sale or, as in the case of taxis, the amount of an essential input, cars, which will be allowed. Supply management belongs to this subset.

While strict import barriers, like the cheese quota of 20,412 million kilograms, are more likely under a supply management system, they are not a corollary of it. A supply management system was put in place for the Canadian poultry sector in the late 1970s and 1980s. The evolution of poultry import barriers has, however, been quite different from that of the cheese import quota; over the period 1980–91, for instance, the global import quota for chicken has nearly doubled, from 22.1 million kilograms to 42.7 million.[30] What singled out dairy supply management is its rigidity and allergy to any form of competition; until recently, margarine sellers were still victims of legal harassment by dairy interests in the Province of Québec, although the Supreme Court of Canada had declared in 1948 that margarine was legal.

Because of the level of task specialization in the contemporary world, exchange rates, as well as consumer prices, have become important determinants of living standards. The value of a currency relies on sales to foreigners, either by exporting or by bringing visitors into the country. The Canadian dairy sector has isolated itself over the twentieth century, moving from the status of an export leader to ultimate protection and minimal exports. In a historical perspective, this evolution relates to the conditions under which farm milk is moving to the processing sector.

While the pricing of milk at the farm gate can be, and has been, a source of conflict, the development of farmers' cooperatives active in dairy processing can mitigate these conflicts as transactions on unprocessed milk become internal to vertically integrated organizations. In some countries, such as the USA, particularly in Wisconsin, cooperatives seem to have been strong enough to oppose dairy farmer cartelization and extensive regulation of dairy production and transactions.[31] Times have changed. Agropur, a cooperative

---

30 CITT, *Rapport descriptif du personnel*, *op. cit.*, p. 100.

31 Opposition, in the Province of Québec, between dairy farmers organized in cooperatives owning processing plants and the union of dairy farmers has been part of history for many decades (see MAPAQ, *Chronologie de l'industrie laitière*, *op. cit.* and Vigneault J., "Repères historiques dans l'organisation de la mise en marché du lait", *Le producteur de lait québécois*, 1990, October, pp. 18–20).

gathering 3,500 Canadian dairy farmers, operating forty processing plants in Canada and elsewhere in North America, has now become a major player in the dairy industry; given the protection enjoyed by Canadian dairy farmers, this processing activity could be criticized by other dairy processors in the same way that land purchases by dairy farmers have been criticized by cash crop farmers and other non-supply management sectors of Canadian agriculture. However, for dairy farmers, who are vulnerable given the perishability of their product, large cooperatives as Agropur are an alternative to the sale of milk to pure processors who might have been tempted to take advantage of their vulnerability.

After nearly half a century of very high consumer prices, Canadians, in their diet, have not completely moved away from dairy products; this reveals a very strong preference for these products among the population. As comparisons with France and other Northern European countries indicate, per capita consumption in Canada could, however, be much higher. The growth potential of the Canadian market is enormous if consumer prices are significantly reduced. The realization of this potential is a win-win situation for many: it will raise the quality of alimentation, particularly in low-income groups, as well as generate jobs at the processing and distribution levels of the dairy sector and bring expansion of dairy farming in rural areas. Real income of families will also increase, since dairy products are not a negligible portion of their budget.

With the very high level of protection it enjoys, the Canadian dairy sector has become unique and somewhat isolated. The days when dairy farmers could dictate the trade policy agenda to federal cabinet ministers, as at the time of the Uruguay round, are gone. It is not politically feasible any more, for maintaining their protection, for them to bar greater access to foreign markets for non-dairy farmers and non-agricultural producers of the Canadian economy. As this protection is gradually reduced, Canadian dairy farmers will be tempted to fight for compensatory payments as if the commercial value of production quotas were a vested right – but it never was. Or alternatively, they could choose a pro-active and socially beneficial approach to the realization of the enormous growth potential of their sector.[32]

[32] A pro-active approach could ask for the annual issuance of new milk production quotas, representing a fixed percentage of the existing quotas, conditional on the remittance of these new quotas to existing dairy farmers, proportionally to current holdings. The realization of this growth potential will be demanding in terms of coordination within the dairy sector, between anticipated consumer demand and decision in farming and processing. The Canadian Dairy Commission and provincial marketing boards have a very good understanding of the Canadian dairy sector. They have the knowledge needed to institute and operate price mechanisms as adjustment devices to equilibrate, with spot as well as future contracts, quantity supplied and quantity demanded of milk and dairy components, in a context of continuous changes, rather than constraining themselves in a laborious, inefficient and expensive central planning approach, which requires collecting large amounts of information of questionable reliability, since it is not binding, and then making predictions of equally questionable reliability.

Contemporary Challenges of the Brazilian Dairy Industry

Hildete De Moraes Vodopives

Sorbonne University

Valter Galan

MilkPoint Inteligencia

1. Introduction

Europe and North America have been the largest producers of milk in the world. However, since 2003, an Organization for Economic Co-operation and Development (OECD) report recognized a shift in production leaders to Asia, Oceania and Latin America.[1] Brazil ranks as number one global supplier of beef and as having one of the largest dairy herds in the world; its dairy potential was considered in a PricewaterhouseCoopers (PwC) study as having "the natural resources to dwarf most competitor exporters in any field".[2] Nevertheless, this alleged agribusiness powerhouse remains a net importer of milk, with productivity and quality standards far below international benchmarks. What is missing? In order to understand the competitiveness challenges of Brazil's dairy industry, we will look into different aspects of the country's farm potential and the industry's main milestones. Regulation, technological advancements and international trade are the fundamental elements in this plot.

A number of factors work in favor of Brazil as an agriculture producer. First of all, the association of a temperate climate and abundant water makes the country the largest in terms of arable land, with around 264 million hectares, of which 74% is committed to meadows and pastures. By way of comparison with another important dairy producer: New Zealand's entire territory is 27 million hectares, including mountains, lakes and rivers.[3] Two specific

1 OECD, "Analysis of International Dairy Trade Liberalisation", in *Dairy Policy Reform and Trade Liberalisation*, OECD, 2006.

2 PwC, *New Zealand Agribusiness Opportunities: South America and China – Background Analysis*, 2012.

3 *Ibid.*

aspects concerning the dairy industry reinforce the country's advantage. The climate is a decisive productivity factor, ensuring low-cost, high quality forage production. Another positive factor is Brazil's strong position as a producer of animal feed, in particular soybean meal and corn.[4]

2. Historical perspective on regulation in the twentieth century

Regulation of the dairy industry in Brazil is threefold. It concerns quality, pricing and tax barriers. Regulation of the sector began in 1934, when animal health service regulation improved standards regarding hygiene for milk production on the farm, product transportation and industrial processing. Although there were records of previous legislation, it was at this point that animal products' sanitary certification was implemented. The next legislative milestone was reached on 29 March 1952, when President Getúlio Vargas approved the creation of the Animal Products Industrial and Sanitary Inspection Regulation (RIISPOA). This defined the form of inspection and quality standards, rules for producers and industrializers, marketing conditions, packaging and labeling. For the milk industry, the new legislation made pasteurization mandatory. M.C. Barcellos[5] highlights the fact that the Vargas "populist and authoritarian" ethos formulated strict legislation[6] aimed at the agrarian sector. Moreover, milk was to be sold exclusively through cooperatives, which hurt farmers' profitability.

In the late 1970s, the Brazilian government decided to implement price discrimination. The effect of these policies was dubious and most of the time they were actually facilitating imports and hurting the Brazilian dairy industry's ability to compete internationally.[7] Brazil was entering a period of high inflation. In an attempt to contain these rising prices, the government imposed price regulation on a number of products; this was the case for soy, corn, beef and cotton, but also for milk. Unfortunately, the growing gap between revenues and costs discouraged new investments at a time of technological advancements in agriculture. Rather than improving productivity and building competitiveness, producers' focus was practically reduced to negotiating with central government for better price conditions

---

4 According to PwC, animal proteins are the main driver of the Brazilian leadership in pork, poultry and beef exports.

5 Barcellos M.C., *São José do Campo Bello: povoamento e pecuária*, Instituto Campo Bello, Série Origens – Caderno 3, Itatiaia Brazil, 2012.

6 See for example Decree Law No. 7,449, of 9 April 1945, which provides for the organization of rural life and Decree Law 8127 of 24 October 1945 in Barcellos M.C., *São José do Campo Bello, op. cit.*

7 Martins P., "Políticas públicas fazem mal ao Leite", Milk Point Inteligencia. In *Juiz de Fora*, Brazil, 2013, pp. 3–6.

and import barriers in order to guarantee their margins. A perverse effect was that the industry was not motivated to invest in technology.

An informal market of milk not channeled through cooperatives surged close to milk districts. One explanation for this informal market is that farms sell milk to local consumers at more competitive prices, and also keep some for their own consumption. While circumventing the formal marketing channels, farmers were naturally avoiding taxes. Informality also reflects the inability of some producers to meet sanitary legislation as well as testifying to poor regulatory oversight. This dual marketing system, formal and informal, has hindered productivity in the field, allowing the coexistence of highly professionalized farmers and artisan producers, who operate within outdated technical standards.

The liberalization wave of the 1990s

This highly regulated sector changed only at the beginning of the 1990s, with the shock of the liberalization promoted by President Fernando Collor. Although the long period of price control had had a negative effect on the industry's margins and competitiveness, the sudden overture of markets scared producers. The end of protectionism followed by an overvalued exchange rate exposed local producers to a competitive international environment. The industry had to adapt in order to meet growing demand, larger diversity of products and smaller margins. Despite initial fears, the dairy industry eventually initiated a process of accelerated investment, which resulted in increased productivity, better logistics and improvement of the entire production chain.

Chronic export difficulties, 1990s-2000s

Although the agribusiness sector accounts for approximately 40% of Brazilian exports, the dairy industry has historically had a negative trade balance. The liberalization of the sector, aggravated by the overvaluation of the Brazilian currency, boosted imports. According to a report of the Brazilian diplomatic department, the first significant exports occurred in 1996, when shipments valued at approximately $100 million were sent to more than 70 countries. Condensed milk, which Brazil first exported in 1997, remains the leading dairy commodity, exported to over 50 countries.[8] Contradicting the official point of view, PwC infers that Brazil has developed few export markets, due mostly to the lack of trade agreements.[9] Instead of encouraging exports, Brazil created trade barriers to protect local companies from imports.

8 Itamaraty, "Brazilian Milk Is Successful in International Markets", Brasilia, Itamaraty, n.d.

9 PwC, *New Zealand Agribusiness Opportunities: South America and China – Background Analysis*, *op. cit.*

Dairy coming from Mercosur[10] is taxed at 28%, while products coming from the European Union incur an extra 14% anti-dumping tax and products coming from New Zealand pay 3.1%. Besides, Brazilian productivity is inferior to those of other Mercosur countries, especially in Argentina.[11] As a way of limiting this competition, Brazil established an import quota from Argentina of 3,600 tons of powdered milk per month.

Technology affecting the industry's geography

While regulation and trade barriers "protected" the industry, the private sector was unable to make necessary investments to boost productivity. PwC remarks on poor record of technology adoption in areas such as bovine genetics and disease prevention.[12] However, other important technological advances that would affect logistics and dairy geography were under way. First, there was the bulk milk cooling tank system, which introduced the concept of integrated logistics, with substantial gains in transport productivity. Although introduced in 1976, in the state of São Paulo,[13] the cooling tanks were expensive for small producers and the technology only became operational on a large scale 25 years later. Its implementation was only possible thanks to a combination of factors. According to Rubez, the end of hyperinflation in 1994 forced companies to shift from financial speculation to operational profitability. Reducing costs was a matter of survival. This scenario, associated with the influence of industry leaders such as Nestlé, was the trigger to the dissemination of cooling tanks. Several multinational manufacturers of bulk tanks began their operations in Brazil. Acquisition was also facilitated by credit lines.[14] The cooling tanks adoption at farm level subverted the hierarchy of milk districts. Once cooled at farm level, the fresh milk could be collected every other day, rather than daily. As a result, the transport scale changed. A truck that carried a maximum of twenty 50-liter milk cans was limited to a two-thousand capacity. The truck was replaced by a system made up of articulated trucks, incorporating an eight-thousand-liter truck and a trailer that could carry 12–15,000 liters.[15] Additionally, the cooled fresh milk could be transported much longer distances. Companies

---

10 Mercosur (or Mercosul as it is called in Brazil) is a customs union of four Southern-cone countries (Argentina, Brazil, Paraguay and Uruguay) established under the 1991 Treaty Of Asunción. In 1996, Bolivia and Chile joined as its associate members. Read more: <http://www.businessdictionary.com/definition/MERCOSUR.html#ixzz3wNmx9hJi>.

11 Gomes S.T., "Cadeia agroindustrial do leite no Mercosul", in *Agronegocios e Desenvolv. Econômico*, 1997, pp. 1–30.

12 PwC, *New Zealand Agribusiness Opportunities: South America and China – Background Analysis*, *op. cit.*

13 Rubez J., *O leite nos ultimos 10 anos*, Associação Brasileira dos Produtores de Leite, São Paulo, Brazil, 2003.

14 *Ibid.*

15 MilkPoint Inteligencia data.

from different districts could purchase fresh milk from different regions, whereas before competition had been limited to a smaller area.

Another technology that revolutionized the consumption end was UHT (ultra high temperature) milk. Easier to transport over longer distances, UHT milk facilitated the dairy production expansion to distant regions, such as the Midwest. The markets of São Paulo and Rio de Janeiro, which were previously supplied by pasteurized milk produced in the states of São Paulo and of Minas Gerais, started to buy UHT milk produced in Goias state (Midwest) and the southern states of Rio Grande do Sul and Santa Catarina. These new milk districts started to grow rapidly while "traditional" milk districts, such as São Paulo's main milk production regions, showed difficulties in adapting to this new competitive environment, and consequently lost market share. Another consequence was the shift in market control among cooperatives. The "traditional" milk districts lost their relevance in the milk manufacturing chain. UHT eventually replaced pasteurized milk as the leading type of liquid milk (Fig. 1).

Figure 1. Brazil – Fluid milks share of category total sales

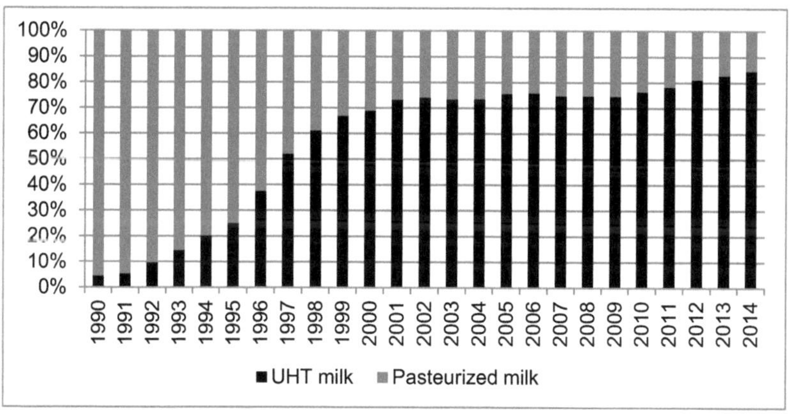

Source: Brazilian UHT Milk Association (ABLV).

The 2000 quality regulation and industry voluntary initiatives

In the interests of understanding the recent regulation on quality, one needs to mention the Milk Quality Program of 2002.[16] The program put together policies to improve milk quality standards and promote milk quality research and better control and inspection services, along with encouraging

16 Regulatory directive No. 51/2002 determined that milk delivery standards measure included: bacteria counts, cryoscopy, acidity, reductase and somatic cells. Standards at point of origin regulated: registered animals, refrigeration facilities, bulk collection centers and mechanical milking.

workforce qualification programs. But initiatives to boost milk quality were not exclusive to the public sector. In a like manner, the private sector was implementing similar efforts. However important these measures were in establishing an institutional parameter, practical results were compromised by the difficulty of controlling Brazil's immense territory. On top of the quality regulation, leading corporations adopted voluntary pricing composition parameters as part of a broader strategy to improve competitiveness.[17] The milk pricing criteria were based on two axes. The first is the criterion of quality parameters. The price varies according to the bacterial count (TBC or TPC – Total Plate Count) and the Somatic Cell Count (Somatic Cells Count). The second parameter relates directly to the percentage content of protein and fat, or milk composition.

Despite some positive results, Brazilian fresh milk quality standards remain far from international best practices, namely the ones adopted in New Zealand, Europe and the USA. As Fig. 2 shows, average results for milk solids in Brazil have been practically the same for the last five years, while New Zealand improves. Milk solids are important because this component has a direct impact on milk powders and cheese manufacturing costs, eventually compromising their international competitiveness. Besides, since the global dairy market is based on these kinds of products, Brazilian competitiveness is compromised at the bottom of the agri-chain, at farm level. The difficulty in improving milk solids' content arises mainly because this depends on bettering cows' nutrition, environment and even comfort, not to mention herd genetics. This point is relevant for the discussion regarding the future of the Brazilian dairy sector and its growth potential.

17 It is important to emphasize that the milk pricing criteria is a specific concept of industry. It differs from those normally used in the economy, where price discrimination can refer to the anti-competitive practices of oligopolies or monopolies.

Figure 2. Milk solids average. Comparing Brazil and New Zealand (Milk solids %)

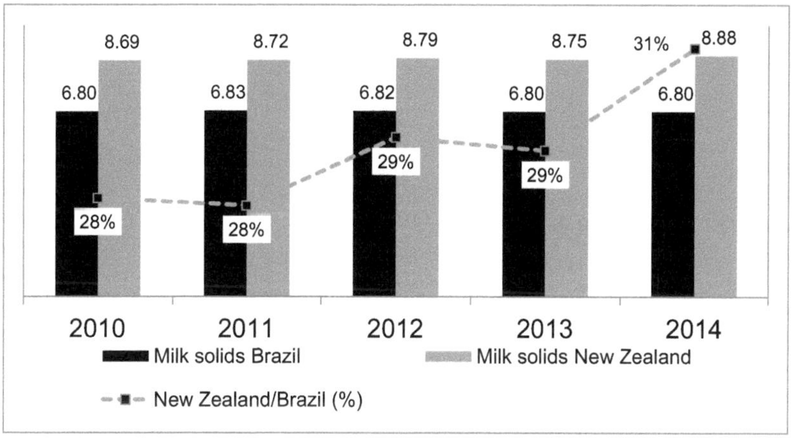

Source: Clinica do Leite (Brazilian results) and Dairy Companies Association of New Zealand.

Demand accelerates, and so does the dairy industry

At the beginning of the twenty-first century, a growing population and increasing per capita income boosted milk demand, making Brazil's production grow by 40% in the last ten years. Milk consumption per person increased from 120 liters per annum to 175 liters.[18] Following this change of scale, demand has been stable, while energy and land costs have increased. Between 2000 and 2008, while world production of cow milk grew on average by 2.1% per annum, Brazil's growth was 4.0% per annum.[19] Fig. 3 shows that, in recent years, fresh milk production in Brazil grew at very high annual rates, compared to the export driven agri-chains: soybean, poultry and swine. From 2005 to 2014, the soybean complex (soybean grains, oils and soybean meal), which accounts for more than 30% of Brazilian agribusiness exports, grew by 5.7%. In the same period, fresh milk fuelled almost exclusively by internal consumption went up by 4.8%.

18 MilkPoint Inteligencia data.

19 Carvalho G., *A Indústria de Laticínios No Brasil: Passado, Presente E Futuro*. Gado de Leite. Circular Técnica 102, Embrapa, Brasilia, 2010.

Figure 3. The Brazilian Agri-chain's Compound Annual Growth Index (CAGR)

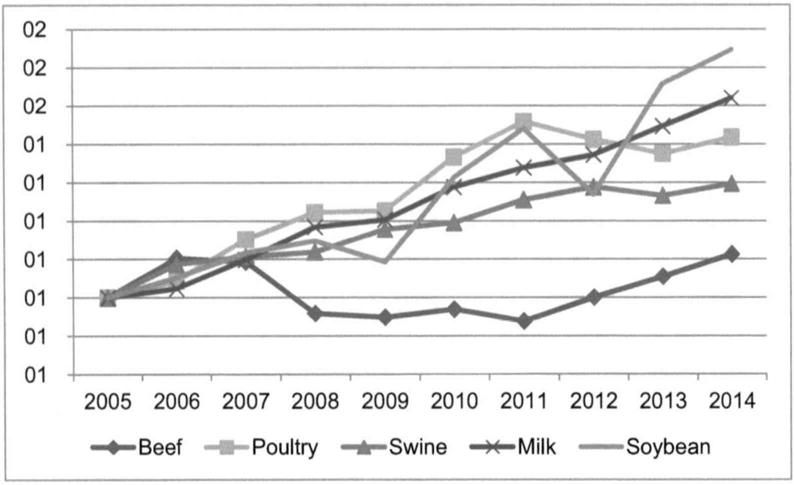

Source: MilkPoint Inteligência, from different sources.

This performance eclipsed other animal protein champions such as poultry (4.0%) and swine (2.9%). It is important to mention that both poultry and swine sectors are controlled by large corporations. BRF (Brasil Foods) accounts for more than 40% of swine and poultry production in Brazil. The agri-chain level of concentration has been a determinant factor in its productivity and profitability. The ability to influence the suppliers' production and specifications allows poultry and swine manufacturers to better adjust to demand.

Dairy's low consolidation level and consequences for the industry's profitability

An analysis of the level of consolidation of the Brazilian dairy industry indicates that the degree of concentration at the manufacturing stage is still lower than in other sectors. Using Leite Brasil data, we estimate that the top twelve Brazilian companies processed around 38% of fresh milk production in 2014, while in 1999 they accounted for 48% of total fresh milk industrialization.[20] Retail companies, however, experienced a very strong consolidation in the same period. In 1999, the top four retail companies accounted for 31% of the sector's revenues, jumping to 65% of retail revenues in 2014 (Fig. 5). One can still observe small family dairy farmers, with low productivity levels and low quality standards, selling their production to small/medium sized dairy industries operating at commodities and low margins markets.

[20] Associação Brasileira dos Produtores de Leite database, <http://www.leitebrasil.org.br/maiores%20laticinios.htm>.

Figure 4. Annual per capita milk consumption in Brazil (liters)

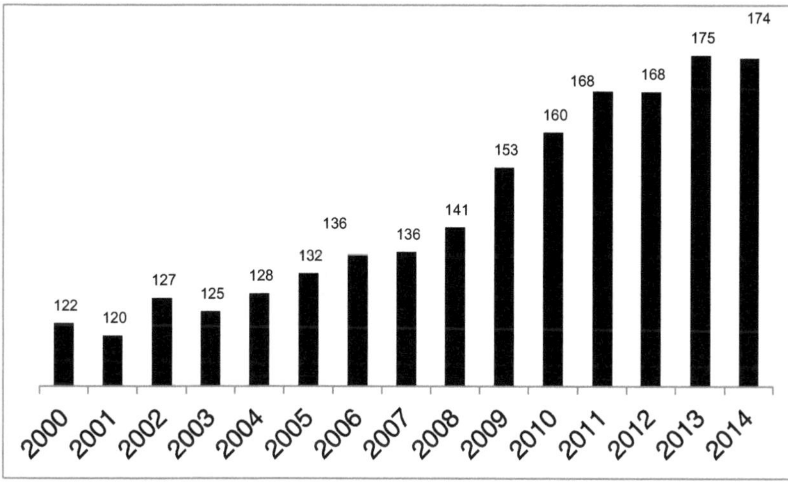

Source: MilkPoint Inteligência, with data from Instituto Brasileiro de Geografia e Estatística (IBGE).

At the same time, the industry has a group of large and professional farmers, who are investing in world-class technology and obtaining higher productivity yields. These large players correspond to 20% of the market and account for 80% of the fresh milk production. In between these two groups, the medium scale farmers struggle to survive. Since their production scale is not high enough to dilute labor costs, their profit margins are always squeezed and their relevance as dairy farmers tends to be reduced. Probably due to a still permissive quality control system and to persisting high fresh milk price levels, the diversity of milk producers can coexist and even grow in the Brazilian market.

Figure 5. Concentration level in the dairy industry and retail channel

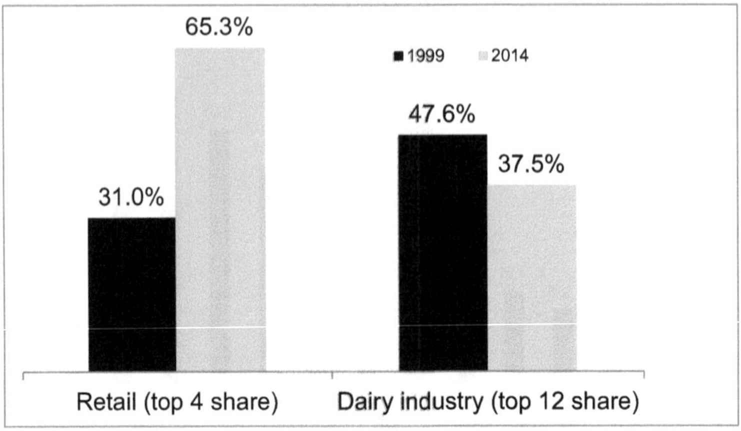

Source: MilkPoint Inteligência, with data from Leite Brasil and ABRAS.

Growing importance of retailers' buying power

Dairy farmers' low consolidation levels contrast with the high level of concentration of the retailers' sector. The result is that retailers' negotiating power has risen significantly in the 2000s, causing low profit margins and low business attractiveness. According to a Rabobank study (Fig. 6), the Brazilian dairy industry's EBITDA margins are among the lowest in the world. The main reason for this is fragmentation level.[21]

Figure 6. Dairy industry margins worldwide

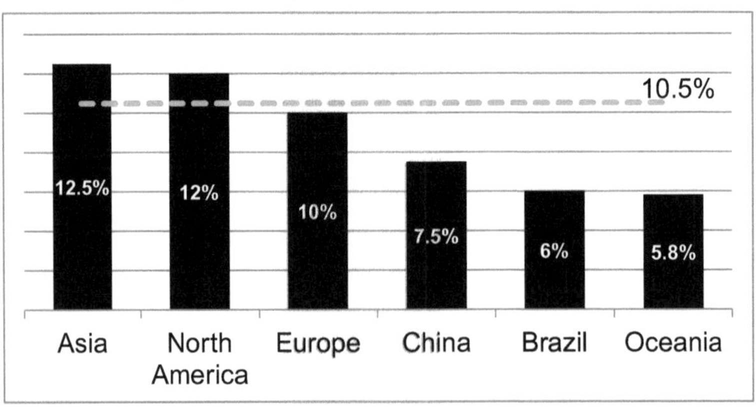

Source: Rabobank Industry Note #467, December 2014.

[21] Padilla A., "Underweight Latin American Dairy. The right strategy for global players?", São Paulo, 2014.

The effect of industry fragmentation: Low coordination capability

The dairy industry's fragmentation translates to what we call the "agri-chain's low coordination capability". In practical terms, it affects the industry quality standards and general competitiveness, simply because it is easier to apply health and tax enforcement to the larger companies while it is almost impossible to control the smaller ones. The higher accountability of corporations such as Nestlé, Danone and Itambé imposes strict standards regarding sanitary and tax issues, standards that are generally more "flexible" in the medium sized or small companies. The other aspect deriving from the industry lack of consolidation is the delay in making adjustments in order to match demand needs. This influences pricing signals, adequate production levels, quality standards, consumption trends, on-farm production technologies and benchmarks. In 2015, for example, instead of reducing production to adjust to a weaker economic pace, small and medium sized regional industries elevated their production in order to make economies of scale. However, they ultimately overloaded the market with milk when demand was not sufficient in the first place. Regarding quality, while the large industries buy milk within the IN62 specifications, the small and medium sized industries suffer from a lack of supervision of quality standards. They operate in a "competitive fringe" supplying price-oriented markets such as areas in the countryside where quality is subject to lower standards. The bottom line is that the whole supply chain competitiveness ends up lower than the international standards.

Figure 7. Milk production in Brazil in 2015

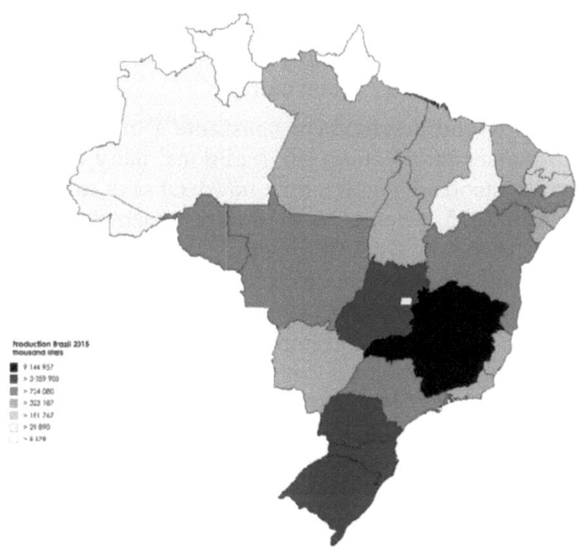

Source: MilkPoint Inteligência, with data from Instituto Brasileiro de Geografia e Estatística (IBGE).

Conclusion

During the twentieth century, the Brazilian dairy industry evolved in terms of regulation and quality standards, but the agri-chain remains below its potential, presenting a low level of international competitiveness. Although the food industry has historically played an important role in the Brazilian economy, dairy fell into a secondary category where policies were not always beneficial to the sector, and regulation eventually became a tool of macro-economic policy. In fact, the Brazilian dairy industry remains extremely strongly protected against international competition.

The role of the public sector in promoting the dairy industry is ambiguous. Looking back, we recognize that dairy has been object of macro-economic policies where the interests of the industry were not a priority. Despite the adoption of the regulation implementing higher quality standards in 2002, fresh milk quality remains below international best practices and Brazilian dairy struggles as a fragmented industry, with inadequate herds and the spread of the informal market. Additionally, government support in research remains timid. The recent technological innovations, such as the cooling tanks, have been led by the private sector.

Observing the past decade, we identify a trend of production growth and demand stabilization. One can presume that Brazilian exports did not grow since 2000 because the priority was the local market, which was blooming. Although production is likely to continue growing at a fast pace in Brazil, consumer spending tends to grow at a slower pace. As a result, there could be an oversupply of milk over the next 5–10 years. The concerns expressed by analysts (MilkPoint) as well as by investors' advisors (PwC) fall into three categories: product quality and development, productivity, and international trade policies. The challenges for 2016 are to address these fundamental issues and compensate for lost opportunities.

Product quality and development constitute a marketing opportunity. According to market information, the traditional dairy categories' (UHT, pasteurized milk, standard yogurts, milk powders) sales volumes contracted by around 5% in 2015, whereas "new" products such as Greek yogurt and lactose-free products showed a 22% growth.[22] This aspect also concerns costs and price reduction. There is consensus in the industry that standard dairy product lines, production costs and retail prices should be managed and reduced, in order to maintain and improve consumption. This issue relates to the whole supply chain and is connected to on-farm productivity and the distribution structure. Productivity is a complex issue. In addition to the low average productivity on the farm, the industry faces other problems in the supply chain. The most important of these are: high manufacturing costs, low quality and short shelf-life of the finished product, and the tax

22 Partial estimates from November 2015.

burden. Industry costs in Brazil remain high, and productivity levels, both per cow and per hectare, are lower than international benchmarks. Finally, tax reductions, along with improvements to the trade agreement strategy, would allow faster access to important foreign markets. Brazil's most sensible alternative would be to export dairy products in a sustainable way, not just as opportunistic exporters.

The complexity of the Brazilian dairy industry allows multiple possibilities in terms of historical research. The role of macroeconomic policies in conjunction with market factors and technological advances represent dynamic forces subject to constant change. Different periods suggest different approaches. The 1990s liberalization started an economic cycle in which the international questions became more relevant from the point of view of study of the industry. The comparison between Brazil and other leading dairy producers is a straightforward one. In this global context, dairy's low consolidation and technology competitiveness are even more relevant. The degree of industry fragmentation versus retailers' high purchasing power can be studied by comparing dairy with different industries' market dynamics. The agri-chain optimization and technological advancements are two other pressing subjects for research. They can be broken down into a series of different categories, including: fresh milk composition, manufacturing economies of scale, and logistics optimization.

Studying the challenges and accomplishments of Brazil's dairy industry in today's competitive environment leads to three main conclusions, which are in fact valid for other industries as well. The first is that natural advantages are not an indicator of automatic success. In other words, the country's ideal climate and abundant water supply are not sufficient to make Brazil an international leader in this industry. The second is that Brazil's population, no matter how huge, is not capable of consuming the country's production. Exploring Brazil's potential ultimately depends on the effective interaction of the private and the public sectors. Companies excel in innovation and marketing when provided with adequate legislation and policies. The case of the Brazilian dairy industry reinforces the necessity for a public sector with a holistic approach and a long-term strategic plan.

New York State Cheese

Nineteenth and Twentieth Century Transformations

Julia Lapp

Ithaca College, Ithaca, New York

1. Introduction

When most people think about cheese production in the USA, New York State probably does not come to mind. Wisconsin, California and Idaho collectively produce more than 50% of American-made cheese,[1] with almost half that total coming from Wisconsin alone. New York ranks third in the nation for overall dairy production and much of that milk goes to the production of cheese, placing New York fourth after Idaho in cheese production. In 2014, 6.9% of US cheese came from New York.

But the current ranking of New York State in the production of cheese today belies the story of the North American cheese industry, the early evolution of which largely occurred in the state. Over a span of approximately a hundred years, between the early nineteenth and early twentieth centuries, New York State was the leading American producer and exporter of cheese. The initial growth of the industry in the state, the industry developments that occurred there, and the later accelerated production in western states provide an informative example of how shifts in technology, tastes and demographics shape food systems.

This chapter provides an overview of the story of cheese production in New York, and by default the USA, spanning the early nineteenth to mid-twentieth centuries.

2. Nineteenth Century Evolution of Cheese Making in New York State

New York lies at the heart of the climatic Humid Continental region of the USA and within the "American Dairy Belt", described by the

1 "Cheese Statistics". *Wisconsin Milk Marketing Board*, 2015, 25 October, <www.http://media.eatwisconsincheese.com>.

nineteenth-century dairy expert Xerxes Willard as "between the fortieth and fifty-fifth parallels of latitude, from the Atlantic to the Mississippi, and possibly to the Pacific".[2] Willard described the region as including "New England, New York, Pennsylvania, the Northern parts of Ohio, Illinois and Indiana, the greater portion of Michigan, Wisconsin, Iowa and Minnesota, and a part of the Canadas". In 1871, when Willard wrote that description from his desk in Herkimer County, New York, only an estimated third of this area was developed for dairying, most of which was in New York State. At that time, New York was the premier cheese-making state in the nation, with more than 800 factories and annual production of 98 million pounds, 65% of the nation's total.[3]

The Erie Canal

Agriculture boomed in New York State after the 1825 completion of the Erie Canal. The canal spanned the length of the state from Buffalo to Albany and provided the nation's first structural transportation system linking the Great Lakes interior with the eastern seaboard. New York became known as the "Empire State" of the USA: the gateway to domestic and international markets as the canal opened channels for the marketing of goods all across New York and into Ohio. At a time before refrigeration, cheese and butter were the primary market products of fluid milk. With the canal, these could now be shipped to New York City, other urban centres, and the eastern seaboard at a fraction of the time and expense.[4] With the promise of healthy markets and means of transportation, more farm families arrived in the state, established dairies and began making cheese. Dairying and cheese production climbed during the 1830s and expanded into western parts of the state and the Midwest, where land was cheap. The first recognized agricultural census in the USA in 1840 shows New York as the leading dairy state with almost 2 million "neat cattle", more than any other state, and "products of the dairy" valued at $10,496,021, slightly less than one-third of the US total of $33,787,008.[5] Cheese-making was established across New York, and in 1846 the canal carried almost 35 million pounds of it to eastern markets.[6]

---

2 Willard X.A., *Willard's Practical Dairy Husbandry*, 2nd edn, New York, D.D.T. Moore, 1872, p. 17.

3 *Ibid.*, p. 20.

4 Bernstein P.L., *Wedding of the Waters: The Erie Canal and the Making of a Great Nation*. New York, Norton, 2005, p. 27; McMurry S., *Transforming Rural Life: Dairying Families and Agricultural Change, 1820–1885*, Baltimore, MD, Johns Hopkins University Press, 1991, p. 100.

5 United States Department of Agriculture, *Compendium of the Enumeration of the Inhabitants and Statistics of the United States*, Washington, DC, Thomas Allen, 1841, p. 359.

6 Durand L., "The Historical and Economic Geography of Dairying in the North Country of New York State", in *Geographical Review*, 1967, Vol. 57, No. 1, p. 31.

At this time, farmers were improving husbandry techniques to optimize the health of cows and enhance productivity.[7] Seed shipments along the Erie Canal indicate changes in cattle feeding, away from cows foraging on their own, towards improvement of pastures through seeding with clover and alfalfa. Emphasis was on quantity of milk production and quality, mostly with the goal of producing cheese. McMurry writes:

> In their deliberations about grass, American cheesemakers lacked the sophistication of their English counterparts. Yet, they recognized that feeding was a key influence on the cheese's quality. [Cheese-makers] were correct to associate clover pasture with rich cheese, for clover provides the nutrients necessary for conversion to milk with high butterfat content. And while Yorkers valued timothy for its high yields of both hay and milk, this plant did not dominate the most-mentioned mixtures. This is important to note, because it indicated that household cheesemakers placed a high value on quality, downplaying the sheer quantity in their plans.[8]

Agricultural and dairy educators, such as Willard, wrote extensively on the importance of husbandry, cleanliness and care in manufacturing high quality products to capture the market. In 1843, Willis Gaylord wrote for the New York State Agricultural Society that "great neatness is required in every department of the dairy, from the milking of cows to sending the products to market".[9] Others, like the award winning cheese-maker Alonzo Fish of Herkimer County, New York, advocated the production of a premier state product.[10] At that time, Herkimer county was the nation's leading cheese-producing county, producing around 10% of the nation's cheese in 1849.[11]

Marketing Farmstead Cheese

By the late 1830s, the state's cheese production was reaching a level that attracted special cheese buyers or "factors".[12] These middlemen travelled throughout the state, visiting cheese dairies to buy directly from farmers, transport the cheese to cities such as New York or Boston, and sell to exporters or cheesemongers. Factors usually visited farms in the springtime, prior to

---

7 McMurry S., *Transforming Rural Life*, *op. cit.*, pp. 27–9; New York State Agricultural Society (NYSAS), *Transactions of the New York State Agricultural Society*, Albany, New York, E. Mack, 1852, Vol. XII, pp. 412–14.

8 McMurry S., *Transforming Rural Life*, *op. cit.*, p. 28.

9 NYSAS, *Transactions*, 1844, Vol. IV, No. 84, p. 119.

10 *Ibid.*, pp. 220–7.

11 Durand L. "The Migration of Cheese Manufacture in the United States", in *Annals of the Associate of American Geographers*, 1952, Vol. 42, No. 4, p. 270.

12 Gilbert B.D., US Department of Agriculture, Bureau of Animal Industry, in *The Cheese Industry of The State of New York*, Bulletin No. 15, Dairy No. 6, Washington, Government Printing Office, 1896, p. 11; McMurry S., *Transforming Rural Life*, *op. cit.*, pp. 46–51.

the production of the year's "make", to negotiate with the farmer for a price based on the market value and the quality of the product. As transportation by canal and rail systems improved, cheese production increased and competition among factors and producers grew with the saturation of local markets. By the late 1840s, factors were beginning to offer "cash on delivery" to farmers for their product. This system placed the majority of risk on the factors and dealers, who sought to enhance the uniform quality of products they sold. There were many variables in the unsystematized "farmhouse" method of cheese-making, that gave unpredictable results. Cash on delivery meant that buyers would have the ability to assess the quality prior to purchase. This increased competion among cheese-makers to produce better products.

Dealers, factors and leaders in the dairy industry continued to encourage farmers to take steps to enhance the quality of their cheese, primarily to appeal to foreign markets, mostly England. In 1844, the cheese dealer T.C. Peters wrote for the New York State Agricultural Society that "Our home markets are already glutted and we must look abroad for an outlet that will carry off the excess of our products", adding that "Cheese has already become a very important item in the list of our exports; and with proper care, all that we can make may find a ready sale in the English market".[13]

The Factory System

In 1850, New York produced almost 50 million pounds of farmhouse cheese, nearly half the US total.[14] But the big advance came in 1851, when a dairyman and cheese-maker named Jessie Williams, in Oneida County, New York, negotiated a plan with his son to pool the milk of both their farms for Jessie to make cheese. As a result, Williams and his son are credited with creating both the nation's first "associated" milk cooperative and its first "cheese factory" for large quantity production.[15]

The Williamses' factory was located along the banks of an Erie Canal tributary for easy shipment to markets and for hydropower (Fig. 1). Its success facilitated wide interest from as far away as Ontario, Ohio and Wisconsin, and Mr. Williams became a respected educator and promoter of the factory system.[16] In 1864, he was elected first President of the newly formed New York State Cheese Makers Association, which today is the New York State Cheese Manufacturers Association.[17]

---

13 NYSAS, *Transactions*, 1844, Vol. IV, No. 84 pp. 238–9.

14 Gilbert B.D., *Cheese Industry*, *op. cit.*, p. 14.

15 NYSAS, *Transactions*, 1868, part II, Vol. XXVII, No. 84, pp. 885–6; Rahmer, F.A., *Jessie Williams: Cheese Maker*, Holland Patent, New York, Steffen, 1971.

16 *Ibid.*, p. 10.

17 *Ibid.*, p. 18.

Figure 1. The first American Dairy in Oneida County, New York State

Source: Jesse Williams' Cheese Factory (circa 1851).

The advent of the factory system brought several advantages to the state's cheese industry and farmers. Primarily, it relieved farmers, mostly farm women, of the laborious task of cheese-making, thereby allowing more time and attention to be given to other aspects of dairying, such as enhanced milk production.[18] It provided a local market for farmers to sell their milk and it consolidated the infrastructures for cheese production, distribution and marketing.[19] The factory system also introduced specialization into the industry workforce; whereas once the farmer had been the milk producer, cheese-maker and cheese dealer, now these were three separate professions.[20]

The number of cheese factories grew throughout the New York countryside and, because farmers could not transport their milk long distances via horse and wagon, factories were built at "cross-roads" and "wherever and whenever there were a few farmers owning 200 or more cows with a radius of two or three miles".[21] Many factories were originally cooperatively owned by farmers with hired cheese-makers, and they were linked to the seasonality of milk production.[22] Later, when significant strides were made in cattle feeding to extend the milking season, cheese factories grew in size.[23]

---

18 Mcmurry S., *Transforming Rural Life*, *op. cit.*, pp. 138–9, 143–4, 152.

19 Gilbert B.D., *Cheese Industry*, *op. cit.*, pp. 36–9; Durand L., "Historical and Economic Geography of Dairying", *op. cit.*, pp. 31–3.

20 Brunger E., "Dairying and Urban Development in New York State", 1850–1900, in *Agriculture History*, 1955, Vol. 29, No. 4, p. 173.

21 Bakken H.H., *American Cheese Factories in Wisconsin*, Bulletin 156, Madison, University of Wisconsin Agricultural Experiment Station, 1930, p. 5.

22 Brunger E., "Dairying and Urban Development in New York State", *op. cit.*, p. 173.

23 Mcmurry S., *Transforming Rural Life*, *op. cit.*, p. 32.

Between 1861 and 1865, 461 factories were built in New York, that produced 72 million pounds of cheese in 1865.[24] By 1870, the census bureau reported 1,313 cheese factories in the USA, with 818 in New York, 195 in Ohio and 54 in Wisconsin.[25]

New York State dairymen also played a role in the rise of cheese factories outside the state. For example, dairyman Chester Hagen of Denmark, New York built Wisconsin's first associated cheese factory in 1864.[26] William Dempster Hoard, called the "the father of Wisconsin dairy", migrated to Wisconsin from Stockbridge, New York during the Civil War and later became a leading promoter of dairying to Wisconsin farmers.[27] Both men contributed to the creation of the Wisconsin Dairyman Association in 1872.[28] Similarly, Canada's first cheese factory was established in Oxford County, Ontario 1864 by a former Herkimer County, New York dairyman named Harvey Farrington, who was instrumental in the development of both Canadian and American cheese export.[29]

By 1866, New York's cheese exports to Britain reached 45 million pounds.[30] That same year, Willard reported that England imported 80 million pounds from Holland and manufactured 179 million pounds domestically. Although competition with producers in western states and Canada was growing for the New York industry, his address to the New York State Cheese Makers Association urged that "Our struggle for supremacy must now be with the European dairymen", referring to the English producers themselves, as well as Dutch and Swiss cheese-makers who were engaging in factory-based cheese production.[31]

In 1880, there were 1,652 New York State factories, which produced nearly 130 million pounds of cheese, totaling 56.5% of US output.[32] Production peaked in 1881 with 148 million pounds.

24 Durand L., "Migration of Cheese", *op. cit.*, p. 272.

25 United States Census Bureau, *Ninth Census Vol. III, Table X Special Statistics of Manufactures*, Washington, Government Printing Office, 1872, p. 594.

26 Wisconsin Cheese Board, *History of Wisconsin Cheese*, Web, 28 Oct. 2015.

27 Hoard's Dairyman, *Hoard's History*, Web, Nov., 1, 2015.

28 Watertown Historical Society, *Wisconsin Dairyman Association*, 30 Sept. 2015.

29 American Dairymen's Association, *Fourteenth Annual Report*, 15 Jan. 1879, Utica, New York, Curtiss & Childs, pp. 172–3.

30 NYSAS, *Transactions, Part II*, Vol. XXVII, 1867, p. 886.

31 Willard X.A., *Address Before the Cheese Maker's Association*, Albany, New York, Van Benthuysen, 1865, p. 13.

32 Brunger E., "Dairying and Urban Development in New York State", *op. cit.*, p. 173; Gilbert B.D., *Cheese Industry*, *op. cit.*, p. 13.

Cheese Boards

With the replacement of farmhouse cheese by factory-made, the old system of travelling "factors" disappeared and factory-men sold directly to cheese dealers. In May 1871, in Little Falls, New York, the American Dairymen's Association created the "Dairymen's Board of Trade", which became the nation's first network of "cheese exchanges" or "cheese boards", providing a system by which producers could be informed of market values for their negotiations with dealers.[33] Cheese boards were established in many towns across the state, "wherever a large cheese interest" existed.[34] Rather than buyers going to the factories to make purchases, the cheese exchanges occurred at weekly meetings "where the latest quotations were posted in plain sight" and factory-men were "prepared to act with intelligence and in accordance with the latest news of the market".[35] Between 1871 and 1894, cheese exchanges were established throughout New York and into Ohio, most often near railroad yards for expedient shipping. In the 1870s, with Britain being a primary market, through use of the telegraph, cheese producers could follow the price quotations from Liverpool, England and exchanges became the sites at which national and international prices were established. In 1875, approximately forty British buyers were headquartered in New York.[36] Montreal was also a center for British buyers as Canadian production increased.

3. Pressures on the Production of New York State Cheese

East Coast Urban Milk Sheds

Competition with Midwestern and Canadian cheese-makers was increasing. Additionally, pressures on the New York cheese industry were stemming from the expansion of milk sheds of the eastern urban centers. In 1850, the total population of New York State was slightly over 3 million people, with less than one third living in the state's urban centers.[37] Over the next fifty years, the state's population more than doubled to 7.2 million people and urbanization resulted in more than half of the population – approximately 4 million people – living in cities.[38] The resulting effect

33 Wickson E.J., "Butter and Cheese", in *Harpers New Monthly*, Vol. 51, New York, Harper, 1875, p. 824.

34 Gilbert B.D., *Cheese Industry*, *op. cit.*, p. 34.

35 *Ibid.*, p. 39.

36 Durand L., "Historical and Economic Geography of Dairying", *op. cit.*, p. 35.

37 United States Census Bureau, *The Seventh Census of the United States, 1850*, Washington, Robert Armstrong Printer, 1853, pp. 90–111.

38 United States Census Bureau, *The Twelfth Census of the United States, 1900*, Vol. 1, part 1. pp. lx-lxi, Washington, Robert Armstrong Printer, pp. 31–2.

on cheese production was increased competition with the growing urban demand for fluid milk.

Prior to the advent of rail transportation, milk in cities could only be transported from rural farms located within a small radius. For the majority of the state's dairy farmers, milk production for market centered on the manufacture of less perishable cheese and butter. In the early 1800s, urban demand for fluid milk, most often fed to young children, could be procured from urban dairies tied to distilleries. This "distillery" or "swill" milk came from cows stabled in unhealthy conditions that were fed the mash residual from distilleries and breweries. The swill milk industry was known to be a public health concern, and in 1858 the *New York Times* wrote that an estimated 8,000 children in the city had died as a result of drinking adulterated milk.[39] Temperance and health advocates spoke out against the "swill milk" dairies and promoted the consumption of "pure, unadulterated country milk" from rural farms.[40]

In 1842, the New York and Erie Railroad in Orange County, New York, approximately 65 miles north-west of New York City, carried the first rail shipment of 240 quarts of milk from Orange County farmers who speculated that they could get a price for their "country" milk that was higher than that paid for butter and cheese and successfully compete with the "swill milk" producers. Ten years later, farmers were shipping nearly 15 million quarts annually to the city milk dealers via the Erie, Harlem and Hudson River railroads.[41]

As railway systems expanded throughout the state, the interest in fluid milk led to a shift in economic strategies for dairy farmers.[42] By the 1860s, butter and cheese were increasingly manufactured in factories, rather than at farmhouses, and prices paid for fluid milk were often significantly higher than those paid for butter and cheese. In 1879, milk was being transported to New York City from as far away as 250 miles, and by 1900 dairies at the furthest reaches of New York and eastern Ohio were supplying milk to New York City and other urban centers.[43]

Milk depots or "creameries" sprung up across the rural New York landscape along rail lines to provide milk collection points. Some of these were owned by producer cooperatives, but most were owned by milk dealers. These rail-side milk depots offered several advantages to upstate dairy farmers

39 "How We Poison Our Children", in *The New York Times*, 13 May 1858; "Swill Milk and Infant Mortality", in *The New York Times*, 22 May 1858.

40 Dupuis E.M., *Nature's Perfect Food: How Milk Became America's Drink*, New York, New York University Press, 2002, pp. 21–5.

41 Mullally J., *The Milk Trade in New York and Vicinity*, New York, Fowlers & Wells, 1853, pp. 24–31.

42 Brunger E., "Dairying and Urban Development in New York State", *op. cit.*, pp. 169–74; Durand L., "Migration of Cheese", *op. cit.*, pp. 263–82.

43 Durand L., "Historical and Economic Geography", *op. cit.*, p. 33.

that created competition with cheese factories. First, like the cheese factories, creameries and depots were built in or near farm communities, thereby opening urban milk markets to rural farmers. Given the perishable nature of milk, the proximity of dairy farms to railways or cheese factories (usually within five miles) was a determinant of where farmers would haul their milk. Second, the creameries offered the same benefits to farmers as the cheese factory, but, for the most part, paid a price above that paid for cheese. Like the cheese factories, creameries contracted with farmers for the season, and the creamery man, like the cheese-maker, assumed responsibility for the milk once it was purchased. The only advantage left to the cheese-maker was his provision of whey back to farmers for use as livestock feed, predominantly for pigs. Over time, urban demands for fluid milk diverted significant amounts of milk away from the production of cheese. Between 1864 and 1927, the proportion of the state's milk that went to fluid sales (as opposed to cheese and butter) went from 9% to 72%.[44] However, the expanding urban centers also had a positive effect on cheese and butter production, in that the overall domestic demand for these goods increased. In addition to cheese exports, this growth fuelled the spread of the factory system.

---

[44] Spencer L., Blanford C.J., *An Economic History of Milk Marketing and Pricing*, Columbus, Ohio, Grid, 1973, p. 375.

Figure 2. Counties in New York State

Source: Internet Archive Book Images 2016.

Loss of the British Market

Even with the more consolidated production of factories, the specialized work of cheese-makers, and the disseminated efforts by educators and experts to improve quality, the ability of New York's industry to produce a reliably premium cheese continued to be a problem. There were several reasons for this. High market prices frequently inspired over-production and marketing of hastily made cheese without sufficient whey extraction and cheddaring, leading to "lack of keeping quality" during shipping and storage.[45] Factories paid farmers for milk by volume, enticing some dishonest farmers to add water or skim the cream off of the milk when butter prices rose. Factory-men also engaged in this controversial practice, thereby producing skim milk cheese or "filled cheese" by adding oleomargarine or lard to skim milk. The cheese looked like whole milk cheese and was often fraudulently labeled as such.[46]

The taste and texture of skim milk cheeses were not what British consumers demanded, further damaging the reputation and market value

45 Gilbert B.D., *Cheese Industry*, *op. cit.*, p. 22.

46 *Ibid.*, pp. 23–4.

of New York's cheese. In 1878, dairyman J.M. Peters exhorted the American Dairymen's Association and New York cheese-makers that skim milk cheese was "still the greatest curse of the cheese trade and that your attention should be promptly given to finding some more profitable use for your skimmed milk than that of cheese-making. If you throw it away, it is certainly better employed than in the manufacture of that pestiferous abomination".[47]

Stephen Babcock, of Oneida County New York, was a Cornell-educated Agricultural Chemist at the New York State Agricultural Experiment Station in Geneva, New York, who, in 1888, began working at the University of Wisconsin-Madison Agricultural Experiment Station. While there, he invented the first affordable system for assessing butterfat content and detecting adulteration of milk, called "Babcock's Test". This allowed for enhanced quality control in the industry, as well as component-based milk pricing strategies.[48]

While New York producers grappled with quality control, the export trade was slipping away as other nations were filling Britain's demand for cheese.[49] Because of US protectionism and import duties on cheese, Canada sought export markets elsewhere for its growing production. And having witnessed the New York experience, Canada put in place strict regulatory laws to ensure buyers that the product contained no additives and was made with whole, unadulterated milk. The Canadian cheese industry had other advantages over the Americans in the form of government grants, oversight by a dairy commissioner, and a national program of training for cheese-makers to see that "every cheese maker in the Dominion is well instructed before he begins his work and that he receives additional instruction each year to enable him keep up with the march of improvement".[50] By comparison, Gilbert describes the American cheese-makers as being "too apt to go on in their old and often careless methods, with no one to instruct them or to suggest a better method".[51] It wasn't long before British interest in Yankee cheese began to wane. By the early 1890s, Canadian cheese production rivaled that of New York State and Canadian cheese captured the British export market, with twice the amount of exports to Britain as those from New York in 1895.[52]

Problems in the domestic market also played a role in the downturn of production by New York's cheese-makers. The growing Temperance Movement led to the New York State legislature's 1896 passage of the Raines Law, with the provision to ban saloons from offering "free lunches" as an

47 American Dairymen's Association, *Fourteenth Annual Report*, 1878, Utica, New York, Curtiss & Childs, p. 182.

48 Wisconsin Historical Society, *Babcock, Stephen, 1843–1931*, Web, 29 Oct. 2015.

49 Gilbert B.D., *Cheese Industry*, *op. cit.*, pp. 21–4.

50 *Ibid.*, p. 40.

51 *Ibid.*

52 Durand L., "Historical and Economic Geography", *op. cit.*, p. 35.

enticement to buy alcohol. This law resulted in domestic demand decreasing by around one sixth of the annual production (15,000,000 pounds). In 1896, Gilbert described the law as "one of the severest blows yet inflicted on the cheese industry".[53] Given also the competition with fluid milk markets, and quality control issues, Gilbert described New York cheese producers as being "weighed down by these unfavorable circumstances, as well as by the faulty manufacture of much of our own stock".[54]

4. New York State Cheese Enters the Twentieth Century

Throughout the end of the nineteenth and into the twentieth century, the expanding rail system into upstate New York, coupled with dropping cheese prices and increased fluid milk prices, compelled more dairy farmers to enter the fluid milk markets. Cheese manufacturers had to compete for a decreasing supply of raw material and factory building and production was on the wane as the New York City milk shed expanded into the northern reaches of the state.[55]

In 1881, there were 1,652 cheese factories in New York, which produced nearly 148 million pounds of cheese, totaling 56.5% of US production and the highest amount produced in the state before the 1950s.[56] In 1900, the number of factories in operation dropped to 1,151 and production decreased by nearly 10% to 127 million pounds, still more than a third of the total US production of 324 million pounds.[57] By this time, the number of cheese factories in Wisconsin had increased to 1,500, 1,227 of which exclusively made cheese, producing a total of 77,748,600 pounds.[58] Only three other states had more than 100 cheese factories: Ohio, with 221; Michigan, with 130; and Pennsylvania, with 124. That year, the number of US cheese production facilities totaled 3,883 – 61% of which were found in Wisconsin and New York State.[59]

By 1905, British exports were almost totally absent from New York's cheese market and ongoing urbanization increased domestic demand.[60] Even so, between 1900 and 1910 cheese production declined 19% in New York

---

53 Gilbert, B.D., *Cheese Industry*, *op. cit.*, p. 42.

54 *Ibid.*, p. 41.

55 Brunger E., "Dairying and Urban Development in New York State", *op. cit.*, pp. 169–70.

56 *Ibid.*, p. 172; State of New York, New York State Department of Agriculture and Markets, New York State Agriculture Statistics Service (NYSDAM, NYSASS).

57 Durand L., "Migration of Cheese", *op. cit.*, p. 272; NYSDAM, NYSASS.

58 Durand L., "The Cheese Manufacturing Regions of Wisconsin, 1850–1950", in *Wisconsin Academy of Sciences, Arts and Letters*, 1953, Vol. 42, p. 113.

59 Durand L., " Migration of Cheese", *op. cit.*, p. 272.

60 Durand L., "Historical and Economic Geography", *op. cit.*, p. 36.

State while fluid milk sales more than doubled.[61] In 1910, when New York's production slipped to under 100 million pounds, Wisconsin rapidly scaled up manufacture to 148 million pounds.[62] Wisconsin remains the leader in US cheese production to this day.

In 1918, the largest cheese exchange in the nation, located in Cuba, New York, was moved to Plymouth, Wisconsin. Here it became the Wisconsin Cheese Exchange, later becoming the National Cheese Exchange.[63]

The 1920s and 1930s brought the advent of improved roads and truck transport, and in the 1940s improvements in refrigeration for both rail and truck were made. These factors facilitated growth of urban milk sheds with wider collection areas for fluid milk from farms that once produced milk primarily for cheese. Cheese producers found themselves with even fewer dairy sources for their raw material. Only those few, large manufacturers who could compete for milk with the fluid market or who were able to hold on to their patrons out of sheer loyalty were able to stay in operation.[64]

In addition to the effects of expanding milk-sheds, the significant trends in US cheese production of the early twentieth century were those of increased consolidation in production and distribution and the development of processed cheese.[65]

Consolidation in Cheese Production, Distribution and Retail

Closures or mergers among factories led to increased consolidation that mirrored the overall trend across the food system. Businesses that had previously operated only at specific points in the food system (producers, distributors or retailers) now had control over multiple points. For example, distributing companies took ownership of retail facilities or retailers acquired production sites. Consolidation in cheese began with the ownership of multiple rural cheese factories by fewer and more successful milk dealers.[66]

Consolidation across industries also occurred. For example, integration of the meat industry resulted in a handful of the large meat packing corporations diversifying their product lines to include cheese. In 1914, only 11.1% of US cheese distribution was controlled by large corporations. By 1918, 48.5% of US cheese distribution was controlled by the "Big Four"

---

61 Roswell D.C., *Origin and Development of the Dairymen's League*, 1938, no pub., p. 1.

62 Durand L., "Cheese Manufacturing Regions", *op. cit.*, p. 121.

63 Hamm L.G., March R., "The National Cheese Exchange: Impacts on Dairy Industry Pricing", in *Dairy Markets and Policy Issues and Options*, Feb. 1995, p. 1.

64 Durand L., "Migration of Cheese", *op. cit.*, pp. 272–3.

65 Novakovic A. Professor of Agricultural Economics, Dyson School of Economics, Cornell University. Interview by Julia Lapp, Ithaca, New York, 3 Sept. 2015.

66 Dupuis E.M., *Nature's Perfect Food*, *op. cit.*, pp. 152–8.

packers: Cudahy, Wilson, Armour and Swift Corporations.[67] Cudahy and Wilson Companies were of great importance in Wisconsin, while Armour and Swift Companies handled a large share of New York state cheese. Of that total, 55.6% was coming from Wisconsin and 25% was supplied by New York. That same year, New York was cited as "ranking first [in the USA] in the total value of its dairy products",[68] a fact that underscores the important place of fluid milk production in the state's dairy economy.

Processed Cheese

As "standardization" was becoming the industrial norm in US food production and distribution, cheese, with its protean ripening and ageing process, became a focal point for systematization by producers. The development of "processed" or "prepared" cheeses and the consolidation and control gained by large-scale cheese processors – namely Kraft – had significant effects on the national industry.

Methods for processing cheese to stop ripening and extend shelf-life had been developed in Switzerland in 1911.[69] In 1916, the first US patent for "processed" cheese went to James L. Kraft, a small-scale cheese distributor in Chicago, Illinois. Kraft had an interest in addressing the problems that occurred with natural cheeses such as spoilage and changes in flavor or texture. Kraft speculated that natural cheese could be pasteurized to increase its stability.[70] In 1912, J.L. Kraft & Brothers established a New York City business office. By 1915, the company was producing and selling the first commercially successful, tinned, pasteurized product made of blended cheeses.[71] The timing of Kraft's patent in 1916 coincided with US entry into the First World War. The company sold over 6 million pounds of tinned, processed cheese to the US military, boosting its commercial success.[72]

Other producers were also trying to establish similar methods of processing. In 1915, Elmer E. Eldredge of the Phenix Cheese Company in South Edmeston, New York applied for a patent on his product "Phennett".[73] And in 1918, a similar product was under development by Swiss cheese-maker Emil Frey of the Monroe Cheese Company in Monroe, New York.

---

67 Nichols W.H., *Post-War Developments in the Marketing of Cheese*, Research Bulletin 261, Agricultural Experiment Station Iowa State College of Agriculture, Ames, Iowa, June 1939, p. 54.

68 *New York Produce Review*, weekly publication of Urner Barry Company, New York, 14 May 1919.

69 Gerber Company web page, *About Gerber*, Web, 30 Sept. 2015.

70 United States Patent Office, *Patent Number US1186524*, Web, 10 June 2015.

71 Kraft Foods Group, *Kraft Foods Corporate Timeline*, Web, 3 Oct. 2015.

72 Kraft Foods Group, *History of James Lewis Kraft, 1874–1953*, Web, 3 Oct. 2015.

73 United States Patent Office, *Patent Number US1389577*, Web, 10 June 2015.

Frey sought to incorporate broken, unsaleable Colby and Cheddar cheeses and Swiss-style cheeses with defective eye generation into new products. He experimented with a method established in 1895 for the manufacture of "cheese food", which was made by grinding naturally aged cheese into paste and mixing it with equal amounts of evaporated whey.[74] Frey called the product "Velveeta" in reference to its smooth texture when hard or melted. He founded the Velveeta Company in 1926. It was purchased by Kraft in 1927, and has been under the Kraft label since that time.[75]

In 1928, Kraft purchased the Phenix Cheese Company, which gave Kraft control over the most significant patents in the industry and monopolistic control.[76] In 1929, Kraft-Phenix had become the largest cheese manufacturer and procurer of natural cheeses (to use in its processed products) in the country, with an annual distribution of 205.1 million pounds, 42.4% of the nation's total.[77] Even though many described processed cheese as an inferior product, its popularity skyrocketed due to its reliably mild taste, its shelf-life and versatility, and the elaborate marketing strategies of the Kraft-Phenix Company, such as the Kraft Music Hall radio show.

Up until the early 1930s, processed cheeses sold for a higher price and profit margin to manufacturers than were natural cheeses. By 1937, with falling prices and profit margins (3.5 cents per pound), processed cheeses accounted for 51.1% of all domestic Cheddars and 39.5% of total US cheese production.[78] By this time, total US cheese production was over 600 million pounds, of which approximately 49 million pounds were coming from New York State and 353 million pounds from Wisconsin.[79]

As milk continued to be diverted away from cheese, fewer, larger factories had the competitive advantage in the consolidated food system by becoming "multi-purpose" in their manufacture of dairy products.[80] In addition to cheese, these factories produced an assortment of products including powdered milk, butter, powdered whey and condensed milk. As a result, the bulk of production through the 1930s in New York state was achieved by fewer than 300 factories centered on northern and western regions of the state.[81] The remaining small, "specialized" cheese-producing facilities made an assortment of natural cheeses, such as Limburger, mozzarella, ricotta, Muenster, Colby

74 Geiling N., Smithsonian Institute, *There Is No Shortage of History When It Comes to Velveeta*, Web, 20 June 2015.

75 Monroe County Historical Society, *History of Cheese Making in Monroe*, Web, 15 Sept. 2015.

76 Nichols W.H., *Post-War Developments in the Marketing of Cheese*, *op. cit.*, pp. 75, 77.

77 *Ibid.*, p. 80.

78 *Ibid.*, pp. 87–8.

79 United States Department of Commerce, United States Census Bureau, *Biennial Census of Manufactures*, 1935, Washington, United States Government Printing Office, 1938, p. 74.

80 Nichols W.H., *Post-War Developments in the Marketing of Cheese*, *op. cit.*, pp. 73–85.

81 NYSDAM, NYSASS.

and Cheddar.[82] For the most part, these "cross-roads" factories continued their operations as in earlier days, with milk delivery coming from nearby farms, on average around twenty-four farms per factory, and they annually produced 10–15 million pounds of cheese throughout the 1920s.

5. Italian Cheese Manufacture in New York State

Reports of Italian cheese manufacture in New York state extend back to 1864, when a factory owner in Holland Patent, New York reported to the New York State Cheese Manufacturer's Association on his process of heating and salting the whey by-product of American cheese to produce "ricata" cheese, which "in Italy is sold in market as a great luxury".[83] He argued that the production of ricotta added value to an otherwise problematic waste product from cheese manufacture and hence wished to "see if we can make *ricata*" an article of commerce in the USA.[84] But it wasn't until many years later that Italian cheese production notably began in the USA, with the arrival of immigrants to east and west coast cities throughout the Italian diaspora; specifically between 1880 into the 1930s, bringing both the know-how for production and the market demand.[85] Most Italians arrived in New York City through the gates of Ellis Island, totaling over 4 million people by 1920, 10% of the total immigrant population.[86] Early consumer demands for Italian cheeses were met by imports of up 45 million pounds annually.[87] As Durand notes, "The first generation of Italian immigrants seems to have preferred the brands of their former homeland. They were familiar with them from custom in Italy, even though the majority of immigrants came from Southern Italy or Sicily rather than from the dairy regions of the Po Valley". As a result, the USA imported between 80 and 90 million pounds annually of Italian cheeses, half the yearly export volume of Italy.[88]

The large populations of first and second generation Italians in east coast cities, particularly New York, provided nearby local markets for domestic production. Early on, this was more the case for New York State than for Wisconsin, where the population of Italians was only 11,000 in 1920.[89]

---

82 Durand L., "Historical and Economic Geography", *op. cit.*, pp. 39–40.

83 De Angelis W.W., "Salting the Whey", New York State Cheese Manufacturer's Association (NYSCMA), *Extracts from the Report*, Albany, New York, Van Benthuysen Printing, 1864, p. 25.

84 De Angelis W.W., "Salting the Whey", *op. cit.*, p. 25.

85 Durand L., "Italian Cheese Production in the American Dairy Region", in *Economic Geography*, 1948, Vol. 24, No. 3, pp. 218–19.

86 United States Library of Congress, *Immigration-Italian*, Web, 9 Aug. 2015.

87 Durand L., "Italian Cheese Production", *op. cit.*, p. 219.

88 Cinotto S., "The Italian American Table: Food, Family and Community in New York City", Chicago, IL, University of Illinois Press, 2013, pp. 147–8.

89 University of Wisconsin, *Center for the Study of Upper Midwestern Cultures*, 1 Nov. 2015.

Metro New York City still accounts for approximately 30% of the Italian cheese consumed in the USA.[90]

Early manufacturers were cheese-makers of Italian training and descent, as opposed to dairy farmers, who either purchased or rented small "cross-roads" manufacturing facilities with nearby supplies of milk. Cheeses traditionally made from sheep or water buffalo milk in Italy, like ricotta, Pecorino romano and mozzarella, were made from more abundant cow's milk in the USA.[91]

Over the twentieth century, manufacture of Italian cheeses was carried out by an array of businesses, some of which went on to become leaders in US production. Most of these companies were started by first generation Italian immigrants. Among these businesses are such iconic brand names as Alleva, Polly-O and Sorrento, all of which continue production in New York State today.

Giuseppe Pollio arrived in 1899, began making ricotta on Coney Island, and later established a retail outlet in Brooklyn, New York.[92] The company expanded manufacture to provolone and mozzarella, changed its name to "Polly-O" so as not to be associated with the polio outbreaks of the late nineteenth and early twentieth centuries,[93] and eventually opened factories in upstate New York. Polly-O became one of the first large-scale ricotta and mozzarella manufacturers in the USA, famously known for its popular mozzarella snack-food Polly-O "string cheese". In 1986, the company was acquired by Kraft Foods.

Sorrento cheese was started by Louis Russo in 1947 in Buffalo, New York and named after Russo's Italian birthplace.[94] The company manufactured mozzarella, provolone and ricotta, and expanded to include Spanish, French and grated cheeses. In 1992, the company was acquired by the French Groupe Lactalis, the world's largest dairy products corporation. In 2013, after 66 years under the Sorrento name, the company was rebranded as Galbani, a brand name with global recognition, and already owned by Groupe Lactalis. Production of Galbani products has continued in the Buffalo, New York plant, making it the largest ricotta production site in North America, with an annual manufacture of 75 million pounds of ricotta and 68 million pounds of mozzarella.

Domestic production of Italian cheeses in New York state represented only a fraction of imports up until the late 1920s. In 1921, production of all Italian-style cheeses (primarily mozzarella, provolone and ricotta) was at

---

90 Tagliaferro L., "Q&A: Mark Pettie; Keeping Alive 100 years of Cheese Making", in *The New York Times, N.Y./Region section*, 8 March 1998.

91 Durand L., "Italian Cheese Production", *op. cit.*, pp. 212–20.

92 Tagliaferro L., *The New York Times*, 8 March 1998.

93 Mongiello A., Founder and Owner of Formaggio Cheese Company, Hurleyville, New York. Interview by Julia Lapp, 20 Aug. 2015.

94 Robinson D., "Sorrento Cheese Name Ages Out as Owner Rebrands Product as Galbani", in *The Buffalo Business News*, 6 Sept. 2013.

2.3 million pounds, and by 1929 there were twenty-six production sites in the state, many located in southeast New York close to the city market, and almost all under operation by cheese-makers of Italian descent.[95] Throughout the 1930s and 1940s, Italian cheese manufacture continued to grow in importance in the New York industry as a result of more Italian cheese-makers establishing production, established American cheese producers taking on the manufacture of Italian-style cheeses, a greater reception for domestic-made products among second and third generation Italian immigrants, and a tapering off in the number of new immigrants.[96] By 1939, production of Italian-style cheese had reached 5.1 million pounds in New York State, while direct imports from Italy declined throughout the 1930s.[97] Import tariffs imposed by the 1930 Hawley-Smoot Tariff Act, which added 7 cents per pound tax to imported cheese, also affected the affordability of imported versus domestic cheeses.[98] Despite the push by importers to promote Italian imports to immigrants as "essential links with their distant homeland and consuming them [as] an act of patriotism",[99] by 1940 national production of Italian-style cheeses exceeded imports and more than doubled during the years of Second World War.[100]

In New York state, production of Italian cheeses rose to over 16 million pounds in 1949, nearly 20% of the state's total cheese production of 85 million pounds.[101] Into the 1950s, with increased interest in Italian-style cuisine, particularly from returning US soldiers, the prevalence of Italian restaurants and pizzerias rapidly rose, accompanied by increased manufacture of Italian-style cheeses.[102] In 1963, the total production of Italian cheeses (mostly mozzarella) in New York state – 35 million pounds coming from 32 of the state's cheese plants – surpassed production levels of American cheeses, at 33 million pounds coming from 31 cheese plants. By the end of the twentieth century, there were twelve plants producing more than 400 million pounds of Italian-style cheeses, 248 million pounds of which were mozzarella.[103]

---

95 Durand L., "Italian Cheese Production", *op. cit.*, p. 220.

96 *Ibid.*, pp. 221–2; Cinotto S., "The Italian American Table", *op. cit.*, pp. 147–8.

97 NYSDAM, NYSASS.

98 Cinotto S., "The Italian American Table", *op. cit.*, p. 166.

99 *Ibid.*, p. 155.

100 Durand L., "Italian Cheese Production", *op. cit.*, p. 220.

101 NYSDAM, NYSASS.

102 Novakovic, interview, 3 Sept. 2015.

103 NYSDAM, NYSASS.

Part II

Terroir, Typicality and Market Openness in the Dairy Sector

Parmigiano-Reggiano Cheese

The Industrialization of Typicality

Stefano Magagnoli

University of Parma and Food Lab
(Research Laboratory on the History of Food)

1. Parmigiano-Reggiano cheese: the building of a reputation

The reputation of Parmigiano-Reggiano cheese outside its production area is documented at least as far back as the fourteenth century. In the whole of Northern and Central Italy, it was one of the most widely appreciated and expensive products, so much so that it was mentioned by Boccaccio as a symbol of opulence and good living.[1] In the renowned *Libro De Arte Coquinaria* written in the fifteenth century by the Italian cook Maestro Martino de Rossi, Parmesan cheese is the only food product identified by its origin. Likewise, two centuries later, in Vincenzo Tanara's *L'Economia del Cittadino in Villa*, Parma cheese is among the few place-based foods that are mentioned with a specific territorial patronymic.[2]

Clearly, being an expensive product, Parmigiano-Reggiano cheese needed some sort of guarantee regarding its origin,[3] and, by the same token, this task was fulfilled by the middlemen: it was the importer who assured buyers regarding the cheese's origin.[4] In turn, the importer had to rely on trustworthy local merchants from Parma, who needed to guarantee

1 Giovanni Boccaccio, *Il Decamerone, Ottava giornata, Novella Terza.*

2 Maestro Martino, *Libro de arte coquinaria*, digital edn by C. Martinelli: <http://www.uni-giessen. de/gloning/tx/martino2.htm>, p. 42; Tanara V., *L'economia del cittadino in villa*, Venezia, 1674, p. 169, pp. 172–3.

3 Guarantees about the origin the quality of products are peculiar to every luxury good (food and non-food ones), which are constantly exposed to attempts of imitation and counterfeiting; see Magagnoli S., "Made in Eataly: identità e falsificazione", in Belfanti C.M. (ed.), *Contraffazione e cambiamento economico. Marche, imprese, consumatori*, Milano, Egea, 2013, pp. 71–97. The *ancien régime* guild system was for many centuries an effective response to such risks. See for instance Guenzi A., Massa P., Moioli A., *Guilds, Markets and Work Regulations in Italy, 16th-19th centuries*, Aldershot, Ashgate, 1998.

4 Giagnacovo M., *Formaggi in tavola. Commercio e consumo del formaggio nel basso Medioevo*, Roma, Aracne, 2007.

not so much the origin (which was obvious considering where it had been purchased) as the quality of the cheese, which had to be equal to its fame and price. Consequently, from the Middle Ages onwards, "designation of origin" became – although not officially – a necessity for food products, such as Parmigiano-Reggiano cheese, which were consumed outside the production area. This was an issue involving traders in a game of mutual guarantees that obviously had costs, which could, however, be easily shifted onto the end consumers, who showed little or no sensitivity to price.

This strong identification between product and area of origin – from which stems a real *reputation* of the product followed by the *reputation* of the place that produced it – imposed very strict controls over production systems. With regard to fabrics or other manufactured goods, this control was performed by urban guilds, whereas for food products, which were almost always produced in the countryside, it was traders who selected the manufacturers who were more attentive to quality. Essentially, only those who produced high quality cheese and cured meat could have access to the rich international markets that yielded higher profits. It was therefore the opening up to foreign markets which obliged an increase in quality standards. The trader could not afford to place on the market any cheese from Parma which did not meet the standards of very demanding consumers, who were willing to pay very high prices. When this type of consumer bought Parmigiano-Reggiano cheese, they wanted to be sure of product quality. Selling on markets distant from the production area required great attention to product quality. This kind of "quality control" was the main duty of the middlemen and, since manufacturing systems were still archaic and extremely empirical, it could be performed only *ex-post*, that is, at the end of the production and maturation process, just before shipment to the end market.

Given the reputation enjoyed by Parmigiano-Reggiano cheese in the Late Middle Ages,[5] it is likely that this process of quality improvement had begun between the thirteenth and fourteenth centuries. It is therefore probably true that Parmigiano-Reggiano cheese producers were pre-eminent among cheese-makers. The strong geographical rooting also is a fact. The dairy industry in this part of the Po Plain, which covers roughly the present-day provinces of Parma, Reggio Emilia, Modena, Bologna and Mantua, certainly saw great economic activity. Large land estates, sharecropping farms, convents and monasteries were all involved in this production, which ensured large profits. In its crucial phase, however, it was concentrated in the so-called "*caselli*" (generally small family-run dairies).[6]

---

5 Iotti M., *Storia del formaggio di grana Parmigiano Reggiano (1200–1990)*, Modena, Aedes Muratoriana, 1991.

6 Ammerman A.J., "The Growth of Parmesan Production in the Nineteenth Century: The View from the Land", in C. Chang, H.A. Koster, *Pastoralists at the Periphery. Herders in a Capitalist World*, Tucson and London, University of Arizona Press, 1994, pp. 122–39 (126).

It is possible that the "designation of origin" was the result of a historical process involving trading channels and communication routes, rather than on the actual production techniques used. Parma was in fact the place where dairy production from a much wider area was concentrated for sale, and the place from where it was subsequently shipped. This process was also fostered by the higher commercial value that a cheese "from Parma" (i.e. shipped from Parma) enjoyed on the market. Clearly, this enlargement process had its limits, beyond which it was no longer profitable for manufacturers to have their products shipped from Parma rather than from other trading or shipping centers.

2. From Reputation to Designation

From a quantitative point of view, the "great leap" supposedly took place between the end of the eighteenth and the beginning of the nineteenth century. As a result of the expansion of the markets and the developments in transport, the demand for products like Parmigiano-Reggiano cheese increased significantly.[7] At this stage, individual producers tried to exploit the reputation of Parma and its surroundings as a marketing tool. Several advertising boards from the nineteenth and early twentieth centuries show that this process was fairly consistent. The few producers who succeeded in exporting directly tried to enhance their brand without distancing themselves too much from the "collective" image of a product that coincides with an area. On the other hand, the middlemen, who amassed and sold products from hundreds of micro-units, were very interested in eliminating individual brands in order to sell a product which was as homogeneous as possible, and whose price was easier to control. These two trends eventually converged towards a solution more logical and profitable, including for those producers in a position to export directly, who, by giving up their brand, faced fewer problems in trading and, above all, became part of a much more controlled market.

This process was confirmed in the most complex period in Italy's commercial history. Indeed, in the second half of the nineteenth century, because of the prolonged instability in government and legislation,[8] Parmigiano-Reggiano, together with Grana Padano and Fontina, was the only dairy product sold on international markets without interruption.

---

7 In this regard see Levati S., "'Cibo sano, comodo a conservarsi e al trasporto, di squisito gusto': il commercio del Parmigiano nello Stato di Milano tra Sette e Ottocento", in Battilani P., Bigatti G. (eds.), *Oro bianco. Il settore lattiero-caseario in Val Padana tra Otto e Novecento*, Lodi, Giona, 2002, pp. 65–98.

8 Chiapparino F., "Tra polverizzazione e concentrazione. L'industria alimentare dall'Unità al periodo tra le due guerre", in Capatti A., De Bernardi A., Varni A. (eds.), *Annali della storia d'Italia*, Vol. 13, *L'alimentazione*, Torino, Einaudi, 1998, p. 239.

Especially for Grana and Parmigiano, exports were clearly supported by the important role that these cheeses played in Italian and international cuisine. They were often considered ingredients rather than products, and thus essential for a number of recipes.[9] To some extent it was the same mechanism which, for example, supported exports of certain wines from the South. These were rarely consumed "pure" but were nonetheless considered indispensable for the wine industry, in particular the French industry, because they were used for blending and adjusting the alcoholic content of local wines.[10]

During the nineteenth century the Lombard system, industrially more advanced, realized the importance of a strong and recognizable brand, and the Chamber of Commerce of Milan adopted the following resolution in 1895: "Parmigiano cheese designates only the cheeses made near Lodi and Milan, while the cheese made in the Emilian provinces is designated Yellow Reggiano cheese".[11] Parma reacted immediately: in 1896 the Chamber of Commerce launched a marketing campaign to communicate to markets that the "true" Parmigiano cheese was that made in the area of the ancient Parmesan State. And in order to underline the fact that the pre-eminence of Parmigiano was the result of long-standing tradition, the Chamber of Commerce of Parma commissioned the respected scholar Carlo Rognoni to write a book about the history of the cheese.[12]

All this explains why Parmigiano-Reggiano was the first typical product to be protected by a designation, and why its producers were the first to join into a Consortium, although the process that led to the creation of this collective institution was anything but straightforward. In 1901 it was the Chamber of Commerce of Reggio Emilia which proposed the setting up of a trade union for cheese producers and traders to provide a guarantee of authenticity of products destined for export. Thus, right from the start, the project was closely linked to a strategy aimed at international markets. The Reggio-Emilia Consortium of 1901 was too weak and too limited geographically, since it excluded *de facto* the producers from the provinces of Parma, Modena and Mantua and, for this reason, had to deal with the hostility of the economic and political organizations of these areas.

---

9 This had been the only use for Parmigiano-Reggiano since the late Middle Ages. See Capatti A., Montanari M., *La cucina italiana. Storia di una cultura*, Roma-Bari, Laterza, 2006, pp. 57–67. See also early 19th-century French and English cookbooks: Charles-Yves Cousin D'Avallon, *Nouveau dictionnaire de cuisine, d'office et de pâtisserie*, Paris, 1826, pp. 234 and 326; Duncan Macdonald, *The new London family cook: or, Town and country housekeeper's guide*, London, 1808, pp. 102, 132, 169 and 237.

10 Unwin T., *Storia del vino. Geografie, culture e miti dall'antichità ai giorni nostri*, Roma, Donzelli, 1993, pp. 297–9.

11 Sani E., *Il "Parmigiano Reggiano" dalle origini ai giorni nostri*, Reggio Emilia, 1958, p. 136.

12 Rognoni C., *Per la storia del formaggio grana. Documenti*, Parma, 1896.

In any case, the producers from Reggio-Emilia clearly considered their cheese to be decidedly, if not substantially, different from the cheese produced in Parma and in the other provinces, which form today's area of origin. These distinctions and terminological confusions are also found in other sources. The renowned economist Stefano Jacini pointed out that the name "Parmigiano" was used to indicate "grana-type" cheeses produced in Lombardy and, in turn, this cheese was often very different in quality levels.[13] Some time midway through the nineteenth century, the name seemed therefore to indicate a type of cheese more than its origin in a strict sense, but it also proved to be an effective marketing tool for markets far away from the production area. At this stage, the name "Parmigiano" appears to have almost totally absorbed other designations (Lodigiano, Lombardo, Milanese, etc.), which, however, had enjoyed good reputations since the Middle Ages, in Italy and beyond.[14]

An additional step towards the birth of the Consortium was taken in 1909, when, in an attempt to find a solution to the problem of authenticating the origin, representatives from the Chambers of Commerce of Parma, Reggio, Modena and Mantua met to find common ground for a hypothetical branding of the grana-type cheese produced in these four provinces. There were protracted discussions about the name that the cheese should use in national and foreign markets. But until the end of the First World War, there was no solution to the problem.

As noted above, the Grana cheeses produced in the four provinces undoubtedly had different characteristics. On the other hand, even though standardization attempts had been going on for some time, the production processes were still essentially empirical, and those few practical manuals which existed were badly written and far from reliable.[15] The production facilities, mostly based on small-sized dairy farms, furthered this sort of "qualitative fragmentation" of the product. At the end of the nineteenth century, we can say that although Parmigiano did not exist, there were several, very similar, grana-type cheeses. Local traditions or the know-how of the individual cheesemaster were able to introduce primary qualitative elements.

In the theoretical paradigm of this chapter, the establishment of the Consortium and the standardization of denomination could thus be interpreted as a first attempt to standardize a product which, being totally

13 Jacini S., *La proprietà fondiaria e le popolazioni agricole in Lombardia*, Milano, Borroni e Scotti, 1854, p. 65. Similar statements can be found, around 1840, in Valéry's celebrated travel guide: "Fromage célèbre, dit de Parme, quoique fait dans la campagne de Lodi"; see Antoine-Claude Pasquin (Valéry), *L'Italie confortable manuel du touriste: appendice aux voyages historiques, littéraires et artistiques en Italie*, Paris, s.d., p. 83.

14 Giagnacovo M., *Formaggi in tavola*, *op. cit.*, pp. 74–8.

15 Cavazzoli L., "Agricoltura e alimentazione nelle campagne mantovane: 1901–1914", in N. Azzi (ed.), *Sapere scientifico e questione sociale tra '800 e '900*, Mantova, Istituto per la storia del movimento di liberazione nel Mantovano, 1984, pp. 115–53.

artisanal, was highly differentiated: a first attempt to standardize Parmigiano cheese through the making of an *Industrial Avatar*.[16]

One of the decisive thrusts in this direction was probably determined by the strong expansion of cooperative movements.[17] Farmers' dairy companies and cooperative dairies profoundly changed the organizational structure of the sector in this area. At the end of the nineteenth and beginning of the twentieth centuries, cheese production had in fact been almost entirely separated from both the farm and its sale. The dairy was often organized as a service provided by a third party, similar to the milling industry. As a result, producers sold their cheese only locally, whereas sales outside the production area were entrusted to middlemen, thus reproducing the organizational model of the Middle Ages. Very few producers were able to cross province borders using their own name and brand. In the second half of the nineteenth century, even with very high export volumes, the production of the cheeses which were to become Parmigiano-Reggiano and Grana Padano was fragmented into small or very small units.[18]

The problem of "designation of origin" affected not only Parmigiano-Reggiano, but also other traditional cheeses which were having to tackle a market with new characteristics. The First World War, which represented a great opportunity for growth in production volumes driven by wartime demand, momentarily set these problems aside. But the difficult post-war situation persuaded operators and organizations to take up the same agenda interrupted by the war. Again, the decisive impulse came from the cooperative movement, which, especially in the provinces of Mantua and Reggio, promoted a powerful process of production innovation and rationalization.[19] Between 1920 and 1921 the relaunch of the dairy sector became a necessity for administrative institutions and for the economic organizations of the four provinces. A permanent inter-regional committee was established to support the dairy sector in September 1921, based at the Chamber of Commerce of Reggio Emilia. The Committee was made up of eleven members: four representatives from the Chambers of Commerce of Parma, Reggio, Modena and Mantua, four representatives of producers from the various provinces, and three dairy technicians.

---

16 Ceccarelli G., Grandi A., Magagnoli S., "The Avatar: An Economic History Paradigm for Typical Products", in *Id.* (eds.), *Typicality in History. Tradition, Innovation, and Terroir/La typicité dans l'histoire. Tradition, innovation et terroir*, Brussels, P.I.E. Peter Lang, 2013, pp. 69–86; Grandi A., Magagnoli S., "Contrefaçon ou démocratisation du luxe? Les Avatars du Vinaigre Balsamique de Modène", in Sougy N. (ed.), *Luxes et internationalisation (XVIe-XIXe siècles)*, Neuchâtel, Editions Alphil-Pressses Universitaires Suisses, 2013, pp. 231–47.

17 Castronovo V., Galasso G., Zangheri R., *Storia del movimento cooperativo in Italia 1886–1986*, Torino, Einaudi, 1987, pp. 180–2.

18 Chiapparino, "Tra polverizzazione e concentrazione", *op. cit.*, pp. 235–6.

19 Romani M.A., Fanin E. (eds.), *Nel solco della terra le radici dello sviluppo: il sistema agro-zootecnico-alimentare a Mantova 1860–2000*, Venezia, Marsilio, 2001, pp. 131–3.

The main functions of the Committee included identifying Reggiano cheese, as it was called, through an analysis of physical characteristics and taste, and through a definition of the production area. The Committee also had commercial and technical functions, such as supporting dairies in production and maturation phases, as well as exports. At the seventh Milk Processing International Conference in 1926, the "Interprovincial Consortium" promoted a proposal to define the "names of types of cheese deriving from their region of origin" in order to prevent fraud at the expense of buyers in collaboration with some French producers. The "designation model" adopted was essentially copied from the AOC system for wines (Appellations d'Origine Contrôlées), which had been under development in France since the first years of the century.

In 1928, the time was also ripe in Italy. At the request of the "Association of Entrepreneurs" and of the "Provincial Council of Corporate Economy" (i.e. the Chamber of Commerce) of Reggio-Emilia, the "Voluntary Consortium for the Protection of Grana Reggiano" was founded. This was basically a commercial initiative, although production issues were also involved. The Consortium in fact had the function of guaranteeing that its members' cheeses met minimum requirements in terms of maturation, fat percentage and origin of raw materials, etc. But its main aim was to promote the product in domestic and overseas markets.[20]

This time, the operation was immediately welcomed by almost all the producers from Reggio Emilia. In view of the success and undoubted advantages that the Consortium granted producers from Reggio, after several meetings, and after accepting the membership of producers from the neighboring provinces, the Consorzio Volontario Interprovinciale ("Interprovincial Voluntary Consortium") was set up in 1934, with the inclusion of producers from Parma, Modena and Mantua. Finally, in 1937, the production area was enlarged to include that part of the province of Bologna to the west of the River Reno (Fig. 1).

[20] Zannoni M., *Il Parmigiano Reggiano nella storia*, Parma, Silva Editore, 1999, pp. 132–3.

Figure 1. The zone of production of PDO Parmigiano-Reggiano cheese

Elaboration by author.

Only in the following year, "protected designation" of the cheese produced in the area covered by the Consortium was established by decree as "Grana Parmigiano-Reggiano", a designation which is still in use today.

The enlargement of the production area from the province of Reggio-Emilia to include areas further afield should probably be interpreted as a response to the increasingly pressing need for a stronger critical mass in a period of convulsive transformation of markets, as well as an enduring consumption crisis. The operation was immediately seen as a loosening of traditional limitations, or more precisely the tie between the cheese and its production area, which still existed. The political and cultural climate of the period was in fact favorable to references to tradition and the centuries-old stability of "peasant civilization", authentic or otherwise. Consequently, Consortium promotional activities were up against certain dead ends. On the one hand, Parmigiano-Reggiano cheese was strongly characterized by its artisan dimension, but on the other, volume of output and standardization were becoming increasingly more important. While Parmigiano-Reggiano was associated with a specific area of the Po Plain, whose extraordinarily rich soil had long been recognized, it was also known abroad as a classic ingredient of Italian cuisine in general,[21] and thanks to its "universality" enjoyed a big advantage in terms of image.

The Consortium was inactive during the Second World War, but resumed after it. Overall production in the Parmigiano-Reggiano district showed a rapid increase, and reached 630,000 quintals in 1954. This, however, led to a crisis of over-production, which was accompanied by growing competition with Grana Padano cheese.[22] So it is not surprising that in aligning with the pursuit of modernity which swept over consumers throughout the Western hemisphere, in a very different social and cultural post-war climate, the Consortium tried to do away with its history and to make its product "less provincial".

3. The Marketing of Parmigiano-Reggiano Cheese between Tradition and Innovation

A true communication strategy for Parmigiano-Reggiano Cheese appeared after the Second World War, in the early 1960s, in parallel with the expansion of markets during the economic boom. Nevertheless, in the first decades of the twentieth century, significant investments were made in posters, which were occasionally genuine works of art, such as those by Achille Luciano Mauzan for the Bertozzi company (Fig. 2) and by Gino Boccasile for Tavella (Fig. 3). The *affiches* made for Pelagatti, a pioneering company in modern Parmesan marketing, are also worthy of note. They depict a close connection between Parmigiano and luxury (Figs 4–5).

Boccasile's work was characterized by what was almost an obsession with smell, and Fig. 3 shows a pig-faced chef enraptured by the aroma of a wheel of Parmesan cheese. Shutting his eyes to give absolute priority to smell, the chef seems to be abandoning himself to his craving for Parmigiano, anticipating with his nose the flavors which would then enchant his taste buds. These are metaphorically represented by his pouting lips, appealing for the long-awaited kiss of the product. In the poster by Mauzan, Parmesan is the olfactory focus of three men with huge noses. As evocative as the previous image, this scene is characterized by an increased "materiality". The three men look hot and excited by their fragrant proximity to the cheese. Ready to eat, they first fulfill their collective olfactory desire, which only heightens their wish to devour the cheese.

In 1961, one of the first radio commercials associated Parmigiano-Reggiano with the "ballad of an ancient and noble cheese, that has been

21 The spread of Pellegrino Artusi's book *La scienza in cucina e l'arte di mangiare bene* (Firenze, Landi, 1891) drove the use of Parmigiano-Reggiano in numerous recipes.

22 Guenzi A., "Il sistema agroindustriale", in *Storia di Parma*, Vol. I, *I caratteri originali*, Parma, MUP, 2008, pp. 453–79 (472).

famous for seven centuries [...] with two buckets of good milk you make one kilo of genuine and delicious cheese [...] those who eat it are wise; these are the wheels guaranteed by a hot-iron brand, which is the most creditable brand of all, as every good chef tells us".[23] Apart from the somewhat wobbly poetry, there are four elements to point out: the reference to its ancient lineage as a luxury product; the underlining of its genuineness and the artisan production techniques; the product being guaranteed by the Consortium through hot-iron branding; and the fact that, as well being eaten by itself like any other cheese, Parmigiano-Reggiano can also be grated and thus used in different dishes.

In the 1963 campaign (Fig. 6), the reference to tradition ("for seven centuries, a great cheese") is more explicit. It was not yet a complete narrative code, but the gaps from the past had been filled. "Modernity" was being flanked by "Tradition", for the first time indicating a *syncretism* between tradition and innovation.

Overall, the poster was very elegant. The text had a single headline: "Parmigiano-Reggiano: for seven centuries, a great cheese". There was no other text element to divert attention from this statement. The elision of the final syllable of the Italian word "*grande*" is also significant; it not only endowed the sentence with musicality and rhythm, it also played with an allusion to "*Grana*", the historical designation of Parmigiano as well as the name of the product that was to become its chief competitor, Grana Padano.

The 1964 message, significantly resorting to black and white, is, on the other hand, remarkably complex. There are several elements to note (Fig. 7):

1) The pleasure of buying is distinct from the act of consumption ("buying it means tasting it"). Even before using taste, consumers are delighted by the act of buying it.
2) The recalling of a memory ("you taste that flavor [...] which is also enticing to your memory"). There is a clear reference to Proust in this statement. Once the consumer has nibbled at a piece of Parmigiano-Reggiano, the flavor triggers a *madeleine* of the past[24] hinting at something or somebody. It levers the "memory" of the consumer, attracted not only by the product's savory taste, but also by an association with a past event connected to positive sensations

---

23 Archivio del Consorzio Parmigiano-Reggiano, fondo comunicazione pubblicitaria, spot radiofonici, 1962.

24 This is a clear allusion to Proust's "madeleine" and refers to "unintentional memory" triggered by an unexpected sensation, which sets off on a journey into the past. The past is re-experienced as though it were the present. The concept of the "madeleine" is a trigger for unintentional and evocative re-creation of the past. See Benjamin W., *Per un ritratto di Proust*, in *Id.*, *Avanguardia e rivoluzione*, Torino, Einaudi, 1973). On the power that taste has to evoke personal memories, see also Barolini H., *Festa: Recipes and Recollections of Italian Holidays*, San Diego, Harcourt Brace Jovanovich, 1988, p. 52.

and emotions. Individual consumers are left to make their favorite connection.

3) The product "doubles up" ("as a dressing or as a table cheese"): at the same time, Parmigiano-Reggiano is both food and dressing; especially for consumers geographically more distant, this is a short "Instruction Manual".
4) It is a familiar product ("because you know it well"): the statement suggests great familiarity, that is, it proposes the image of a well-known product that resides in everyone's memory.
5) The reference to naturalness ("it is a natural product, made today using the same artisan systems as seven centuries ago"): this says not only that the product dates back seven centuries, but that the processing techniques have also remained unchanged.
6) The underlining of uniqueness ("unique in the world"): Parmigiano-Reggiano is "unique" and "exceptional". There is also the intention to suggest this competitive factor: other cheeses (the competitors) may be good, but they are not unique.
7) The price factor ("and you pay with pleasure"): the product costs more than its competitors. Consumers are asked to make the effort to pay a "premium price" because the product not only gratifies them but grants them social status.

The headline of the poster used in 1966 is very clear-cut (Fig. 8) and introduces some new elements. With its statement that "Parmigiano-Reggiano is the husband of Italian cuisine" the product becomes gendered and dons a precise identity associated with the social role of "husband". Playing on the evocative function rooted in traditional culture, the product thus assumes the important role that in those years was attributed to the "male householder". This is a mimesis based on the metaphorical message that represents Parmigiano-Reggiano as the "husband" associated with a well-groomed female figure, the smiling housewife of an urban family getting to grips with the meat sauce crock.

The text and image are intimately linked, inspiring a "reassuring" message which revolves around traditional family values and thus underlines a reference to tradition, embodied in the product being advertised. The Parmigiano/husband represents traditional equilibrium in defense of a social model which is gradually crumbling away.[25]

Two years later a further leap forward is recorded (Fig. 9). "This is the table cheese for the man who deserves the best". The headline is tightly linked to the graphics, which illustrate two central elements: the torso of

---

25 On changes occurring in Italian society during the 1960s see Crainz G., *Storia del miracolo italiano. Culture, identità, trasformazioni fra anni Cinquanta e Sessanta*, Roma, Donzelli, 1996.

a man sitting in front of a wheel of Parmigiano, on top of which there is a plate with a wedge of cheese and the special knife to break off small chunks, and the arms of a woman – most probably his wife – who is affectionately embracing him. What is unusual is that the two characters' faces are missing, that they fill only a small part of the scene, and only with certain details.

Despite the incomplete representation, their roles and identities are clear. That the man belongs to the urban middle class is shown by his jacket and tie. In fact, this is nothing more than another way of representing the "husband" who appeared in the 1966 text. By his side is his doting wife, sketched in with just enough details to identify who she is; the wedding ring testifies her marital status, the bracelet on her right fore-arm is that of a woman from a moneyed family, while her well-groomed hands with their lacquered nails tell of someone who does not concern herself with housework.

The communicative innovations may be plentiful, but the cultural groove they lie in is comparable. The evocation of the family dimension remains explicit, even though the communicative blueprint appears more "refined", leveraging the more "subtle" visual elements (the wedding ring, the tie, etc.). Moreover, the association Parmigiano/husband, which had been explicit in the previous campaign, is now simply entrusted to the opening lines of the advertisement ("for him").

The "closure" of the advert, concentrated in the *pay-off*, is worthy of interest, as it comes across as somewhat subversive compared to the past. The product is no longer suggested to the consumer exclusively as a "dressing", but as food to be consumed as a main course. The product is riding the crest of the wave of "modernity", which, especially for people no longer engaged in arduous, taxing manual labor, entails lighter and lighter lunches, suitable for tackling professional or intellectual tasks during the afternoon.

In 1969, Italy was swept by an unprecedented wave of workers' demonstrations, following the massive student protests of the year before. The 1969 Parmigiano publicity machine (Fig. 10) appears to be colluding with the "silent majority", who in Italy were unfavorable to both protest movements.

"A meal in itself": this single sentence conveys the whole message, which marks a radical departure from the past. The family unit has been banished, as has the centrality of the male husband figure. Here we see a complete change of direction, right in the middle of the social protests and upheaval, with the appearance of a group of men at a formal dinner table lorded over by a wheel of Parmigiano-Reggiano. There are several elements to note:

1) These are adult men in formal attire: businessmen, educated and prosperous, who find themselves at a gastronomic feast during a business meeting.
2) The iconography hints at tradition, reproducing the classic image of weddings at that time.

3) The wheel of Parmigiano holds the place of the most important guest at the head of the table and the product is given an "authority" never previously associated with it.
4) A precise consumer audience is targeted: upwardly mobile professionals, totally insensitive to the price factor.

4. Treasure and Pirates: the 1960s from Poster Design to "Carosello"

The year 1971 saw the first advertising campaign which unequivocally leveraged the "virtue" of the *terroir* to promote the product. For several complementary reasons, the campaign formed an important break with the previous forms of communication. First, there was a need to impress the cheese's "reputation" on expanding markets by linking it to its area of origin, which was becoming more and more distant in both geographical and cultural terms. A second factor was the Italian political agenda of that period: in 1970, the administrative subdivision of the country into Regions, set up by the 1948 constitution but never in force, was finally implemented.

The whole issue was widely discussed in the public arena and the ten years of debate covered political aspects as well as the distinctive physical-geographical traits of the various regions of the country, where eno-gastronomical differences played a major role.[26] Two further interrelated elements probably came into play here. On the one hand, the French AOC model was finally applied to Italian wines, through the Denominazione d'Origine Controllata ("Protected Denomination of Origin") system. On the other, these years saw a growing interest in "regional cuisine", linked to a significant increase in domestic tourism boosted by mass car ownership. This rapidly developing demand did certainly not go unnoticed by marketing departments.[27]

The communicative blueprint is simple but at the same time sophisticated (Fig. 11): the wheel of Parmigiano-Reggiano, already sliced to show the

26 On the role that administrative subdivisions play in creating a framework favorable to the rise of food typicality see Csergo J., "L'emergere delle cucine regionali: la Francia", in Flandrin J.L., Montanari M. (eds.), *Storia dell'alimentazione*, Roma-Bari, Laterza, 2007, pp. 643–55.

27 The first "Denominazione d'Origine Controllata" (DOC) was decreed in 1963 for Marsala wine, shortly followed by several other wines, like Chianti, which had a long-standing connection with a specific area of production. Among the most prominent opinion-makers who guided the rediscovery of regional cuisine in Italy was the gastronomist, oenologist and writer Luigi Veronelli; in 1966 he co-authored the first systematic book on this topic; see Carnacina L., Veronelli L., *La cucina rustica regionale*, Milano, Rizzoli, 1966, 4 Vols. For the interplay between gastronomy and tourism see Csergo, "L'emergere delle cucine regionali", *op. cit.*, pp. 652–4.

crystalline texture of the cheese, lying on a parchment map of the place of origin, comprising the provinces of Parma, Reggio Emilia and Modena, plus a part of those of Mantua and Bologna.

These few elements strengthen the advert's communicative power by transmitting the product's sense of "tradition" and "lengthy history" to the consumer. Is not simply a matter of promoting a cheese, but of celebrating a unique product in which the very best virtues of a wealthy and prosperous *terroir* are embedded.

The sentence in the headline ("Treasure Island: Parmigiano-Reggiano's Place of Origin") can only be understood if compared to the image underneath:

1) Note first and foremost the map: it is a roll of parchment on which the production area of Parmigiano-Reggiano is drawn in the strokes and colors of an ancient survey maps. This has a definite emotional impact: Treasure Island – clearly read between the lines – is not a component of the advertising message, but represents an objective, irrefutable historical fact, deeply rooted in memory.
2) The contrast created by the muted, slightly uneven color of the parchment and the equally uneven crystalline texture of the Parmigiano is intended to evoke a *chromatic assonance* in the consumer's unconscious. The triangulation is simple. The map bears witness to the product's roots in a given *terroir* which has a "long history" behind it, made up of symbols and traditions, confirmed by the ancient parchment the Parmigiano is lying on. The product, destined to gratify the consumer's taste, is seen to have the same colors – and indeed seems to be made of the same material – as the parchment. Or more precisely, in a mimetic process, it is the parchment that seems to be made of Parmigiano. At the end of this mental process, the circle closes and the goal of selling history as a "procreator of typicality" has been fulfilled. From this point of view, it is history (and its use) that builds on "tradition" rather than the product itself.[28]

From 1969 to 1972 the Parmigiano/Treasure analogy was to appear in commercials shown on "Carosello", the Italian state television's main advertising "container". A series of cartoons, "Briganti Mattacchioni" (lit., "The Joker Bandits"), shows four unlikely bandits in search of treasure to steal.[29] The bandits use a variety of methods to reach their goal, from infiltrating a soldiers' camp to steal their wages to trying to open the gates of

28 Our methodological reference is to Hobsbawm E.J., Ranger T. (eds.), *The Invention of Tradition*, Cambridge, Cambridge University Press, 1983; Hobsbawm's approach is applied to typical food in Magagnoli S., "L'invenzione 'industriale' della tradizione: il cartello dell'Aceto balsamico tradizionale di Modena", in *Food & History*, 1, 2005, pp. 225–63, in which the "invention" of Modena's balsamic vinegar is analyzed.

29 For an analysis of this series, see Giusti M., *Il grande libro di Carosello*, Milano, Sperling & Kupfer, 1995, p. 170.

a castle, and even trying to burgle an Egyptian pyramid and force open the secret treasure vault.

However, the four bandits are the classic anti-heroes of these cartoons, and, like the more famous Wile E. Coyote, they inevitably end up victims of their own escapades. Flattened, battered or knocked out by unexpected explosions, the four jolly fellows quickly take to their heels, leaving room for the tail end of the commercial. This extols the real virtues of the "Treasure Island" of the Po Plain, which for many centuries has kept embedded the only real treasure worth knocking oneself out for, Parmigiano-Reggiano.

The series was very effective, thanks above all to the charm of the music, the originality of the bandits' adventures and the overall brilliance of the communicative idea. The general narrative standard is powerful and explicit, appealing as it does to the ideals of history and tradition. Without rhetoric or artificiality it maintains irony and a light touch, and succeeds in conveying the tradition of the product through the characters' misadventures and jocularity.

The series was extremely successful, but showed no clear link with the place of origin or the long history of the product. The link, however, was clearly shown in the next series (starting in 1973), set in the castles of the five provinces of the Consortium production area. The series, symbolically, started in the castle of Torrechiara, in the province of Parma. This series, in many ways, was very pedantic, probably unsuitable for attracting or involving TV audiences in the 1970s. This was relatively unimportant. The campaign aimed to show a strong link between product and place. And although the link was historically weak and somewhat artificial, it appeared here very realistic.

The link is realistic, although the nature of the Grana cheese was no longer the same. By the 1970s, the raw material had changed and Friesian cattle were being milked instead of the traditional breed of Red Cows. Manufacturing and maturation techniques had also been modernized. The territory or area of production itself had changed. This had little impact on marketing. The typicality of the Parmigiano cheese was perceived as very high, and this was the important thing for consumers. The typicality of Parmigiano-Reggiano conveyed emotions. On an international scale, Parmigiano cheese came to be synonymous with quality.

5. *Terroir*, Industry and Typicality

In ancient Rome, goods arriving to and from all corners of its boundless empire were indeed marked by their intrinsic properties but, above all, by qualities related to their place of origin.[30] In Renaissance Florence, as in sixteenth-century Venice, it was possible to find oriental fabrics or precious

---

30 Carcopino J., *La vita quotidiana a Roma all'apogeo dell'Impero*, Roma-Bari, Laterza, 1993, pp. 269–70.

artefacts coming from every part of the Mediterranean, as well as rare wines from Spain and Sicily and cheeses from Parma and Lodi.[31] Only with market integration and the increase in food consumption, in particular among the upper classes, did typicality become a key feature of distinction.[32] Karl Marx himself explains the difference between absolute and monopoly rent using the example of superior quality wines.[33] Marx obviously could not take into account that monopoly rent can be "artificially" determined by laws such as those that legally protect place-based food and wine following the French AOC system. The case of Parmigiano-Reggiano is similar in this since, as we have seen, its place of origin underwent a process of historical construction (or reconstruction) whose contours are blurred if not questionable.

Terroir is the protagonist of this study; a concept which is sometimes hidden and sometimes clearly visible. Although geographical and historical references to the place of origin were explicitly used as promotional marketing tools only in the 1970s ("Treasure Island"), the various literary, historical and iconographic sources described here all tend to emphasize and even idealize it. Even if it is true that "terroir et typicité ne sont pas réductibles l'un à l'autre. La typicité peut avoir d'autre fondement que le terroir", as Madeleine Ferrières argues, the bond between typicality and *terroir* is clearly the simplest to distinguish and thus to define.[34] For Parmigiano-Reggiano, even though the process of definition is today firmly part of its marketing, it appears to be a historically weak, if not actually artificial, process. For instance, the area in which Parmigiano-Reggiano is currently produced was divided into five distinct political entities until the end of the eighteenth century, and covered three different states until 1860.

But the typicality of Parmigiano-Reggiano was, and is, taken almost for granted, and this distinctive feature has enhanced its reputation since the end of the Middle Ages. The physical borders of its typicality, beyond which Parmigiano-Reggiano becomes today a different cheese (i.e. Grana Padano), have nevertheless changed over centuries and the boundaries of the current area of origin were established only shortly before the Second World War. If it is accepted that a place and its history are able to convey emotions and influence consumer preferences, the success story of place-based foods can be better explained. The reason typicality started to be manipulated, expanded, reduced and even transfigured right at the beginning of the Retailing Revolution becomes clearer. In addition, twentieth-century developments in place-based food legal protection systems can be better explained, typicality having become

---

31 See for instance Malanima P., *La decadenza di un'economia cittadina*, Bologna, il Mulino, 1982.

32 On this topic see *Typicality in History*, *op. cit.*

33 Marx K., *Capital*, Vol. III, Chicago, Charles H. Kerr and Co., 1909, p. 900.

34 Ferrières M., "Terroir: jalons pour l'histoire d'un mot", in Ceccarelli G., Grandi A., Magagnoli S. (eds.), *Typicality in History*, *op. cit.*, p. 26. For a wider concept of *terroir* see Trubek A.B., *The Taste of Place. A Cultural Journey into Terroir*, Berkeley and Los Angeles, University of California Press, 2008.

a plastic material which is partly shaped by "a narration of memory".[35] Among place-based foods Parmigiano-Reggiano is perhaps a borderline case. Its typicality proved to be so effective in transmitting emotions to consumers that it has been, and still is, used internationally as a synonym of quality, opening the way for a multitude of unlikely imitations: products that do not even try to copy the original except in the evocative force of its name.

After the Second World War, the Parmigiano-Reggiano production system in Emilia had two options. One was to become more industrial, through producers combining and merging into larger firms, product diversification and the integration of manufacturing, preservation and retailing. The other was to preserve the existing organizational model and enhance the value of typicality and quality. The second option was of course chosen, but the Consortium introduced innovations such as selection of cattle breeds, specification of cattle diet, and cattle disease prevention. The role of the Consortium was strengthened in the protection of quality, and in 1954 it became the "Consortium of Parmigiano-Reggiano".

Subsequent history shows that the complex and segmented model of organization was unchanged over time, although the number of producer firms fell. The model is the only one of its kind in Italy and overseas: it is specialized, localized, and closely reflects its socio-economic and institutional context, as well as elements of nature. Thanks to its size and the type of its members, it is a model which tends to reward labor rather than capital.[36] It is also extremely unusual, in that it is highly specialized in producing a "typicality" which is not a niche product. It is a system based on the advantages of artisan production, such as a focus on quality and traditional methods, and at the same time on the advantages of industrial production such as high output, a good quality-price ratio, and sales through different channels. So the system, like the product, is original.

Figure 2. *Advertisement poster, Parmigiano-Reggiano* Bertozzi [Italy, 1930, by L. Mauzan].

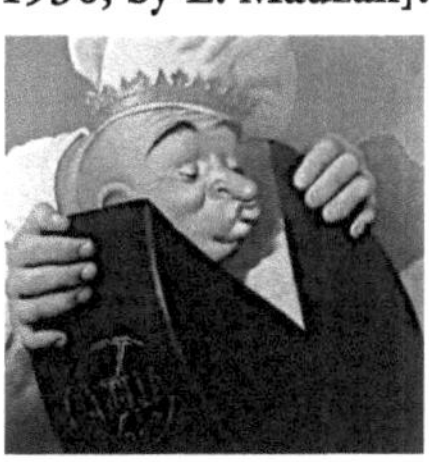

Figure 3. *Advertisement poster, Parmigiano-Reggiano* Tavella [Italy, ca. 1940, by G. Boccasile].

Figure 4. *Advertisement poster, Parmigiano-Reggiano* Pelagatti [Italy ca. 1930].

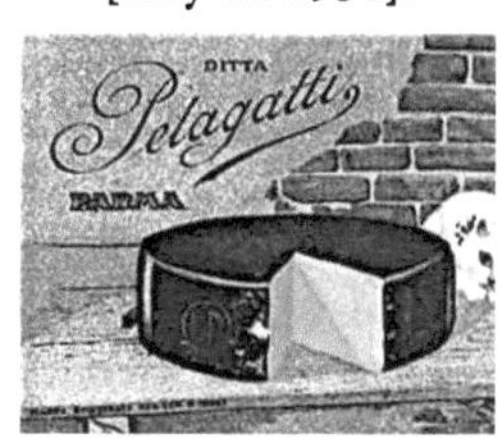

[35] Franchi M., "The Contents of Typical Food Products: Tradition, Myth, Memory. Some Notes on Nostalgia Marketing", in Ceccarelli G., Grandi A., Magagnoli S. (eds.), *Typicality in History*, *op. cit.*, pp. 45–68.

[36] De Roest K., *The Production of Parmigiano Reggiano Cheese. The Force of an Artisanal System in an Industrialised World*, Assen, Van Gorcum, 2002.

Figure 5. *Metal plate, Parmigiano-Reggiano* Pelagatti [Italy ca. 1930].

Figure 6. *Advertisement poster* [Italy, 1963, by Consortium Parmigiano-Reggiano]

Figure 7. *Newspaper advertisement* [Italy, 1964, by Consortium Parmigiano-Reggiano].

Figure 8. *Newspaper advertisement* [Italy, 1966, by Consortium Parmigiano-Reggiano].

Figure 9. *Newspaper advertisement* [Italy, 1968, by Consortium Parmigiano-Reggiano].

Figure 10. *Newspaper advertisement* [Italy, 1969, by Consortium Parmigiano-Reggiano].

Figure 11. *Newspaper advertisement* [Italy, 1971, by Consortium Parmigiano-Reggiano].

Roquefort: a Large Company Creates its Territory

Sylvie Vabre

Toulouse Jean Jaurès University

1. Introduction

"Seven steam engines with 130 horsepower are used for starting the machines (for brushing, piercing, lifting) in cheese ripening cellars or in cold rooms or for refrigeration".[1] These remarks about the Roquefort industry are excerpts from the first survey on the dairy industry in France at the turn of the twentieth century. They highlight the peculiarities of the cheese industry in the village of Roquefort. There, machines from the first and second industrial revolution were used in establishing a model for cheese production that was drawn from that of heavy industry. In fact, in 1901, nearly five thousand metric tons of cheese were produced by fourteen cheese-ageing companies, all located within an 80 km radius from the village of Roquefort.[2] The concentration of this cheese industry is all the more remarkable because a single company, the Société des Caves et des Producteurs Réunis de Roquefort, a limited company, controlled three quarters of the cheese-making industry there.[3] Since the company's creation in 1851, it had accounted for more than half of all production, but in 1911, it bought its main competitor, the Société Rigal et Cie. In so doing, the company consolidated its dominance over production. On the eve of the First World War, the company had entered into the select ranks of the top one hundred French companies.[4]

1 "Sept moteurs à vapeur d'une force de 130 chevaux sont utilisés dans les caves d'affinage ou dans les réfrigérants pour la mise en marche des machines (brosseuses, piqueuses, monte-charge, etc.) ou pour la production du froid". Ministère de l'agriculture. Offices de renseignements agricoles. Service des études techniques, *Enquête sur l'industrie laitière*, Paris, 1903, p. 39.

2 The village had nine Roquefort ageing companies or affineurs. *Enquête sur l'industrie laitière*, cit., p. 40; Marre E., *Le roquefort*, Rodez, E. Carrère, 1906, pp. 150–3.

3 This amounted to 3,822 metric tons, Archives Départementales (henceforward AD) de l'Aveyron, 54 J 2692. Today the company's name is Roquefort Société. In this article, I refer to this company by the term "Société".

4 Kogut B., "The large firm in France", in *Entreprises et Histoire*, 1989, No. 19, pp. 113–51; Vabre S., *Le sacre du roquefort. L'émergence d'une industrie agroalimentaire*, Rennes, Presses Universitaires de Rennes, 2015.

At first glance, such entrepreneurial dynamism in the cheese industry is surprising. In fact, the growth of the French cheese industry overall remained modest. With 3,330 establishments recorded in the 1901 survey, cheese-making in France was dominated by small manufacturers.[5] Cooperatives, whose number multiplied in the second half of the nineteenth century, accounted for 46% of all cheese production. However, as Claire Delfosse has pointed out, this industry attracted considerable interest from scientists and the French State, even though the sector was poorly industrialized.[6] Moreover, although France was the fourth largest European cheese exporter on the eve of the First World War, far behind the Netherlands, Switzerland and Italy, it had a trade deficit in cheese, with imports much higher than exports.[7] In fact, the cheese industry is similar to other food industries, in that it is so strongly tied to agriculture that it is often difficult to distinguish between them.[8] In addition, historians have described pre-war France as having a "fragile" economic balance and as being a "semi-industrialized" nation.[9] When the key industrial sectors in the second industrial revolution are compared, France is considered as being behind England and Germany. In this somewhat pessimistic landscape, the results achieved by the Roquefort industry are rather unexpected.

In the second half of the nineteenth century, Roquefort was the first cheese to enter the industrial era by creating national and international demand. The various characteristics mentioned above were decisive in creating a unique case in the food and dairy industries, and presented a new model in French industry before 1914. Against all odds, this creation, combining a search for profit and capital accumulation, ended up leading to the establishment of the first *Appellation d'origine* [designation of origin]. In 1925, for the first time, a law defined the territory within which a product could be collected and produced. Today, however, territorially identifying food appears to be "naturally" attached to craft production targeting a small market. Yet the Roquefort model, the first to be certified by the French Parliament, shows that in fact there was nothing natural about it – quite the contrary. In this chapter, I will show how the anchoring of this model in the territory through the law was in fact driven by the company's strategy of reaching national and international markets. Breaking into these markets and staying in them were the reasons for the company's choices, especially with the radical shift caused by the First World War.

---

5 Some 64,500 metric tons of cheese were produced, amounting to around 19 metric tons on average per company. *Enquête sur l'industrie laitière*, *op. cit.*, p. LXXX.

6 Delfosse C., *La France fromagère*, Paris, La boutique de l'Histoire Éditions, 2007, p. 36.

7 23,525 metric tons in 1913 and 16,958 metric tons exported, *ibid.*, p. 50.

8 Verley P., "Les industries alimentaires au XIXe siècle: croissance industrielle et consommation", in Marseille J., *Les industries agroalimentaires en France – Histoire et performances*, Paris, Le Monde Editions, 1997, p. 40.

9 Asselain J.-C., *Histoire économique de la revolution industrielle à la Première guerre mondiale*, Paris, PFNSP/Dalloz, 1991, p. 112.

To explain this apparent paradox between rhetoric and reality, we must examine the drivers of growth before the Great War. First, I will analyze the commercial efforts of the Société Company and show the decisive role they played. Second, to understand the business model created, I will show how the cheese was linked with the local territory in which it is made. Finally, the First World War created new circumstances that forced Société to make certain choices, which guided the future of the cheese until very recently.

2. Conquering Distant Markets

On the eve of the First World War, Société's Roquefort was sold in all French regions and around the world. How was this achieved, since the village of Roquefort is located on the edge of the Larzac, a vast limestone plateau far from the main roads and the large cities in which this cheese sold the most? – Paris is 800 km away.

Figure 1. From Paris to Roquefort

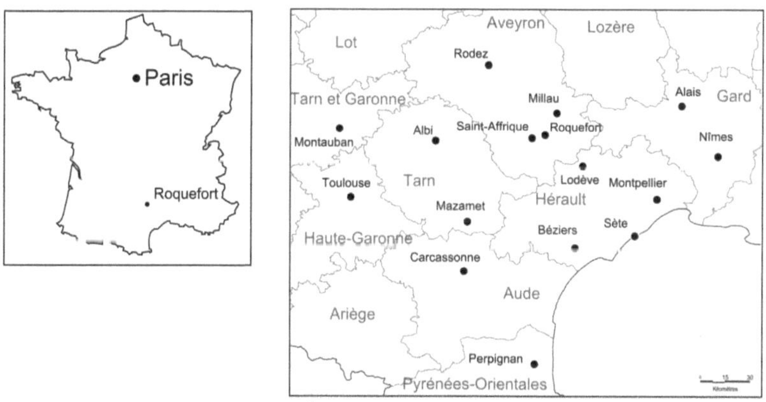

Elaboration by author.

Tackling the cheese trade head-on reflects the capitalist spirit that drove the founders of the Société de Roquefort. Starting from the company's creation, sales representatives were recruited. They were first located in Paris and then paired with the expansion of railway networks, assuring the increasing dissemination of the product. Since the village is located far from the railways, the carters were organized to arrive right on time at the shipping station.[10] By 1881, all French regions had at least one point of sale; at the same time, major French commercial centers like Paris, Toulouse and Marseilles saw consumption of Roquefort increase. In 1914, there were

[10] The railway arrived in Tournemire in 1874, a village located 3 km from Roquefort.

146 sales representatives, 31 of whom were located abroad.[11] The company's organizational capabilities can also be seen in its constant adjustment of product and demand. Indeed, how can one sell an unstable and evolving food product whose characteristics cannot be ensured once they leave the cellars? In the middle of the nineteenth century, around 30% of the company's cheese could be marketed as *surchoix*[12] for the Parisian market, since the ageing process enabled them to withstand a long journey. Other cheeses classified as first or second choice were offered at lower prices in French provincial towns and were selected according to the tastes of the consumers there. For example, the characteristics and taste of the Roquefort sold in Toulouse were different from those of the Roquefort sold in Marseilles, because they had a stronger taste. This kind of marketing according to the location continued to be used by Société, despite the various technical innovations adopted at the beginning of the twentieth century that enabled a more standardized product. Thus the Roquefort sold in Lille was not quite the same as that in Marseilles, but both carried the same brand image that had been printed on the cheese since 1858: the word "Société" surrounded by an oval.[13]

The most remarkable part of Société's marketing achievements was its convincing of foreign palates, as can be seen in the appearance of the accounting item "exports" in the company's account books from 1854. The cheese having previously been sold through specialized dealers, the company now decided to take over this role. "Remember to export; we think this has a great future before it", prophesied Etienne Coupiac, the company's Managing Director, to Florent Roucoules, the son of one of the partners, on a sales tour in 1864.[14] The director spared no effort to break into foreign markets, and the company's much-talked-about stand at the Universal Exhibition in Paris in 1867 increased the company's exports. Spurred by this success, the share of export revenue leapt from 1.8% to 2.5% in one year. The company's account books also show the arrival of new wholesale and semi-wholesale customers. Milan, Seville and Ostend were now listed as new cities with Roquefort consumers, and in cities where Roquefort was already consumed both the number of customers and sales increased. The case of Belgium is particularly telling: with eleven Roquefort dealers in 1867, the Société Company had a turnover in that country of 2,519.05 francs, a growth of 9.25% over the previous year. In 1868, the company added 21 more traders and increased turnover by 114.61%. No doubt many Belgians took advantage of the proximity to Paris to come to the exhibition,

11 Vabre S., "Le représentant et le roquefort de la Société des Caves (1851–1914)", *Entreprise et Histoire*, April 2012, pp. 131–46.

12 *Surchoix* is often thought to refer to best quality, but in the case of Roquefort it refers to the cheeses that were best able to travel.

13 The brand was registered with the Commercial Court at Saint-Affrique in 1863.

14 "Rappelez vous de l'exportation; nous pensons que c'est d'un grand avenir". AD de l'Aveyron, copie de lettres, 54 J 3356, f° 493, 23 avril 1864, Étienne Coupiac directeur à Florent Roucoules.

where they had been able to taste the cheese. The Universal Exhibition, a world event, accelerated this trend of growth over the long term. Between 1854 and 1868 the company's exports grew by 31.24% per annum, while overall turnover grew by only 4.9% in the same period.[15] This dynamism did not stop, and foreign sales reached 25% of total sales in 1913. The Société Company had customers in 36 countries, the USA then accounting for 33% of the company's total turnover.

Tab. 1. Sales in 1913

(%)

Paris	13.31
Other French territories	61.18
Exports	25.51
TOTAL	100.00

Source: AD Aveyron, 54 J 2537, 54 J 2564, 54 J 2605, 54 J 2621.

Tab. 2. The ten first cities buying Roquefort in 1913

(%)

	% exports	**% sales**
New York (USA)	33.07	8.44
Copenhague (Denmark)	8.51	2.17
Buenos Aires (Argentina)	7.49	1.91
Constantinople (Ottoman Empire)	5.81	1.48
Vienne (Austrian Empire)	4.24	1.08
Londres (Great Britain)	3.02	0.77
Saint Petersburg (Russia)	2.99	0.76
Bruxelles (Belgium)	2.70	0.69
Frankfurt (German Empire)	2.11	0.54
Holstebro (Denmark)	1.82	0.47
Autre pays	28.24	7.2
TOTAL	100	25.51

Source: AD Aveyron, 54 J 2621.

In fact, being present at such major world events was not enough to create success. It was also necessary to know how to communicate to the public, to make them interested in a product that was at the very least exotic and by some even considered toxic. For example, to avoid any problems, the experts at the 1862 Universal Exhibition in London thought it best to exclude Roquefort cheese on the grounds that the samples sent, despite being selected as being of the "extra" best quality from the cellars in the village, were liable to go off and smelled bad! This refusal reveals consumers'

15 Sales from exports rose from 786.16 francs in 1854 to 46,422.05 in 1868, while sales from all business rose from 901,351.40 francs to 1,863,952.90 francs over the same period. The percentage of turnover from exports rose from 0.8 to 2.5% of the total.

difficulties when faced with a product whose taste was unknown to them. While Roquefort certainly enjoyed a reputation for excellence, promoted by Diderot and Enlightenment philosophers, by the middle of the nineteenth century this fame had somewhat diminished. Roquefort, with its natural color and appearance displayed to a knowing connoisseur visiting the Paris Exhibition of 1855, was not seen in the same way by a large majority of visitors to London and elsewhere, despite the excellence of the cheese.

In this sense, the 1867 Paris exhibition signaled a radical change. From that time on, it was the history and geography of Roquefort, and the expertise involved in cheese-making, that took center-stage. The cheese itself was no longer displayed; it was explained, but not seen. Instead, introducing the cheese by a reconstruction of caves, and the presence of women working in them along with some animals, was a success. The picturesque Roquefort pavilion received accolades from the press: they felt the company knew how to pique the public's curiosity, in perfect adherence to the objectives of the fair's organizers. Both entertainment and education were involved, and the mysteries of how a cheese ages as well as the mastery of skills needed were the highlight of the spectacle. To make people remember, an informative brochure was distributed to those interested, in particular reporters, thereby offering them everything they needed to write their articles. This new strategy was complemented by advertising posters in several languages as well as large billboards at the Universal Exposition in Chicago in 1893.[16]

From a production point of view, the growth in exports was a triple advantage. Exports were young cheeses, made in the months of March/April, which had no buyers in France at that time of the year. Indeed, British, American and Scandinavian customers preferred a cheese with a lighter taste, only slightly aged, and these young cheeses were only aged a short time in the cellar, no more than a few weeks or even days, while it took four to six months of ageing for French customers. Promoting exports thus meant a greater rotation in the cellars. In other words, the development of exports amounted to increasing production capacity without investing in the expansion of the premises. To this double bonus from the perspective of production management there was also a commercial benefit: cheese destined for export was considered to be in the category of luxury goods characterized by high prices, with an extra profit margin for the business. The company thus acquired international stature, thanks to a product that had been renowned since the eighteenth century. Although the company certainly contributed to its own success, it benefited from the special distinction that comes from being a French *luxury* product, as described in a report from the Universal Exposition in London.[17]

---

16 Vabre S., "Constructing Roquefort Cheese: Universal Exhibitions 1851–1914", Teughels N., Scholliers P. (eds.), *A Taste of Progress: Food at International and World Exhibitions in the 19th and 20th Centuries*, London, Ashgate, 2015, pp. 183–201.

17 In this report, the English market was found to be tailored towards mass-market production, while French industry was found to be based on small, craft production and luxury goods. Bergeron L., *Les industries du Luxe en France*, Paris, O. Jacob, 1998, p. 21.

There was, however, a downside. The presence of Roquefort on the international stage encouraged other initiatives as well as competition. With the necessary means, the large company Société promoted its own Roquefort, but other Roqueforts as well, as they took advantage of the path opened up by Société to gain a foothold in the market. Similarly, the market for blue cheeses generally was boosted by Roquefort's renown, and other national-level cheeses benefited from a transfer of know-how, as has been observed for other industrial sectors. For example, Gorgonzola benefited from research on the ageing process that was done in the Roquefort village in Aveyron. Claudio Besana has highlighted the adoption of moldy bread powder in the middle of the nineteenth century, yet it had already been used for fifty years to accelerate the ageing of Roquefort.[18] Similarly, according to the newspaper *L'industrie laitière* [The Dairy Industry], a cheese-maker from Lombardy, Antonio Zazzera, was inspired by the village of Roquefort when he chose to establish cellars in the Valsassina Valley to use the coolness of underground waterfalls there in 1880.[19] The journalist writing the article emphasized the similarity of the two cheeses, saying he could not distinguish between them. Yet he forgot to mention that Roquefort is made with sheep's milk and Gorgonzola with cow's milk, which radically changes the taste and price. In fact, the similarity between the products worked against Roquefort in some markets. For example, some English people, thinking they were tasting Roquefort, were in fact eating Gorgonzola, as the disappointed travelling salesman for Société, Miesels, found on a trip to the UK in 1903.[20] Thus, the Italian cheese benefited from the fame of Roquefort. As noted above, Roquefort cheese was celebrated by Enlightenment thinkers, with Diderot decreeing that "roquefort cheese is unquestionably the best cheese in Europe".[21] In fact, this cheese, a product of know-how described as early as the middle of the eighteenth century, became the successful cheese at the best Parisian restaurants in the middle of the nineteenth century, in the wake of the boom in gastronomy and restaurant dining. However, although its name was known in the main markets, the consumer did not know how to identify it properly.

Building on Roquefort's renown, this decision to expand commercially succeeded in creating vast sales territories. At the same time, the company's production grew from 795 metric tons in 1851 when it was founded to 4,000 in 1914. From that date, production by the Rigal Company, controlled by Société since 1911, must be added. In total, the Société group accounted

---

18 Besana C., *Tra agricoltura e industria. Il settore caseario nella Lombardia dell'Ottocento*, Milano, Vita e Pensiero, 2012, p. 45.

19 *L'industrie laitière*, 13 November 1887, "Les caves glacées du Gorgonzole", p. 504.

20 Archives Roquefort Société à Roquefort (henceforward ARS), lettre de Otto Miesels, salesman, to Pierre Massol, director, 19 mai 1903.

21 "Le fromage de Rocquefort est sans contredit le premier fromage de l'Europe", *Enclyclopédie ou Dictionnaire raisonné des sciences, des arts et des métiers*, Vol. 7, Troisième édition, Livourne, 1773, p. 347.

for nearly 75% of production. Roquefort thus became a factory town whose existence was justified by the natural caves, and it was the caves that entrepreneurs used as the basis for their commercial expansion.

3. The Village Caves: Tieing Sales and Production to the Territory

With Roquefort, the people at Société knew how to exploit the curiosities of nature to build an original cheese industry.[22] The village that gave its name to the cheese in the seventeenth century is built on a long, wide limestone scree through which cold, damp air circulates. This air becomes more humid and flows through the *fleurines*, which are corridors of varying widths throughout the scree. This natural mystery, the real secret of the cellars, was the subject of multiple studies in the nineteenth century. While many caves exist on the limestone plateau of Larzac, they are far from offering the same characteristics: from 5 to 8°C and high humidity.[23] For cheese-makers, the secret of Roquefort cheese is in these caves. The colder and more humid air of some caves means that around 30% of the cheese will be qualified as *surchoix* at the end of ageing, and is sold at a higher price thus allowing the cheese-makers to earn a fortune. These natural conditions were the starting point for an industrial adventure begun in the mid-nineteenth century.

The history of the village's businesses is primarily the story of the struggle to control the best caves. This objective was behind the creation of a non-stock partnership company in 1851, but it did not work. Control of the best caves was only accomplished in 1882, when this association was transformed into a limited company. Since then, one sole company has controlled the row of the best caves for cheese-making. This desire for a monopoly created a climate of war in the village, where the saying "Peace on earth, war in Roquefort" was shown to be true numerous times. The presence of one company controlling more than 50% of production, however, also resulted in a certain sort of comfort for all the cheese-makers, since that company could play the role of leader, driving research and paying the costs needed to introduce the cheese into the greatest commercial markets. These actions confirmed the company's role at the top, while at the same time bringing along other cheese-makers in its wake in a virtuous circle.

---

22 The term "industrie" was used for the first time in 1850 in a pamphlet written by the veterinarian at Saint-Affrique: Roche L., "Nouvelle instruction sur l'industrie fromagère du roquefort", in *Bulletin de la Société centrale d'agriculture de l'Aveyron*, Rodez, 1850, pp. 153–73.

23 Girou de Buzareingues Ch., "Mémoire sur Roquefort, ses caves, ses fromages et l'agriculture des environs", *Annales administrative et scientifique de l'agriculture française*, t. VIII, Paris, 1830.

Thus it was the Société that introduced machinery into cheese-making, starting in 1874–5. The *piqueuses* [piercers] reduced the ageing time by half, thanks to the circulation of air created by the needle holes in the rounds of cheese before placing them in the caves. This innovation enabled the use of a larger area in the caves and increased production to meet demand. Meanwhile, the radius for collecting ewe's milk curds was expanded. The milk was processed into rounds of curd on farms and then brought to the town of Roquefort for refining. *Ramasseurs* [pick-up service] working for the companies were responsible for going from farm to farm, so that the transport fees did not burden the sheep farmers, since they were busy: milking 100 ewes required four people milking twice a day, and so making the curds fell to the women of the farms.

Taking charge of the transport of the cheese curds allowed the farmers to devote more time to taking care of the flock, while making the *ramasseurs* part of the sales staff promoted an increase in production. The search for this "white gold" of sheep's milk curd extended to a radius of 80 km around Roquefort in 1880, the maximum distance for transporting curd in horse-drawn carts; by 1913, the region for collecting curds had extended to the Béarn region and Corsica. In fact, the introduction of *laiteries* in the manufacturing process in the late 1880s, combined with rail transportation, allowed an even greater enlargement.[24] The fight for control of the best caves was similar to the fight to control the *laiteries*. But regardless of where the curd was made, ageing the cheese always took place in the caves of the village, which was perfectly incorporated into the marketing rhetoric used by the company: any cave located outside the village of Roquefort was named a "bastard" cave, a term full of meaning.

This situation drove the strategic capital decisions that marked the village permanently by linking cheese-making exclusively to the caves in the village. Grouping the best ageing caves under the same company with a single commercial brand was one of the manifestations of this strategy. Production choices made at the turn of the twentieth century followed the same direction. For example, when Paul Lebrou, the first engineer hired by the company, adapted the cheese-making process to use refrigeration in the 1890s, the Board of Directors chose to integrate cold machinery into the caves, triggering major renovation projects and expenditures reaching almost a million francs. Even modernity still needed the old caves. Taking things a step further, Société managed to bring all the cheese-makers of the village together to drag a young competitor to court. The newcomers had large refrigerators set up near the main road but just outside of the village, to be able to make Roquefort cheese within the city limits. The cheese could then

24 This term is specific to the making of Roquefort. Milk was transformed into rounds of cheese curd in these *laiteries* and not in *fromageries* [cheese-makers] as was the case elsewhere. Only in the caves of the village could white rounds of cheese curd become a Roquefort.

bear the name "Roquefort" by law.[25] The stand-off began in 1904 and ended in 1907 by amicable agreement. Société, as the "guardians of the temple", managed to make passage through the caves of the village obligatory for a cheese to be called Roquefort. Being made within the town limits was not enough, as inscribed in French law. To obtain a Roquefort cheese, it is mandatory to age it in the caves of the village.

All of these strategic choices concentrated expertise in this area of this small village. The presence of a laboratory for research and quality control in the company's caves signaled its entry into a new era of cheese production and further amplified this concentration of expertise. Similarly, technological changes in the manufacturing chain reinforced quality control. At the turn of the century, young couples from the Rouergue region, who perfectly mastered the techniques, were sent to Corsica to establish *laiteries*. The vocabulary used carries this special meaning of the village. Everything begins and ends with the town of Roquefort.

On the eve of the First World War, the village of Roquefort was the seat of a large French company. Linking their product with the caves, and their history and geography, would have seemed to protect these cheese-makers in a lasting way. However, four years of war would radically undermine this model – the war was not simply a break in normal life but a shift in a new direction that had previously been unthinkable.

4. Reinventing a Territory: The AO in 1925

The First World War created a situation that had been unimaginable a few years earlier, both for Roquefort and for the cheese markets. In the village cellars, cheese rounds made from cow's milk were now aged and marketed under the name "Roquefort". The war did not transform industrial techniques but the product itself. How did this happen? Several factors came together and formed a vicious circle.

First was the question of labor. Mobilizing the army swept up 63% of the male labor workforce throughout the country. In the Roquefort industry, not only were farms lacking farmers, but employees and workers were lacking in offices, transportation and *laiteries*. The latter managed to make rounds of curd thanks to the female workforce, but quality controllers were missing. This profession of recent origin conducted regular checks to guarantee quality, which was defined by the absence or very weak presence of cow's milk. In fact, if used in large quantities, cow's milk radically transforms the taste of Roquefort. The lack of controllers encouraged an even greater presence of cow's milk in the cheese, since the price of ewe's milk initially stagnated during the first two years of the war. Société, like many companies

---

[25] The law of 1824 required the mention of the place of production.

facing an uncertain future, chose to lock in prices of raw material since inflation, a quite new phenomenon, rose sharply. These strategic choices reflect the difficulties in understanding inflation, which had been unknown in the nineteenth century. This company policy amplified the consequences of the labor problem in rural areas: with the lack of labor, farmers began selling off their flocks and moved towards the more abundant cow's milk.

This change in the cheese was also promoted for commercial reasons. Against all expectations, the cheese trade in wartime actually flourished. Starting with the declaration of war, the states at war with France and its allies were banned from trading with them. Moreover, the dynamic trade with friendly countries had shut down by the end of 1916, due to difficulties in transportation and communication. But the loss in sales did not appear catastrophic, and French domestic demand was stronger than ever, while production fell by 25%. In fact, within a few weeks the war had reduced the number of cheeses available in domestic markets. For example, Cantal and Laguiole cheeses were absent from Parisian markets. The structure of French companies partly explains this shortage: many family businesses had closed due to lack of manpower. Only large companies, such as Société, had been able to re-organize themselves in order to continue production at all costs. Moreover, certain cheeses had been reserved and purchased by the army from the beginning of the conflict to feed the troops. Gruyère and Dutch cheeses were absent from civilians' tables but ensured part of the soldiers' meals. The consequence of the growing gap between supply and demand resulted in an inevitable rise in the price of Roquefort, amplified by inflation. In constant francs, the increases were quite remarkable: from 185 francs (1910) for 100 kilograms in 1914, Roquefort reached 209 francs in 1915 and 318 francs in 1918! The turnover of Roquefort producers was enormous. Yet, behind this favorable confluence of circumstances, there was a fragile product. Mixing different kinds of milk often gave way to cheeses made exclusively with cow's milk. The cheese-makers gave these products the name "Roquefort" which, a few years earlier, would never have crossed into the town limits. In 1916, management understood the phenomena at work but the decisions they took only slowed this trend.[26]

At the end of the war, the territories of Société and Roquefort had disappeared. The caves, the traditional territory, were now being used to age all kinds of mixed cheeses which then had a hard time being distinguished from other blue-veined cheeses. In addition, on the foreign markets that were now accessible, other "Roqueforts" had gained hold during the war. Danish Blue, a cheaper "Roquefort" made with cow's milk, triumphed on the English market. In California, people were trying to make this kind of cheese from goat's milk. In fact, countries that had been deprived of imported goods during the war started their own cheese-making, while pre-war exporters such as the Netherlands, Switzerland, Italy and France, hard

26 Vabre S., *Le sacre du roquefort*, *op. cit.*, pp. 489–509.

hit by the conflict, saw their share in international markets reduced. Only Finland, and in particular Denmark, further from the front, were rising export powers.[27] In short, the commercial territories conquered in the second half of the nineteenth century were in danger, as war had reshuffled the cards. This new situation encouraged discussion of daring proposals at the international level.

In 1923, at the World's Dairy Congress in Washington DC, a delegation representing the national committees of the northern European countries suggested removing the established industrial property regulations applied to food products as defined by the conventions of Paris in 1883 and then renewed in Madrid in 1892.[28] According to the proposal, the words "Gouda", "Edam", "Gruyère" and "Roquefort" would from then on define the dairy characteristics and chemical properties of the cheese. They could therefore be reproduced exactly regardless of the region they come from. Indicating their place of production would remain mandatory, so that customers would know if they were buying a Danish Camembert or a French Gouda. These proposals were discussed at length, then were put on hold under pressure from representatives of dairy products linked to their production region. At Roquefort, a summary report welcomed the change in the situation.[29] However, observations were all pointing in the same direction: the existing rules no longer worked and the dairy industry was searching for new regulations.

What to suggest for Roquefort? In 1920, a multi-organization regulation was signed, with the president of the *Syndicat Aveyronnais des Fabricants de Fromage de Roquefort* [Union of Aveyron Roquefort Cheese Manufacturers] and the *Union des Associations Agricoles du Plateau Central* [Union of Agricultural Associations of the Central Plateau] under the aegis of the *préfet* of the Aveyron. The goal of this regulation was to increase the controls and checks done by the French State Fraud department that imposed a ban on cow's milk in Roquefort. Yet this was wasted effort, as they did not achieve the desired results. So again, the cheese-makers sought the help of the State and proposed a law defining Roquefort cheese. This proposal was integrated into the legislative framework in place since 1905 to organize the repression of fraud and falsification of foodstuffs and agricultural products.

The law passed in 1905 had been enriched with new provisions about "protected designation of origin" in 1908, and was completed by a law passed

---

27 Institut International d'Agriculture (ed.), *Le lait et ses dérivés, renseignements statistiques sur leur production et leur mouvement commercial*, Rome, Imprimerie de l'Institut International d'Agriculture, 1924, pp. 144–5.

28 This congress has been organized since 1903 by the International Dairy Foundation, a non-governmental organization that promotes the dairy sector internationally; Bonin H., "La contrefaçon et les guerres industrielles – L'union des Fabricants et l'INPI acteurs de la lutte contre la contrefaçon", in G. Béaur G. *et al.* (ed.), *Fraudes, contrefaçons et contrebande*, Genève, Droz, 2006, pp. 776–89.

29 ARS, Roquefort.

8 May 1919 called the "Appellation d'origine". Yet, asking the national government to consider a law on Roquefort cheese is certainly audacious, but implies two prerequisites. First, one must be sure that French law is the right tool for imposing Roquefort's status in the largest international markets, through bilateral negotiations for example. For Société's legal counsel, Mr. Asselineau, using national legislation would have been effective, since international agreements are based on national legislation. The second stage was more difficult to overcome. The agreement of all stakeholders in the sector and locally and nationally elected representatives had to be obtained, which seems a nearly impossible task because there were three main points of contention in the discussions. While the first point, the ban on cow's milk, quickly gained consensus, the second point on the legal definition of the territory for ageing required lengthy negotiations. Where do the caves for ageing Roquefort cheese begin and end? Is there a specific perimeter? Finally, after many disagreements, it was agreed to have recourse to the courts to decide this point. The third point was the focal point of all tension: defining the area in which milk could be collected. How to delimit that area? In other words, can it be said that only the pasture lands of certain areas are appropriate for feeding the sheep? What territory gives Roquefort cheese its unique taste? The cheese-makers say that it is undoubtedly their caves, but sheep farmers say it is their pastures. As we have seen, the zone for collecting milk had continued to expand throughout the nineteenth century following the increase in demand. Since 1900, Corsica had been delivering milk and the end of the war had allowed the resumption of collecting milk in Tunisia and Turkey. Yet, while the stand-off between sheep farmers and cheese-makers dragged on, Roquefort was not selling well internationally. During the final negotiations, the expression "*l'usage de lait provenant de la France continentale ou de la Corse* [the use of milk from mainland France and Corsica]" obtained the agreement of Société and the other cheese-makers.[30] Roquefort cheese therefore became exclusively made from French milk. On 26 July 1925, the first law on the term "*appellation d'origine*" was passed. Thus, the raw material constituting the cheese, place of origin and place of ageing were henceforth defined.

The war created the conditions necessary for collective negotiation. The loss of international markets, and the confusion reigning in Roquefort village, explain why cheese-makers were able to reach an agreement with the sheep farmers. The legislative frameworks used were designed first and foremost for organizing the wine industry, which had been in crisis since the phylloxera years. However, their use for the Roquefort industry raised the question of the definition of *terroir* in the cheese industry. Was it the caves, as the cheese-makers from the village of Roquefort claimed, or the lands the sheep grazed on? For winemakers, this discussion could not take place: undoubtedly, the *terroir* has primacy over the cellar, as the latter does not have any intrinsic

30 ARS, Roquefort. It became "France métropolitaine" in the law.

qualities. The law establishing the protection of origin for Roquefort was thus a compromise between different groups in the industry and the desire of legislators. However, the primacy of Société continued unabated. It managed to work within these new legal frameworks to reconquer the markets that had been lost.

5. Conclusion

The history of the construction of Roquefort cheese territories occurred in three distinct time periods, contrary to a linear history that insists on the continuity of connection with the territory since time immemorial. As I have shown, it was the First World War that triggered this new direction.

Before the war, the company, positioned between the consumer and the rural world, had built a Roquefort cheese market and had managed to bend that market to its advantage. Production was organized as efficiently as possible, with the company providing solutions for each bottleneck. On the eve of the First World War, Société was a large industrial company that explained Roquefort cheese through the town's history and geography. In the first period, this narrative helped win new customers in new territories, and in a second phase, it helped to protect the company's position in the market. This situation contributed to the rapid construction of an industrial sector that was able to protect and further its competitive advantages, with this large company acting as a first mover.

The war radically transformed the cheese supply and disrupted the sale of Roquefort in foreign markets while also transforming the cheese market in France. Strong demand led to radical changes. Production was adapted in order to satisfy demand. New products called "Roquefort" appeared. Although their taste was very different from that of pre-war Roquefort, they resembled it in appearance. The sales strategies created in the nineteenth century were no longer adapted to this new context. Would Roquefort be subject to radical changes in the years following the end of the war? To counteract the trends in motion, the Société Company convinced the entire industry to embark on a rather audacious path: delimiting the geographical area for collecting the milk and for ageing the cheese in order to tie Roquefort to a given territory using legislation. This action allowed Roquefort to be inserted into the network of international relations that were prominent from that time onwards. The law of 26 July 1925 defining the cheese and its production territories was a new interface/interaction between trade, industry and agriculture. Professionals and politicians tried to adjust supply and demand through a new tool: legislation. However, they did not entirely get rid of the past – on the contrary, they brazenly used the past and tradition to justify this new approach of geographical limitation. Their reinvention of tradition demonstrates the agency and cleverness of the people involved and provided a new way of thinking about the entire cheese industry. Industry, territory and tradition were now linked, and because they were linked, they offered a home base for protecting the product as well as conquering new markets.

The Production of Pecorino Cheese in the Roman Countryside from the End of the Nineteenth Century until the 1930s

Rita D'ERRICO

Roma Tre University

1. Introduction

During the first three decades of the twentieth century, Italy rose to become one of the largest international cheese exporters after having been a net importer right up until the middle of the nineteenth century. This achievement was due in part to the contribution of producers in Southern and Central Italy and the Islands, home to the most exported Italian cheese of the 1930s, Pecorino romano.

Before focusing on the manufacture of Pecorino romano, it may be helpful to delineate the overall situation of sheep's milk cheese production in this area as described by the first complete statistical survey on the milk and derivates processing industry, compiled in 1937 by the Italian National Office for Statistics (ISTAT).[1] At the end of the 1930s, dairy production in the vast area stretching from Tuscany down to Calabria and Sicily and west to the island of Sardinia represented barely a quarter of Italy's overall output in the sector. This situation led Paolo Albertario, editor of the aforementioned 1937 survey, to define the manufacture of milk and its derivatives as an essentially Northern industry centered around Lombardy, Emilia Romagna, the Veneto and Piedmont.[2]

Despite a number of differences in context and conditions at a local level, many overall similarities were shared by producers in Central and Southern Italy which set them apart from those in the North. Prominent among these similarities was a common pattern of livestock distribution which was unlike that found in the Northern regions in terms of both numbers and composition.

---

1 For a detailed look at the 1937 survey see the chapter by Claudio Besana in this volume.

2 ISTAT, *L'industria della lavorazione del latte e dei prodotti derivati. Censimento del 25 maggio 1937*, Roma, Tip. Falli, p. 19.

Cow farming was more widespread in the North, especially in Lombardy, Piedmont, Emilia and the Veneto where, in 1937, cows represented between 80% and 90% of milk-producing cattle[3] and were reared on farms that had overwhelmingly begun to implement more rational and advanced working methods. In Southern and Central Italy sheep represented a similar majority, but farming techniques were outdated and based on extensive land usage. By 1937, 90% of all Italy's sheep livestock was located in this area, of which half was concentrated in Sardinia, Latium and Tuscany.[4] This meant that, apart from small quantities of cow's milk cheese (representing barely 4% of all cheese produced in the area), of which the most appreciated were stretched curd cheeses like Provolone, Caciocavallo, Scamorza and Mozzarella, the dairy industry of the entire area was based on milk produced by sheep. With the exception of the milk reserved for lambs, all sheep's milk was transformed into cheese, since its strong taste made it inappropriate for consumption as milk or butter. Sheep's milk provided around 46,500 tons of sheep's cheese, equivalent to 21% of the total volume of cheese produced in Italy in 1937 (Fig. 1).

Sheep's milk cheeses were not highly regarded within the panorama of Italian dairy products, such that the judging committee for the dairy section of the 1881 Italian industrial trade fair in Milan declared that some of the sheep's cheeses exhibited were repugnant in appearance and possessed a taste and smell which it was impossible to define.[5]

Figure 1. Varieties of cheese produced in Italy in 1937

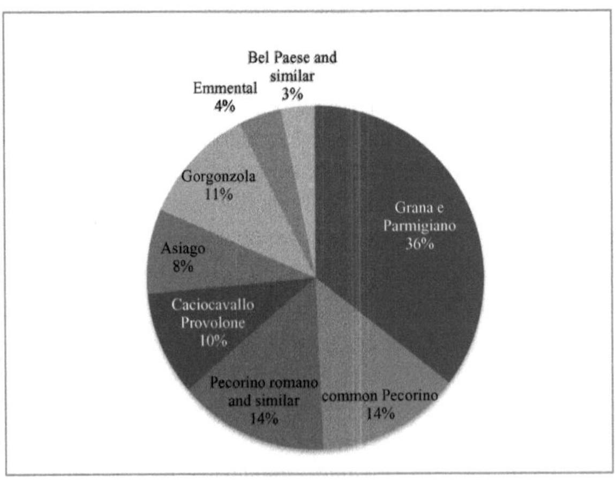

Source: *L'industria della lavorazione del latte, op. cit.*, table 62, p. 104 (our elaboration).

3 *Annuario statistico dell'agricoltura italiana 1936–1938*, Roma, Tip. Falli, 1940, table, 5, pp. 124–5.

4 *Ibid.*

5 *Esposizione Industriale Italiana del 1881 in Milano, Relazione dei giurati, Sezione X, Classe 22*, Milano, Hoepli, 1881, p. 11; see also Besana C., *Sul latte di pecora e sul caseificio pecorino*, Asti, 1892, pp. 4–5.

The many different types of sheep's cheeses fell into two overall categories related to migratory sheep farming on one hand, and stationary flocks on the other. Around half of all products derived from sheep's milk, known collectively as "Pecorino comune" (common sheep's cheese), were produced by stationary sheep farmers, and included various different types of sheep's cheeses that were usually named after their original area of production. Manufacture of these products took place on an extremely fragmentary basis because they depended on small, generally stationary flocks which produced limited amounts of milk; these cheeses were often home-made and their production was integrated with other small-scale agricultural activities: around 70% of common Pecorino cheese was made by domestic producers, though this percentage was as high as 87% in Central Italy and 82% in Southern regions.[6] The manufacturing process of this cheese, then, required a higher level of agricultural participation than any other Italian cheese, meaning that there was no separation between the production of milk and the post-processing phase, whereas the production of cow's cheese, which was typical of the Northern Italian regions, began to assume the characteristics of a fully-fledged industry. These sheep's cheeses were not subject to any particular commercial standard, and their shape, weight and size tended to vary, since they were generally produced for local or family consumption rather than for distant or foreign markets.[7]

The other category of sheep's cheese was produced in large quantities from the milk of migratory flocks, and was superior both in volume and in commercial strength. These flocks often counted several thousand sheep each, and were active particularly in Latium, Abruzzo, Apulia, Sardinia and Tuscany.[8] Unlike other cheeses, their greater volume of production caused them to acquire a certain overall uniformity which made them more appropriate for the wider market.

Pecorino romano fell into this category, and was among the most popular and widely produced cheeses in Central and Southern Italian dairies. After the First World War it became the most exported Italian cheese, a position that it continued to hold until quite recently.[9]

The history of this cheese has so far been relatively unexplored, especially with regard to the early years of its production in Latium and the countryside

6 *L'industria della lavorazione del latte*, *op. cit.*, table 62, p. 104.

7 Besana C., *Il caseificio meridionale e il suo avvenire*, Lodi, Tip. Dell'Avo, 1911, pp. 9–11.

8 The principal varieties, along with Pecorino romano, included Pecorino crotonese (Calabria), Moliterno (Molise), Fiore sardo (Sardinia), Incanestrato pugliese (Apulia), Pecorino siciliano.

9 In 2014, although it had lost its position as most exported dairy product in the 1990s, Pecorino romano was still well-positioned on foreign markets, above all in the USA where it was in third place for volume of exports behind the category of fresh cheeses (Mozzarella, Ricotta, etc.) and those of Grana and Parmesan, see Associazione Lattiero-Casearia Italiana, *Rapporto 2013*, Milano, Editoriale Il Mondo del latte, 2014, table 3.1 pp. 60–74.

around Rome, and offers the opportunity for an interesting case study which makes it possible to focus on specific aspects of the transformation that was taking place during this period in the Italian dairy industry both at a local and at a national level.

2. The origins of Pecorino romano

This cheese's original area of production, as mentioned above, was in the region of Latium and specifically the Agro romano, a mostly uncultivated and unpopulated area of around 200,000 hectares surrounding the city of Rome which, until the Second World War, was one of the most important sites in Central Italy for migratory sheep flocks. For many centuries, shepherds brought their flocks to graze in this vast territory during winter, arriving here from Latium as well as the surrounding regions of Abruzzo, the Marches and Umbria. The system of transhumance had led pastoral sheep farming to become an integral part of the Agro romano countryside since the beginning of the fourteenth century, and was seen as the form of land use most compatible with the environmental conditions of the countryside around Rome. Malaria, the causes of which remained a mystery until the early twentieth century, was at its height between June and late September, making it impossible to establish any form of agriculture requiring the permanent settlement of farm workers. Added to this was the poor agricultural quality of the land, meaning that the safest and most economically viable use for the land of the Agro romano was as pasture for migratory sheep flocks. This solution was found to be of mutual benefit to all concerned, including the owners of flocks, intermediaries, and the landholders, who were mostly aristocrats and ecclesiastical organizations and who enjoyed a regular income by renting their land without having to invest either money or work. Malaria, depopulation and the large estates concentrated in the hands of a few owners formed an unholy alliance which maintained its control until the inter-war years when reclamation operations began to gather pace in the Agro romano.[10]

[10] A vast amount of literature has been produced on the Agro romano, of which the principal works are: "Sulle condizioni dell'Agricoltura e della Pastorizia della Provincia di Roma", in Ministero Agricoltura Industria e Commercio (henceforward MAIC), *Monografia della Città di Roma e della Campagna Romana*, Roma, Tip, Elzeviriana, 1878, 5 Vols., I, pp. LXXXVIII-CXXIII; De Cupis C., *Le vicende dell'agricoltura e della pastorizia nell'Agro romano*, Roma, Bertero, 1911; Sombart W., *La Campagna romana*, Torino, Loescher, 1891; Valenti G., "La Campagna romana e il suo avvenire economico e sociale", in *Studi di politica agraria*, Roma, Athenaeum, 1914, pp. 145–252. A particularly detailed description of the evolution of sheep farming in Latium can be found in Bianchini B., *L'industria ovina nell'economia agraria del Lazio*, Roma, Tip. Le Massime, 1927. A recent publication worth mentioning is: *Antologia dell'Agro romano*, Parisella A., Passigli S. (eds.), 5 Vols., Roma, Regione Lazio – Istituto Nazionale di Sociologia Rurale, 2008.

Each year between autumn and spring, when the area was free of malaria, thousands of sheep poured into the Roman countryside from the Apennine mountains, accompanied by men and animals carrying the equipment necessary for the care of the flocks and for the associated dairy activities. The return trip towards the mountain pasture lands took place between the end of May and mid-June, when heat and malaria made the Agro inhospitable for both men and animals. Cheese-making, then, was a seasonal activity, taking place between October and May in rudimentary mobile dairies travelling with the migrating flocks. The dairies used basic equipment and techniques developed over the centuries in traditional round straw huts topped with a cone-shaped roof, whose smoke-filled interiors also served as sleeping quarters (Fig. 2).

Figure 2. A traditional straw shepherd's hut in the Agro romano

Source: Ministero dell'Economia nazionale, *L'Agro romano nel primo quinquennio fascista*, Roma, Tip. Cuggiani, 1928, p. 8.

Figure 3. The zone of production of Pecorino romano cheese

Elaboration by author.

Shepherds limited their involvement to the first stage of preparation in which a fresh cheese paste was made. Later stages, like salting and ripening, required custom-built facilities and lasted between six and nine months or more. The cheese was then sent to retailers in the city, where it was salted and aged in the cellars below their shops. Agreements between shepherds and retailers were usually renewed on a yearly basis, and required the shepherds to deliver the curds and ricotta (a highly appreciated whey by-product), which were produced between October and May, and carted to Rome two or three times a week.[11] Only a handful of wealthy *mercanti di campagna* (rural entrepreneurs), who lived in Rome and owned land or had long-term leases as well as large flocks of sheep, were equipped to carry out all the phases of production.

Unfortunately there are no complete statistics for this period on Pecorino romano production in the Agro, so figures are generally based on the number of animals and their estimated average annual milk production. At the beginning of the 1880s, yearly Pecorino production in Latium[12] was

11 Camera di Commercio di Roma, *Usi Commerciali*, Roma, Tip. Cecchini, 1927, pp. 57–61.

12 In the division of Italy's administrative territories that was in place between 1870 and 1927, the province of Rome corresponded approximately to modern Latium, with five districts whose capitals were Rome, Viterbo, Velletri, Frosinone and Civitavecchia. Following the

estimated to be around 6,100 tons, most of which came from the Agro romano, where almost half the region's sheep were located.[13]

This volume of production surpassed local needs, representing not only a highly popular and affordable product among Rome's working classes,[14] but also one that was sold in the markets of nearby regions. Company agents from Genoa and Naples made wholesale purchases of the Pecorino salted and ripened in the Agro, Maremma and Viterbo countrysides, which they then sold in Campania and Tuscany, and in Liguria[15] where Pecorino was known as "the Roman cheese"[16] and became a main ingredient in "pesto genovese" sauce. Thanks to its value for money and extended shelf-life, mercantile ships regularly carried Pecorino romano among their provisions on long-range trade routes,[17] and the Italian Royal Navy considered including it among their provisions after complaints about the poor quality of cheese served in their onboard canteens (low quality imitations of Groviere and Dutch cheeses).[18]

The cheese acquired new markets after Rome's annexation to the Kingdom of Italy in 1870, as well as benefiting from a general period of expansion in the dairy sector,[19] as seen in the increasing number of applications received by

---

reform of 1927, the districts were replaced by the four provinces of Rome, Frosinone, Viterbo and Rieti (the latter was annexed to Latium in 1923), with the addition, in 1934, of the province of Littoria (renamed Latina in 1946); see Salvo P., "Vicende territoriali del Lazio la costituzione di nuove provincie", in *Amministrazione Provinciale. Studi in occasione del Centenario*, 2 Vols., II, Milano, Giuffrè, 1970, pp. 29–55.

13 *Atti della Giunta per l'Inchiesta Agraria e sulle condizioni della classe agricola*, Roma, Tip. Forzani, 1883, Vol. XI, fasc. 1, p. 288, fasc. 3, p. 31; of the 708,000 sheep in Latium, 340,000 were present in the Agro romano; other areas of note where Pecorino was produced were the countrysides surrounding the towns of Viterbo and Civitavecchia. It was also made elsewhere, for example in the countrysides of Grosseto and the Maremma in Tuscany, whose environmental conditions were similar to the Agro's.

14 It seems that the wealthier classes preferred the more delicate taste of cow and bufala cheeses like Marzolino and Provola, and the northern Grana and Parmesan cheeses, rather than the somewhat rustic and strong-tasting Pecorino, see Ciarrocca V., "Premesse economiche e indirizzi positivi per la ripresa degli allevamenti ovini", in *Atti del I Congresso Nazionale per la Rinascita dell'Industria Armentizia*, Roma, 1958, Azienda Beneventana Editoriale, pp. 59–71: 63.

15 "L'industria ed il commercio del formaggio Pecorino romano", in *Bollettino della Camera di Commercio e Industria di Roma*, I, 1924, 6, pp. 380–2.

16 Camera di Commercio Industria e Artigianato di Roma (henceforward CCIAR), Secondaria Camera di Civitavecchia, I, 1828–1879, Price lists and commercial reports, b. 696 ter. 1856–1859.

17 "Sulle condizioni dell'Agricoltura e della Pastorizia della Provincia di Roma", *op. cit.*, p. CXV; ricotta's limited shelf-life meant that its principal market was Rome itself, although some was exported to nearby markets.

18 Archivio Centrale dello Stato (henceforward ACS), MAIC, Direzione Generale dell'Agricoltura (henceforward DGA), II vers., b. 165, f. 832, Pecorino cheese for the supply of the Italian Royal Navy, 1876.

19 Fascetti G., "Sulle vicende del caseificio in Italia", in *L'Italia Agricola*, LXIII, 1926, No. 12, pp. 578–82. In particular, the article explains the consequences of the 1870–1 Franco-Prussian war that led to the disappearance from the international dairy market of the two largest producers, leaving the field open for other countries, including Italy.

Rome's City Council for licenses to set up facilities for the salting and storage of cheese.[20] These activities were generally located on the outskirts of Rome where they could benefit from more efficient road and railway networks, particularly in the new Ostiense industrial quarter. This initial modernization process coincided with a period in which the older Pecorino salting businesses were being forced out of the city center after an ordinance published by the City Council on the 29 July 1884 banned cheese salting within the walls of Rome, in response to increasingly vociferous protests from inhabitants forced to put up with the nauseating vapors that rose from the cellars beneath grocer's shops.[21] More advanced levels of uniformity and quality control began to characterize Pecorino manufacturing when purpose-built salting facilities were introduced. These were more spacious and better equipped than the salting cellars used by grocers, some of whom, unable to meet the costs of relocation, had been put out of business by the new legislation.[22]

Businessmen were attracted not so much by the local market as by opportunities that had recently appeared across the Atlantic where there was a growing interest in traditional Italian food products – such as Pecorino cheese, which was used as condiment for pasta – as a result of the influx of Italian migrants to the Americas, which had intensified at the end of the nineteenth century.[23] The first to take advantage of the demand for Pecorino romano were wholesale companies and exporters of food products based outside of Latium. One of these was the Tuscan company Bertolli,[24] which possessed an efficient commercial network and was already exporting oil and wine to the USA when, at the end of the nineteenth century, it also began to export Pecorino romano.[25] Another pioneer in the sector was the company owned by the cousins Giuseppe and Bernardino Castelli, wholesalers from Norcia (Umbria) active in Rome from the 1870s,[26] who sent their first shipment of Pecorino romano to the USA in 1890. More exporters followed in the next few years, and by the beginning of the twentieth century around thirty companies specialized in the salting and export of Pecorino.[27]

---

20 Archivio Storico Capitolino, Titolo 47, "Sanità, licenze e contravvenzioni", b. 16, f. 72; b. 21, f. 209; b. 26, f. 201.

21 Archivio Storico Capitolino, Titolo 46, "Assistenza sanitaria", b. 26, f. 35. The findings of the enquiry commissioned by the Royal Academy of Medicine were published in the *Bullettino della Reale Accademia Medica di Roma*, 1883, meeting of 11 March, pp. 69–89.

22 Tomarelli D., *Il formaggio Pecorino*, Roma, Tip. Agostiniana, 1951 (2nd edn), p. 6.

23 Corti P., "Emigrazione e consuetudini alimentari. L'esperienza di una catena migratoria", in Capatti A., De Bernardi A., Varni A. (eds.), *Storia d'Italia, Annali, 13, L'Alimentazione*, Torino, Einaudi, 1998, pp. 683–719.

24 Segreto L., "Bertolli", in *Dizionario biografico degli Italiani*, Vol. 34 (1988) also available online: <http://www.treccani.it/enciclopedia/bertolli_(Dizionario-Biografico)>.

25 Tomarelli D., *Il formaggio Pecorino*, *op. cit.*, p. 7; Ciarrocca V., "Premesse economiche e indirizzi positivi per la ripresa degli allevamenti ovini", *op. cit.*, p. 64.

26 Biographical notes on the Castelli family can be found in Tomarelli D., "Lo sviluppo industriale del formaggio Pecorino", in *La Pecora*, I, 1927, 20 agosto, p. 2.

27 *Guida Monaci, Guida commerciale di Roma e Provincia*, XLI, 1911, p. 1134.

The strongest of these companies was the Roman Society for Pecorino Cheese (Società Anonima Romana del Formaggio Pecorino, SRFP). This company, founded in 1907 with a capital investment of 3 million lire – almost half of which was underwritten by members of the Castelli family and by the Bertolli company[28] – was, on the eve of the First World War, the second most valuable dairy company in Italy after Polenghi Lombardo. The Society created a vast commercial network, opening offices on the American continent tasked with keeping demand high and managing the supply chain.[29]

After the First World War other businessmen entered the Pecorino romano market; some were from Rome, while others came from other regions, like the Mattia Locatelli company from Lombardy[30] and the Gennaro Auricchio company, which also operated in Lombardy but whose origins were in Campania. In all cases, commercial success was determined above all by companies' ability to maintain a constant and strong position in foreign markets.

Pecorino romano exports, three quarters of which were destined for North America, grew rapidly from the beginning of the twentieth century, surpassing Parmesan and Gorgonzola exports in 1913 in terms both of volume and of value (Tab. 1). The commercial success enjoyed by Pecorino romano soon led to the appearance of the first imitation products, including Vacchino, a cheese produced in Lombardy with characteristics very similar to those of Pecorino, but made from cow's milk rather than sheep's.[31]

A downturn in the market took place during the First World War, due to an export freeze and the consequent loss of foreign markets. The most serious damage, however, was not caused by the closure of the export market, since

---

28 When the company was first set up its main shareholders, who owned 90% of all shares, were: the Castelli and Marinucci company (3,000 shares); the Castelli brothers company (6,500 shares), the Berio, Castelli, Prosperi & C. company (4,500 shares); the Ascarelli, Petrangeli, Ruggeri Italian Cheese Society (4,000 shares); the Ascarelli company (2,000 shares); the Francesco Bertolli company (3,685); the Giovanni Silvestrini company (1,845); Raffaele Piperno (800 shares); and Roberto Castelli (300 shares). CCIAR, *Anagrafe Imprese*, reg. 750/907, SRFP, minutes from management board meeting.

29 See the chapter by Gianpiero Fumi in this volume, in particular Table 4.

30 In 1928 the company also acquired the 1,151 hectare Tragliatella estate near Rome, probably with the intention of beginning the production of cheese curds there, as well as ripening activities; see Mantegazza A., "Locatelli", in *Dizionario biografico degli italiani*, Vol. 65, 2005, also available online: <http://www.treccani.it/enciclopedia/locatelli_(Dizionario-Biografico)/>.

31 *Relazione della Camera di Commercio e Industria di Roma sull'andamento economico del proprio distretto nel 1912*, Roma, Artero, 1915, p. XXI. *Relazione della Camera di Commercio e Industria di Roma sull'andamento economico del proprio distretto nel 1912*, Roma, Artero, 1915, p. XXI. Egisto Castelli set the example for Vacchino producers when, in 1907, he began making this cheese in partnership with the Latteria di Soresina in Lombardy. Castelli, however, soon came to see this as a mistake and the experiment was short-lived, but his example was followed by others, see Federazione Nazionale dei Cavalieri del Lavoro, Archivio Storico, Memoirs of Egisto Castelli; see also "Formaggio Pecorino e formaggio Vacchino", in *La Pecora*, I, 1927, 7, 20 ottobre, p. 1.

companies received compensation in the form of fixed contracts to supply their product to municipal food distribution departments,[32] but rather by the appearance of imitation cheeses produced abroad, like Romanito, made in Argentina by emigrants who had worked in the Southern Italian dairy trade.[33] Romanito became successful thanks to its lower price, and was on the brink of extending its market to the USA. After the war ended, however, and particularly after 1922, Pecorino romano exports began to grow once more as it consolidated its position as the most exported Italian cheese (see Tab. 1).

Table 1. Export figures for Pecorino romano cheese

	TONS	Euro		TONS	Euro
1907	4,565	38,839	1923	8,214	137,595
1910	6,663	60,066	1925	10,882	127,067
1913	9,176	87,088	1928	11,334	159,933
1916	7,921	70,182	1931	9,421	93,974
1919	418	6,171	1934	7,836	63,564
			1936	7,036	67,366

Source: Ministero delle Finanze, *Movimento Commerciale del Regno d'Italia*, Roma, Tip. Civelli, various years. Contemporary lira values have been recalculated to the 2014 value of the euro.

3. From Latium to Sardinia: the divided pathways of Pecorino romano

At the same time as Pecorino romano was conquering foreign markets, its manufacturing base was beginning to expand beyond its traditional area of production, with exporters looking to find more diversified and profitable sources of raw materials; their requests were met mainly by producers in Sardinia, which already had a large sheep population. In the late nineteenth century some exporters began to set up rudimentary manufacturing plants in different parts of the island, designed to carry out the entire Pecorino production cycle from the processing of fresh milk right through to the final product; production techniques were entrusted to experienced dairy operators from Latium and Abruzzo who had been brought in specially. Shepherds in Sardinia traditionally produced a type of cheese known as "Fiore sardo", and had no experience of making Pecorino romano (which had never taken hold among consumers on the island), so after some initial reluctance they realized that they could make more money by selling fresh milk than by transforming it into cheese,

32 Archivio Storico Capitolino, Ufficio XI, *Annona e mercati*, b. 79, Butter, ricotta, cheese, supply contracts, 1917.

33 Bianchini B., *L'industria ovina nell'economia agraria del Lazio*, *op. cit.*, p. 44.

not least because the plant owners were offering to buy their milk at very competitive prices.[34]

The pioneer of this shift was Egisto Castelli, who in 1899 had been the first to set up dairies in Sardinia, and he was quickly followed by other companies from Lucca, Campania and Rome.[35] Fearful of the potential damage of too much competition, the largest companies involved in the export and sale of Pecorino came together in 1907 to form the SRFP, mentioned above, active in the markets of Rome and Sardinia, although it implemented different strategies in each: in Rome it purchased the cheese curd and then salted and ripened it, whereas in Sardinia it bought milk from the shepherds and carried out all successive phases of production.

Industrial-scale manufacture of Pecorino romano took off rapidly in Sardinia, to the point that the 1937 census found that, of the 5,500 tonne total volume of production, 86% came from industrial dairies. The island counted around seven hundred small or medium sized factories which were equipped to carry out all phases of manufacture, employing approximately 3,000 workers,[36] making Sardinia the region with the most industrial dairies in the whole of Central and Southern Italy).[37] The growth of the dairy industry quickly revealed itself to be a precious resource for Sardinia's economy and led to a 123% rise in the sheep population of the island, from 840,000 to 1,876,000 between the two livestock censuses of 1881 and 1908,[38] just when Italy and many advanced North European countries were experiencing a progressive reduction in their sheep populations. This was particularly true in Northern Italy, where pasture land available to sheep farmers was disappearing due to land reclamation and the appearance of intensive cow farming.[39]

Unlike in Sardinia, shepherds in the Agro romano maintained their hold on the curd production phase, a practice which they still carried out in their

---

34 Campus A., *Pecorino romano e sardo. L'evoluzione dell'industria del formaggio Pecorino romano nel Lazio e in Sardegna*, Roma, Arte della Stampa, 1936, p. 12; milk prices increased between 1897 and 1907 from 6 to 25 lire for a hectolitre.

35 Tomarelli D., *Lo sviluppo industriale del formaggio Pecorino*, *op. cit.*; the companies from Tuscany were Bertolli and Silvestrini, De Paoli Luporini and Landucci; Salvatore Cannavale & figli, Colonna & Ciro Piro operated in Campania; Rome was the location of the Castelli companies, the Ascarelli, Petrangeli & Ruggeri Italian Society for Cheese, and the Giovanni & Silvestrini company.

36 Doneddu G., "L'Industria casearia in Sardegna", in *Las industrias agroalimentarias en Italia y España durante los siglos XIX y XX*, Barciela C., Di Vittorio A. (eds.), Alicante, Universidad de Alicante, 2003, pp. 155–73: p. 163; see also the chapter by Giuseppe Doneddu in this volume.

37 *L'Industria della lavorazione del latte*, *op. cit.*, Tav XVI, p. 186.

38 See Ministero di Agricoltura Industria e Commercio, Ispettorato Generale dei servizi zootecnici, *Censimento generale del bestiame del 19 marzo 1908*, Roma, Tip., Civelli, 1910, Table L, p. LXIII.

39 Fotticchia N., "La produzione ovina in Italia", in *L'Italia Agricola*, LXVII, 1930, 6, pp. 309–14; D'Alfonso C., "L'Industria ovina in Italia", in *Rivista di Zootecnia*, I, 1924, 6, pp. 167–74.

traditional huts before selling on their produce to the salters. Only the final phase of production was affected by the transformations taking place in the dairy industry, with the introduction of modern cheese-making facilities equipped for the salting and ripening of large quantities of cheese, but not for the transformation of milk into curds. A 1937 census registered 119 of these facilities in the vicinity of Rome out of an overall total of 205 in the region of Latium (Tab. 2).

Just as Pecorino romano was beginning to achieve popularity, the Ministry for Agriculture, Industry and Commerce appointed Carlo Besana, an eminent scholar and the founder and director of the Lodi Royal Experimental Dairy Laboratory [Regia Stazione Sperimentale di Caseifico di Lodi],[40] to carry out a series of studies of the sheep dairy industry in the Agro, with the aim of improving its production techniques. Besana carried out these studies between 1887 and 1891, and held practical courses on farms in the areas of Santa Maria di Galeria and Prima Porta which were attended by cheese-producing shepherds. The initiative also involved successful trials in the production of other types of cheese, including imitations of French Roquefort and soft cheeses, demonstrating that sheep's milk could be adapted to a variety of dairy products capable of achieving popularity in the domestic market.[41]

Carlo Besana had, nonetheless, insisted ever since his first report to the Ministry in 1887 that the only way that the sheep's milk dairy industry could be modernized was by separating the roles of shepherd and cheese-maker, with the former acting solely as a supplier of milk and the latter taking care of the transformation process.[42]

In the Agro, and in the rest of Latium, shepherds continued to be the main producers of Pecorino curd; fewer than ten dairies equipped for the entire processing cycle appeared in the 1920s and 1930s, and little is known about them,[43] while the 1937 census recorded 993 migratory sheep farmers in the province of Rome, who produced more than half of all cheese curds in the region (see Tab. 2).

---

40 Stoppa A., "Carlo Besana e il rinnovamento del caseificio", in Battilani P., Bigatti G. (eds.), *Oro bianco. Il settore lattiero caseario in Val Padana tra Otto e Novecento*, Lodi, Giona, 2002, pp. 241–70.

41 ACS, MAIC, DGA, V versamento, b. 751, f. 3261, Production of Pecorino cheese in the Agro romano, Carlo Besana's reports detailing the results of the trials were published in the *Bollettino di Notizie Agrarie*, in 1887 (No. 57), 1890 (No. 26) and 1891 (No. 18). The study concluded in 1892 with the creation of the first practical course in sheep's cheese production to be held in Italy, at the Royal School of Agriculture in Rome located in a farmhouse three km from Porta Cavalleggeri, see Besana C., *Sul latte di pecora e sul caseificio pecorino*, *op. cit.*, pp. 14–15.

42 This same opinion was backed over the following years by other scholars and commentators, see Bianchini B., *L'industria ovina nell'economia agraria del Lazio*, *op. cit.*, pp. 46–7.

43 Sparse and fragmentary information can be found on these dairies in "Ancora sui caseifici". *La Pecora*, V, 1931, 14, 31 luglio 1931, p. 5

Table 2. Pecorino romano production figures in Latium from the 1937 census

	Migratory livestock businesses (a)		**Quantity of milk processed each year (b)**		**Production of Pecorino romano (c)**		**Salting and seasoning facilities (d)**		**number of employees (e)**	
PROVINCE	No.	%	hl	%	q.li	%	No.	%	No.	%
FROSINONE	187	9	2,087	1	3,999	7	17	8	26	5
LITTORIA	230	12	14,925	7	3,738	6	11	5	24	5
RIETI	120	6	3	1	3,182	5	1	0	1	0
ROME	993	50	149,076	67	32,428	54	119	58	304	61
VITERBO	458	23	54,837	24	16,835	28	57	28	144	29
LATIUM OVERALL	1,988	100	223,925	100	60,182	100	205	100	499	100

Source: *L'Industria per la lavorazione del latte, op. cit.* (a) Table XXXIX, p. 224; (b) Table XXXIII p. 216; (c) Table XLIII, p. 231; Table XXXIX p. 224; (d) Table XLII, p. 230; (e) Table XLVII, p. 236.

Collective associations were practically non-existent in the sheep's cheese dairy sector of Latium, as was the case in the rest of Central and Southern Italy;[44] Sardinia was an exception, accounting in 1937 for around forty milk-producing collectives that brought together the produce of numerous livestock companies.[45]

In Rome, attempts at forming producer's collectives had been sporadic and were limited to the salting, ripening and retailing sectors, leaving the producers of curds unaffected.[46] The authorities, on the other hand, did little to promote the interests of the sheep's milk dairy industry in Latium. Apart from the Ministry of Agriculture's support for Besana's studies at the end of the nineteenth century, the only initiative in favor of the sheep farming sector was the setting up of the Latium Chamber for Wool and Pecorino (Camera Laziale per la Lana e il Formaggio Pecorino), which nonetheless was only active between 1922 and 1927.[47] Among the aims stipulated by the Chamber's founding statute was the requirement to support and promote products related to pastoral sheep farming by offering consultancy services to producers and retailers, and provide them with assistance in selling their

44 Montanari V., "La produzione di latte e derivati nell'Italia Meridionale", in *Il Latte e i suoi Derivati*, Roma, Arte della Stampa, 1932, pp. 111–27.

45 *L'Industria per la lavorazione del latte, op. cit.*, Tav. 5, p. 170.

46 In 1902, for example, ten wealthy sheep farmers from Abruzzo and Latium set up the Cooperative Society for the Salting and Export of Pecorino (Società Cooperativa per la Salagione e l'Esportazione del Pecorino); nonetheless, it never attracted a great number of participants, and shortly after the war the society was liquidated; see Consiglio Provinciale dell'Economia Corporativa di Roma, *Relazione Statistico – Economica della Provincia di Roma per l'anno 1929*, Roma, Tip. Cecchini, 1933, p. 145.

47 *Statuto della Camera Laziale della Lana e del Formaggio Pecorino e Regolamento della Cassa di Credito Agricolo e Armentizio*, Roma, Tip. Italiana, 1923.

products in foreign markets; plans were made to build storehouses and set up a system of credit secured against commodities (wool and cheese). This credit system was set up in 1923, though it never received adequate resources and after four years went into liquidation together with the aforementioned Chamber for Wool and Pecorino.[48]

Sheep farmers, for their part, were enjoying the benefits of Pecorino's increasing value and growing market, thanks in part to the weakness of the national currency against the dollar, and limited themselves to adapting rather than actually changing their traditional manufacturing procedures. The supply of milk to the cheese-making sector was prioritized, to the detriment of that normally destined for young lambs, highlighting a relative lack of interest for the meat trade which at the beginning of the nineteenth century had been the main source of income for rural businesses in the Agro romano.[49] This new approach affected eating habits in Rome, where "abbacchio" (a new-born lamb around twenty days old) grew in popularity, replacing the previous market for lamb of up to a year old.[50]

Figure 4. Pecorino romano prices from 1910 to 1940*

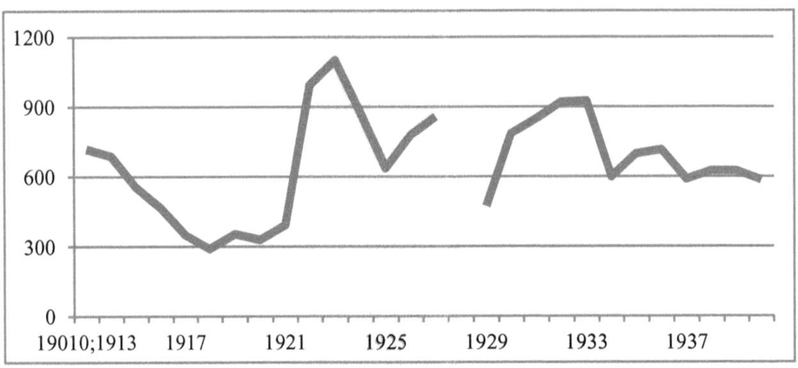

Source: Cianferoni R., *La pastorizia nel Lazio e nell'Abruzzo*, Roma, Istituto Nazionale di Economia Agraria, 1969, table 4, p. 16 (our elaboration). *Figures are based on the 1965 value of the lira.

---

48 Some documentary evidence of the activities of this credit bureau can be found in the Historical Archives of the Bank of Italy, *Vigilanza sulle aziende di credito*, b. 8171, f. 4, Cassa di Credito Agricolo e Armentizio.

49 In a study comparing the financial activities of a number of pastoral businesses typical of the Agro between the beginning of the 19th century and the 1960s, the agrarian economist Reginaldo Cianferoni demonstrated that originally the most profitable product of the pastoral economy was meat, followed by wool and cheese. Between the end of the 19th century and the first decade of the twentieth, however, a change took place and the production of curds together with that of ricotta became the most important sector for sheep farmers, see *La pastorizia nel Lazio e in Abruzzo, op. cit.*, pp. 17–30, in particular table 8. V, p. 22.

50 *Ibid.*, pp. 17–18.

The increased production of sheep's milk was not due to higher productivity, which changed little during this period, but was, rather, a consequence of the growth in livestock numbers. In Latium, the sheep population grew by 72.4% between the censuses of 1881 and 1908 – from 708,000 to 1,220,662 – following, along with Sardinia, the opposite trend to that in other regions. Much of this increase occurred in the Agro, where the sheep population grew from 300,000 to 456,000 between 1881 and 1906,[51] reaching 731,000 in 1924[52] despite the negative effects of the war years, and around one million two years later;[53] Pecorino production also grew, reaching 9,000 tons in 1926.[54]

The period between the end of the nineteenth century and the mid-1920s was a highly favorable one for the sheep industry in Latium, with high prices being commanded by wool and cheese. Sheep farmers, therefore, were not keen to see their profits related to the cheese-making process disappear, profits which came mostly from sales of fresh Pecorino, but also of its by-product, ricotta, which enjoyed much popularity locally.

Local producers of curds tried to obtain official recognition for the unique variety of Pecorino produced in the area of Rome, on the basis of its profound connection with a specific territory. The aforementioned Chamber for Wool and Pecorino Cheese tried for a brief period to distinguish Pecorino produced in Rome from that made in Sardinia with the creation of a special label indicating the region of provenance.[55] This initiative turned out to be unsuccessful because the wholesale and export companies, which sold Pecorino from both regions, preferred to apply their own brand labeling, and considered identification with a specific territory to be more of an impediment than a source of added value.[56]

The businessmen's point of view is clearly expressed by a contemporary commentator:

> The reason it is still called "romano" is because it originated, long ago, in the Agro romano, but now it is produced in large quantities in Sardinia as well as in Apulia and Tuscany, and in Lombardy it is made with cow's milk. Unless producers wake up, this industry will disappear entirely from the Agro romano. To keep it alive they will need much more than a label of provenance, which is what many believe will defend it from their Sardinian competitors. Even if we concede, so as to give the pasture-owners their due, that in some parts of the province of Rome the grass has a flavour which is richer than elsewhere such that it gives cheese a

---

51 "Statistica degli ovini nel Comune di Roma", in *Rivista Agricola Romana*, XXXVII, 1907, 6, p. 128.

52 Lotrionte G., "L'Industria Zootecnica nella Provincia di Roma", in *Roma Agricola*, II, 1926, 3, pp. 43–6.

53 Bianchini B., *L'industria ovina nell'economia agraria del Lazio*, *op. cit.*, p. 23.

54 *Ibid.*, p. 38.

55 "L'industria ed il commercio del formaggio Pecorino romano", *op. cit.*, p. 382.

56 "Formaggio Pecorino sardo", in *La Pecora*, I, 1927, 9, 20 November, pp. 6–7.

unique aroma similar to a bouquet, it must be said that what is needed to achieve and maintain a place in the vast field of international competitors is a uniform, healthy product, unadulterated and with an agreeable appearance, none of which can be obtained as long as this cheese is produced with the same techniques as it was in the times of Galen and Varro.[57]

4. The crisis years and the collapse of Agro romano Pecorino production

What were described as Pecorino romano's "golden years"[58] came to an end in the late 1920s with the appearance of unfavorable conditions, some of a general nature and others connected with the specific local context. These conditions were to challenge the practicality of the traditional manufacturing system described above. The principal cause of this decline was connected to the deflationary effects of the revaluation of Italy's currency against the dollar, which had been initiated in 1927 by Mussolini with his "battle of the lira", better known as the "Quota '90". After a period of rising prices, the value of Pecorino romano began to fall in 1928 in reaction to a stronger lira and the need to remain competitive on foreign markets. The international economy suffered a rapid deterioration after the Great Depression of 1929, which led to a fall in profits and to the adoption of protectionist trade policies by many countries as well as a restriction on new immigrants entering the USA, all of which meant a reduction in the demand for Pecorino abroad. The situation worsened when the dollar was devalued in 1933, causing Italian dairy exports to fall to their lowest ever peacetime level (Tab. 1).

Despite this reduction in export volumes, Pecorino romano remained stronger than other types of cheese on foreign markets, continuing to command the largest portion of total Italian dairy exports. It is therefore understandable that this cheese attracted many imitators in a context of constricting markets and the increasing availability of cow's milk as a result of the numerical rise and higher productivity of milk-producing cattle achieved by Northern Italian producers over the previous thirty years.[59] The search for new commercial opportunities led many cheese producers to create imitations of Pecorino, which appeared not only in Lombardy, where the aforementioned Vacchino was produced from the first decade of the twentieth century, but also in the USA, where the San Francisco-based Kraft Cheese Company began to produce a cow's cheese imitation of Pecorino.[60]

---

57 Sapelli A., "Il formaggio Pecorino romano. Sveglia!", in *Latte e Latticini*, IV, 1927, 1, pp. 5–6.

58 Cianferoni R., *La pastorizia nel Lazio e in Abruzzo*, *op. cit.*, p. 30.

59 Fotticchia N., "I problemi contingenti della produzione zootecnica italiana", in *Rivista di Zootecnia*, VII, 1930, 6, pp. 211–37.

60 "Il Vacchino americano", in *La Pecora*, II, 1928, 4, 20 febbraio 1928, p. 2.

The questionable tactics pursued by competitors who made Vacchino became particularly fierce at the end of the 1920s and its abundance and low cost on the American market was a further cause of a general fall in prices.

The crisis weighed heavily on the companies involved in the sale of Pecorino romano, and their numbers fell by a third in just a few years.[61] Wholesalers, nonetheless, tried to protect their profit margins by passing their losses on to milk suppliers (in Sardinia) and curd producers (in Latium). The balance sheet of the largest cheese producer, the Roman Society for Pecorino Cheese, for example, remained healthy apart from the three years between 1929 and 1931, even though their overall output fell throughout the 1930s due to reduced demand.

Those who suffered most from the crisis were the sheep farmers, since Pecorino prices, like those of wool and meat, were falling while the costs of running their farms remained high; particularly hard-felt was the tax burden, as well as employment costs and rental of pasture lands.[62] Sheep farming, then, began to appear less commercially viable than it had been in previous years, such that at the end of 1929 many livestock owners complained that they were running their business at a loss.[63]

As the depression deepened, cooperative initiatives began to appear in the countryside; in May 1929 newspapers reported the creation of a consortium by shepherds in the Agro romano, which was set up with the support of the National Fascist Confederation of Agricultural Unions [Confederazione Nazionale Sindacati Fascisti dell'Agricoltura].[64] A few months later, a report was presented on the grave situation faced by sheep farming and its products to the Agriculture and Forestry department of the Provincial Council of Corporate Economy (the forerunner of Rome's Chamber of Commerce), whose members included the parliamentarian Annibale Marini, who was also one of the promotors of the Consortium.[65] The department approved the shepherds' request that the government apply regulations to abolish unfair competition from Vacchino and favor the creation in Latium of cooperative dairies for the mutual production of Pecorino and cooperative bodies for its sale and export. Shortly afterwards, requests were made for low-interest loans. The government responded favorably, in part, to the requests put forward by Pecorino romano producers: a positive step was taken with legislation No. 381 of 6 April 1933 ("Disciplines governing the production and sale of

61 Of the thirty or so salting and export companies active in the mid-1920s, only eight were still in business in 1936, see *Guida Monaci*, *op. cit.*, 1936, p. 833.

62 De Matteis M., "Stato attuale dell'industria ovina transumante e mezzi per fronteggiare la crisi", in *La Pecora*, V, 1931, 4, 28 febbraio, pp. 2–4.

63 CCIAR, Verbali Sezione Agricola e Forestale, seduta 12 settembre 1928, p. 41; see also De Angelis A., "Risultati economici di alcuni tipi di imprese pastorali", in *Congresso Nazionale Armentario*, *op. cit.*

64 "Costituzione del consorzio Pastori dell'Agro romano", in *La Pecora*, III, 10, 20 maggio 1929, p. 5.

65 CCIAR, Verbali Sezione Agricola e Forestale, 12 novembre 1929, pp. 4–6.

vacchino and pecorino cheeses"), which obliged producers of Vacchino to apply labeling to distinguish it as a cheese produced with cow's milk.

The government showed much less enthusiasm for supporting the cooperative system of Latium curd producers, as seen in the case of the Consortium for Latium producers of Pecorino, founded in 1931, which set itself the ambitious task of constructing facilities for the production, salting and sale of its members' cheese.[66] The Consortium, however, received no financial support or recognition from government authorities.

Isolation and economic crisis made the sheep farmers vulnerable in a phase of transition characterized by the acceleration of the land reclamation operations that took place in the years known as the "battle of the grain", initiated by Mussolini in the mid-1920s as a counterpart to the "battle of the lira". Land reclamations in the Agro romano had begun shortly after the annexation of Rome to the Kingdom of Italy, but they had gone ahead slowly over the following forty years and did not yet represent a serious threat to the traditional landholding system as long as it remained restricted to the 10 km wide strip (the so-called "decikilometric zone", equivalent to one tenth of the Agro) that surrounded the capital, and as long as the financial resources provided by the state remained inadequate to perform the necessary operations.[67] The traditional extensive landholding system continued to enjoy a positive period between the end of the nineteenth century and the 1920s, proving that it could still be profitable despite its antiquated character.[68] Nevertheless, after the mid-1920s and the order to extend land reclamations to the entire Agro romano territory along with a substantial increase in the low-interest government loans made available to land owners,[69] the land reclamations began to pose a threat to migratory sheep flocks, reducing the land put aside for grazing and leading to a rise in the rental cost of land.[70]

The Ministry of Agriculture's campaign for the transformation of land into cultivated crops, which was one of the conditions that landowners had

66 CCIAR, Anagrafe imprese, fasc. 333/31, Consorzio Laziale Produttori Pecorino Romano. Founding act and statute, 5 marzo 1931.

67 D'Errico R., "La bonifica dell'Agro alla vigilia della Marca su Roma", in Carusi P. (ed.), *La Capitale della Nazione. Roma e la sua Provincia nella crisi del sistema liberale*, Roma, Viella, 2011, pp. 281–300; d'Errico R., "L'Agro romano e il suo settore Ostiense negli anni della grande trasformazione (1870–1930)", in *Roma moderna e contemporanea*, XX, 2012, 1, pp. 11–35.

68 Baldassarre S., Brizi A., *La pecora nell'Agro romano. A proposito del concorso per il miglioramento degli ovini*, Roma, Cuggiani, 1913, pp. 14–18. In 1925, it was estimated that 180 hectares of land set aside as pasture land brought in an income of around 58,000 lire, while the same land, when cultivated with cereal crops, brought in no more than 18,000 lire, see Rivera V., *Battaglie per il grano*, Aquila, Tip. Vecchioni, 1925, pp. 99–100.

69 Ministero dell'Economia Nazionale, *L'Agro romano nel primo quinquennio Fascista*, Roma, Cuggiani, 1928.

70 Romolotti A., "In difesa della produzione ovina nazionale", in *Rivista di Zootecnia*, IV, 1927, 5, pp. 177–80.

to comply with in order to receive the loans, led the land closest to the capital to become dedicated to the production of cereal crops, orchards and cow farms, aimed at satisfying the growing demand in Rome.[71] Permanent pasture lands were to be progressively reduced and replaced with planted grasslands.

The Great Depression of the 1930s, with its accompanying price collapses, closure of export markets and increased operating costs, accelerated the decline of sheep farming and convinced many farmers to cut their losses and leave the industry. In 1930 the sheep population of Latium had reached 1,463,000,[72] higher than it had ever been since records had been kept. The decline experienced by sheep farming became more pronounced in 1936, with a 15% reduction over six years despite the boost received in 1932, when customs duties were applied to wool imports in order to block the falling prices of this raw material.[73] The province of Rome was particularly hard hit, with its sheep population falling by 30% between 1930 and 1936 (from 858,285 to 604,870) and by 22% between 1936 and 1941 (from 604,780 to 472,137) after Italy had entered the Second World War. Pecorino production in the Agro fell proportionately, and in 1930 was estimated at around 3,000 tons, the same as it had been at the beginning of the century. During the war, migratory sheep herding suffered further from confiscation of animals[74] and requisition of farmer's produce[75] between 1942 and 1944, especially during the periods of Nazi and Allied occupation.

The decline in the sheep population of Latium continued during the post-war period, to the point that only 675,000 sheep were present in the region at the beginning of the 1960s, less than half of the total population of thirty years before, representing a structural collapse rather than a short-term crisis. Seasonal transhumance was becoming a thing of the past, as permanent sheep farms began to appear, accelerating the economic decline and depopulation of the small communities located in the Apennine mountains of Abruzzo.[76]

The straw huts where shepherds traditionally produced the Pecorino base product became a typical feature in the repertory of iconic "Roma sparita" imagery, while the Agro territory was being transformed into a

71 Ministero dell'Economia Nazionale, *L'Agro romano nel primo quinquennio*, *op. cit.*, pp. 33–5.

72 *Annuario statistico dell'agricoltura italiana 1936–1938*, *op. cit.*, p. 109.

73 Fotticchia N., "Il problema zootecnico", in *I progressi dell'agricoltura italiana in regime fascista*, Roma, Ministero dell'Agricolrura e Foreste, 1934, pp. 221–37.

74 In Lazio, around 500,000 sheep were confiscated during the war, representing almost half of the total population, Cianferoni R., *La pastorizia nel Lazio e in Abruzzo*, *op. cit.*, p. 46.

75 For example, the warehouses of the SRFP in Rome were raided by German troops on 9 September 1943, and dairies in Latium and Sardinia were damaged during the war, see CCIAR, *Anagrafe inprese*, SRFP, minutes from management board meeting, 12 February 1944.

76 Marcolini E., "L'Allevamento ovino nel Lazio", in *Roma Economica*, XVIII, 1965, 1, pp. 14–20.

built-up periphery of Rome, with road traffic growing ever more intense and the presence in some areas of intense crop cultivation and cow farms. The image of sheep grazing in the fields was fading away from the Roman landscape. They were replaced in the 1950s and 1960s by the first dairy factories designed to cover the entire Pecorino romano production cycle: the new protagonists in this field were companies like Lopez, Ronconi, Brunelli, Fulvi (some of which had begun business in the 1920s and 1930s).

In 1954, producers from Latium and Sardinia came together to form the Consortium for the protection of Pecorino romano (Consorzio per la tutela del Pecorino romano), with the aim of protecting its production and sale in Italy and abroad and of limiting the use of denomination status to members of the Consortium.[77] Its Latium-based membership, however, soon began to decline, leading to the decision in 1978 to transfer the Consortium's operational headquarters to Macomer in Sardinia.[78] By 1994 the Latium membership had all but disappeared, and the head office, too, moved from Rome to Macomer.[79]

At the same time, the manufacture of Pecorino in Latium was declining and Sardinia was becoming the principal center of production. Today it is responsible for 97% of all Pecorino produced in Italy.[80]

5. Conclusions

Pecorino produced in the Agro romano was for centuries the end product of a complex production cycle that began with the requirement for winter pasture lands of migratory sheep flocks from the Central Apennines and ended in the workshops of Roman grocers. Although it was already known locally in remote times, Pecorino only gained international renown at the end of the nineteenth century thanks to a combination of favorable events which were mostly independent of the local context. These events included the growing demand for Italian products in emigrant communities and the ability of wholesalers and exporters to exploit emerging commercial opportunities for a product that until then had been consumed in a more limited area. Wholesale exporters also created new manufacturing opportunities beyond the confines of Latium, benefiting Sardinian producers greatly.

77 CCIAR, *Anagrafe Imprese*, reg. 1954/2277, "Consorzio per la tutela del Pecorino romano", Founding charter, 5 July 1954.

78 CCIAR, *Anagrafe Imprese*, reg., 449/960, "Consorzio per la tutela del Pecorino romano", Founding charter, 28 November 1978.

79 *Ibid.*, minutes of the general assembly, 15 June 1994. Of the Consortium's 151 members, the only company from Latium was Sini Fulvi from Viterbo.

80 According to data provided by the Consortium for the Protection of Pecorino Romano Cheese, total production in Latium during the 2014–15 dairy-making season was 883 tons, compared to 29,285 tons produced during the same period in Sardinia.

The owners of sheep flocks, among whom were the wealthy Roman "rural entrepreneurs", who only had control of the initial phase of the production cycle of Pecorino, were unable to break away from their traditional practices and develop alternative manufacturing techniques, and made little progress in shaking off the individualistic nature of pastoral society; perhaps an even greater cause of the decline was the lack of support offered by public institutions to pastoral trades in a time when the political priorities for the Agro romano were focused on its position as the site of one of the most ambitious campaigns of integral land reclamation. It is, nonetheless, possible to affirm that the extensive pastoral system that dominated the Agro region for many centuries proved itself to be a highly profitable business for landowners and sheep farmers until the outbreak of the Great Depression in the 1930s, despite its outdated nature.

The collapse in prices and the closure of foreign markets that took place from the end of the 1920s were not the only factors that determined the decline of Latium's pastoral system, since these temporary effects were overcome in the post-war period elsewhere, as in Sardinia. The finality of the crisis was, rather, a result of the structural transformation of the areas surrounding the capital, which saw a growth in demographics, property development and manufacturing, all of which were incompatible with the requirements of the pastoral habitat that was fundamental to the sheep's milk dairy industry. Even if they had introduced innovations into their manufacturing practices, sheep farmers would probably not have been able to turn back this process.

Production and Processing of Sheep Milk in Sardinia (1950–2015)

Giuseppe Doneddu
University of Sassari

1. Introductory remarks

Up until the mid-twentieth century, the economy of Sardinia was still largely dominated by the combination of cereal and herding activity.[1] While the former comes to a point of crisis in the 1950s, the second is confronted with the great economic and social transformations of the last part of the century and goes unscathed until the new millennium: a specialization determined by the particular geographical and anthropic situation of the island which exceeds centuries-old vicissitudes and which presents itself as a case study of great interest to historians and economists. In the 1990s, a young French researcher, attracted by its peculiarities, decides to devote his doctoral thesis to sheep farming in Sardinia.[2] On the basis of a detailed analysis of the characteristics of the region and its economic and social issues, he acutely identifies the reasons for the long-term persistence and its further more recent development that finds no equal on either shore of the Mediterranean. On an island where large floodplains are historically bound to cereal cultivation and mountainous areas forced the transhumance[3] of cattle to the plains in winter months, still in the early twentieth century, the hilly areas with abundant natural grassland are the most suitable for the development of an activity run by a class of landowners-farmers also particularly active in the production and marketing of cheese. The production traditionally consists of a soft cheese destined for domestic consumption and a salty, long-life and low cost cheese that is largely directed towards the continental markets with a productivity that in the course of the modern period averaged around

1 Camba R. *et al.*, "Un'industria assistita", in Cannari L., Chiri S. (eds.), *Lo sviluppo economico della Sardegna*, Bologna, il Mulino, 2000, p. 39; Doneddu G., "Il sistema agropastorale sardo nel periodo unitario", in Bencardino F. *et al.*, *Mezzogiorno-Agricoltura. Processi storici e prospettive di sviluppo nello spazio EuroMediterraneo*, Milano, Angeli, 2011, pp. 493–528.

2 Paoli J.C., *Patrons et bergers sardes: origine et transformations d'un élevage ovin méditerranéen*, Institut national agronomique Paris-Grignon, thèse soutenue le 28 mai 1997.

3 Ortu G.G., "La transumanza nella storia della Sardegna", in *Quaderni Bolotanesi*, 1988, 14, pp. 159–73.

2,000 tons per annum.[4] If we calculate that 12 million liters of milk are needed to produce this amount of cheese and that 50 liters on average are produced annually by each one of around one million sheep and goats that make up the cattle at that time, it seems clear that around a quarter of the total Sardinian production is destined for export.[5]

The contemporary age begins in Sardinia around the middle of the nineteenth century, with the end of the feudal system and the "perfect fusion" between the island and mainland territories of the Kingdom.[6] With Italian unification, the national law leads to the privatization of most state-owned land, the place of the traditional activities of the rural subordinate classes.[7] These lands are acquired mainly by notables, already holders of large estates who rent them to small farmers without further investment. With the increase in maritime traffic and trade, Western Europe is overrun by cheap grain from Russia and America.[8] The price of wheat decreases by a third while the price of the cheese remains stable. The Sardinian durum wheat production destined for export is now directed towards a largely grown domestic local population (around 680,000 inhabitants in 1881 and 800,000 in 1901).[9] The tariff war with France ends the lucrative export of livestock from Sardinia.[10] These and additional factors favor the further growth of sheep farming, certainly more profitable, with a development driven by landowners-farmers who take advantage of the presence of cheap labor.

2. The birth of the dairy industry

In this situation at the end of the nineteenth century, a new fact is introduced. For various reasons, not least the considerable increase in demand

4 Di Tucci R., "L'industria casearia sarda sulla fine del Settecento", in *Giornale degli economisti e rivista di statistica*, 1930, 70, pp. 732–46; Doneddu G., "L'industria casearia in Sardegna", in C. Barciela y A. Di Vittorio (eds.), *Las industrias agroalimentarias en Italia y España durante los siglos XIX y XX*, Alicante, 2003, pp. 155–7.

5 Doneddu G., "Il Censorato Generale", in *Economia e Storia*, 1980, 1, pp. 86–7.

6 Sorgia G., *La Sardegna nel 1848: la polemica sulla "fusione"*, Cagliari, Fossataro, 1968.

7 Doneddu G., "La questione della terra in Sardegna tra pubblico e privato", in *Tra Diritto e Storia. Studi in onore di Luigi Berlinguer promossi dalle Università di Siena e di Sassari*, Soveria Mannelli, Rubbettino, 2008, I, pp. 963–73.

8 Sereni E., *Il capitalismo nelle campagne (1860–1900)*, Torino, Einaudi, 1977, pp. 222–369; Porisini G., *Produttività e agricoltura: i rendimenti del frumento in Italia dal 1815 al 1922*, Torino, ILTE, 1971; Lei Spano G.M., *La questione sarda*, Torino, Fratelli Bocca, 1922, p. 318.

9 Corridore F., *Storia documentata della popolazione di Sardegna (1479–1901)*, Torino, Carlo Clausen, 1902.

10 Gerlat S., "Les répercussions de la rupture commerciale franco-italienne de 1887–1888: la crisi économique sarde", in *Cahiers d'Histoire*, 1966, XI (3), pp. 257–65; Coda L., *La Sardegna nella crisi di fine secolo: Aspetti dell'economia e della società sarda nell'ultimo ventennio dell'Ottocento*, Sassari, Dessì, 1977, pp. 158–61.

for cheese coming from Italian immigrants in America, the dairy industry, operating mainly in Central Italy and which find itself in the middle of a production crisis, move large parts of its activities to the island.[11] This is how the first industrial dairies come to life (around fifty large companies with approximately twenty workers each), to which the other small dairies provide the product after the first processing phase (around a hundred and fifty small farms with five to six workers each).[12] The latter are strategically placed in the Sardinian territory to treat the fresh milk and prevent deterioration. A profound transformation of the Sardinian dairy landscape takes place in this way: the breeder/owner, who previously controlled the whole process from milking to marketing, now only supplies the raw material to the dairy where the industry determines the entire route by changing the typology of the product. The traditional Pecorino sardo cheese forms of around two to three kilos[13] (which still continue to be produced, albeit in small quantities, by many breeders) are replaced by Pecorino romano, which has a greater degree of salting and a longer shelf-life, and which is marketed in forms considerably heavier, varying between 20 and 35 kilos. The new production is carried out by master cheese-makers from Central Italy, who employ local labor. These innovations, in a closed world like that of the Sardinian countryside of the era, take place through a series of measures. The work of convincing and intermediating with some local merchants who help overcome mistrust and resistance is fundamental, and as a result these local merchants turn themselves into producers by acquiring modern dairies. Added to this is the availability of cash from the industrial manufacturers: in the early years of their settlement they purchase milk at prices higher than those current on the island, and with immediate payments and advances on future production they inextricably bind breeders to their manufactory.[14] In the space of around a decade, on the eve of the First World War, the situation of the Sardinian dairy sector appears profoundly changed compared to in the past. The rising demand is covered by the exponential growth of the cattle park in the first ten years of the twentieth century and reaches 2 million heads.[15] The increased amount of milk production is also linked to the selection of highly productive animals by breeders making wiser purchases

11 Vöchting F., *La Questione Meridionale*, Napoli, Istituto Editoriale del Mezzogiorno, 1955, pp. 383–91; Bussa I., "L'industria casearia sarda: storia, conseguenze e prospettive", in *Quaderni Bolotanesi*, 1978, 4, pp. 24–7; see also the chapter by Rita d'Errico in this volume.

12 Le Lannou M., *Pâtres et paysans de la Sardaigne*, Tours, Arrault et C., 1941, pp. 304–5.

13 Cerchi Paba F., *Evoluzione storica dell'attività industriale agricola caccia e pesca in Sardegna*, IV, Cagliari, Stef, 1977, p. 403.

14 Sanna L., "Vite parallele. L'esperienza industriale del Nord Sardegna nel vissuto e nella documentazione del sistema bancario e dell'informazione, dell'universo camerale e accademico", in Di Felice M. L., Sanna L., Sapelli G., *L'impresa industriale del Nord Sardegna. Dai pionieri ai distretti: 1922–1997*, Bari, Laterza, 1997, pp. 212–13.

15 Campus A., *L'allevamento della pecora*, Piacenza, Federazione italiana dei consorzi agrari, 1929.

in farm fares and exhibitions[16] (in 1928 the sheep herd book of Sardinian breeds is introduced). Furthermore, in some areas, transhumance livestock starts to become settled, with a positive impact on productivity.

Thanks in large part to the Sardinian dairy industry, the importance of the export of cheeses grows such in the early twentieth century that Italy transforms itself from an importing to an exporting country.[17] On the eve of the First World War, Italy occupies second place behind the Netherlands. During this period, grana-type cheese represents one third and the Pecorino cheese one fifth of the national production while exports are equivalent (over 9,100 tons for each type in 1913 with an increase in the previous years of over 100% for the Pecorino and over 30% for the grana-type cheese).[18] The Pecorino reaches a position of great importance, particularly in trade with the USA. In this country, due to the high demand from our southern emigrants, it is sold at prices 40% higher than Reggiano, Gorgonzola and Pecorino imitations, making it a particularly advantageous export. The residual portion of the production covers the entire island and the traditional markets of central and Southern Italy, where its value is less than 30% of the grana-type cheeses.

Immediately after the war, Alivia, a careful economist, highlighted this Sardinian feature, arguing for the necessity of complete liberalization of trade in dairy products to improve the regional economy.[19] In fact, the international market is far from saturated and further prospects for this increasingly in-demand product in new markets are opening. Alivia affirms in particular that the causes of the economic difficulties of the sector at the time are not so much the natural adversities but the erroneous policies of the national government. The conclusion is that the Italian government must facilitate the export by renouncing directing the product to the domestic market with its laws that are partly a remnant of the rationing restrictions of the First World War and with red tape that negatively affects freedom of trade. Along with the positive aspects, the difficulties that the export of Pecorino romano encounters are also highlighted. The first is caused by the activity of some Italian entrepreneurs who open dairies in Spain which produce cheeses similar to those in Sardinia: in particular Manchego, a seasoned Pecorino cheese that can become a dangerous competitor. Other Pecorino imitations are produced in Argentina and California; but we must not forget that even in the industrial north of Italy they attempt to mimic the Pecorino cheese

16 Falchi E., "Un'azienda agro-pastorale del Marghine. Formazione e prezzi dei prodotti", in *Quaderni Bolotanesi*, 2001, 27, pp. 305–30.

17 Dettori G., *Il movimento economico della provincia di Cagliari e della Sardegna dal 1881 al 1912*, Cagliari, Società Tipografica sarda, 1915.

18 Di Felice M.L., "La rivoluzione del Pecorino romano. Modernità e tradizione nell'industria casearia sarda del primo Novecento", in Mattone A., Simbula P.F. (ed.), *La pastorizia mediterranea. Storia e diritto (secoli XI-XX)*, Roma, Carocci, 2011, pp. 973–6.

19 Alivia G., *Per la libertà economica della Sardegna. L'industria e l'esportazione dei formaggi sardi*, Sassari, Gallizzi, 1921.

through a mixture of sheep's and cow's milk, sometimes even entirely basing it on cow's milk.[20]

The two decades between the wars are characterized by two periods completely different from each other.[21] In the 1920s there was a further increase in production and foreign trade. The 1930s, by contrast, are a time of profound crisis. The domestic policy clearly favors the expansion of cereal production compared to animal farming, which in Sardinia alone sees a loss (1937–8) of around 600,000 heads, of which more than 460,000 sheep and goats.[22] The international economy is characterized by the crisis of 1929 and the years following, and the closure of the US market has serious repercussions for the cheese export. This event clearly shows the danger of producing and marketing the product almost completely towards the North American market: a positive situation when demand is high, but clearly negative when the political and economic conditions and the unfavorable exchange rate for the dollar push towards depression of demand without appreciable alternatives.

3. The time of change

In the second half of the twentieth century, the opening of international markets raises profound difficulties in the cereal industry, still dominant on the islands and Central and Southern Italy, which has to deal with the stagnation of grain prices determined by foreign competition. As for Sardinia, this situation causes strong migration from the countryside.[23] It is a real exodus, that at its peak affected around 400,000 people, a third of the total regional population.

The repercussions for animal farming are significant, with a definite specialization in ovine and bovine cattle in the hills and mountain areas where laborers are no longer dedicated to cereal cultivation. The overall stability of the price of milk and cheese on the market allowed resistance to the obvious difficulties of the countryside until the end of the 1950s.[24] In the decade after the Italian industrial growth and the prospects offered by urban

---

20 Alivia G., *Economia e popolazione della Sardegna Settentrionale. Relazione sulle condizioni attuali (1928–1930) economiche e demografiche della regione*, Sassari, Gallizzi, 1931, pp. 227–32.

21 Campus A., *Il Pecorino romano e sardo: l'evoluzione dell'industria del formaggio Pecorino nel Lazio e in Sardegna*, Roma, Società Anonima Arte della Stampa, 1936.

22 Tognotti E., "Le campagne sarde nel regime fascista (1927–1939)", in *Archivio sardo del movimento operaio, contadino, autonomistico*, 1977, 8/10, pp. 171–3.

23 "L'emigrazione in Sardegna nel decennio 1961–71", in *La programmazione in Sardegna*, 1972, 39/40, pp. 25–8; Ortu L., Cadoni B., *L'emigrazione sarda dall'Ottocento ad oggi*, Sassari, Gallizzi, 1982.

24 Falchi E., "Un'azienda agro-pastorale del Marghine. Formazione e prezzi dei prodotti", in *Quaderni Bolotanesi*, 2001, 27, pp. 305–30.

development, but also through high fluctuations in the commercialization of cheese, conversely, a growing gap between the rural income and the average industry used as a comparison parameter is caused.[25] The problem for Sardinian farmers in this situation is therefore to keep their standard of living comparable to the average Italian non-rural standard of living. In this period, as in the past, control of the herd continues to be entrusted to one shepherd per hundred-average sheep and productivity appears blocked by the resources, by the nourishment of livestock, and the production of spontaneous fodder found during the transhumance.[26] Added to these factors are the cost of labor, often still bound by agistment contracts,[27] and, especially, the cost of renting land, which remains high despite the law on fair rent which remained largely unapplied. The small- and medium-scale shepherds dedicated to transhumance overcome this stage using the opportunities offered by the agricultural crisis. Already by the start of the 1960s many breeders, most of which come from disadvantaged mountain areas, moved with their cattle, attracted by the favorable real estate market conditions, to the (by this time) fallow countryside of Central Italy, especially in northern Latium, Umbria and Tuscany.[28] Others, successively, buy extensive plains, the traditional terminals of transhumance, until then rented at a high price on the island (Fig. 1). In the early 1970s the national De Marzi-Cipolla law automatically extends the precarious annual contracts for twelve years and recalculates the value of the land income, radically changing the relationship between the former owners of pastures and shepherd tenants, and making the rent and the purchase of land particularly favorable to the latter.[29] The regional development policy of the inland areas also allowed for various interventions in support of the tenants, starting with credit facilities for the acquisition of property that according to some calculations reached up to 300,000 hectares in the hill and mountain areas.[30] From this period on, thanks to the availability of land, farmers could devote themselves to the modernization of their companies and to initiating what we call the forage revolution. Strong regional subsidies allowed the purchase of tractors for

25 Olla D., *Il vecchio e il nuovo nella economia agropastorale in Sardegna*, Milano, Libreria Feltrinelli, 1969.

26 Porcheddu D., "L'allevamento ovino in Sardegna nell'analisi economica di Gavino Alivia", in Mattone A., Simbula P.F. (eds.), *La pastorizia mediterranea*, *op. cit.*, pp. 1026–30.

27 Ortu G.G., *L'economia pastorale della Sardegna moderna. Saggio di antropologia storica sulla "soccida"*, Cagliari, Edizioni della Torre, 1981; Ferrante R., "L'allevamento nel diritto civile dell'età moderna: il contratto di soccida e la codificazione ottocentesca", in Mattone A., Simbula P.F. (eds.), *La pastorizia mediterranea*, *op. cit.*, pp. 452–60.

28 Berlinguer L., "Emigrare (con le pecore) in Toscana", in *Ichnusa*, 1982, 1; Meloni B., "Le nuove frontiere della transumanza e le trasformazioni del pastoralismo", in Mattone A., Simbula P.F. (eds.), *La pastorizia mediterranea*, *op. cit.*, pp. 1051–76.

29 Setzi A., Antoniacci E., "L'evoluzione dell'economia pastorale sarda nel periodo 1960–1980", in *Quaderni Bolotanesi*, 1988, 14, pp. 143–52.

30 Idda L., "Realtà e problemi del mercato fondiario sardo", in *Quaderni Sardi di Economia*, 1972, 5–6; Paoli J.C., *Patrons et bergers sardes*, *op. cit.*, pp. 200–4.

planting grasslands, the construction of stables and the predisposition of the various measures necessary to make a herd more productive.

Figure 1. Areas in Sardinia that were once used for transhumance and were purchased by shepherds after 1970

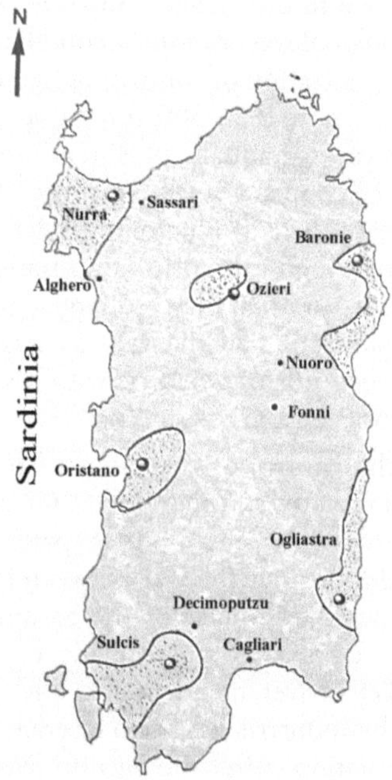

Source: Paoli J. C., *Patrons et bergers sardes*, *op. cit.*, p. 3 (elaboration by author).

4. Private and cooperative industrials

Immediately after the war the situation in the dairy industry is similar to that of the pre-war period: approximately 300 dairies of primary processing and a few dozen dairies concentrated mainly in Macomer, Thiesi and Cagliari, which carry out the complete transformation and preservation of the cheese.[31] Among all these, around a dozen industrials excel and monopolize the export of Pecorino cheese to the North American market. The product, which undergoes profound fluctuations both in production and in market value, is advantageously marketed by these entrepreneurs, who avoid putting it in circulation during periods of price shocks, setting

[31] Pili P., "Industrializzazione dei prodotti del suolo di Sardegna", in *Atti del convegno di studi per l'industrializzazione della Sardegna sotto l'egida della Regione Sarda* (Cagliari, 10–14 aprile 1953), Roma, La Tipografica, 1957, III, pp. 22–3.

up stocks to be used at the appropriate time.[32] They also determine the price of milk by intervening with payments at the beginning of the campaign or at the end of the summer and using the price fluctuation of the cheese according to their convenience. The myriad small farmers who cannot afford maintenance of stocks in anticipation of better times due to the size of their businesses are forced to sell to the bigger companies at lower prices in times of crisis. The private industrial sector in the second half of the century carries with it an increasing concentration of companies that develop alongside their modernization. From the more than 300 companies of the 1950s, with the demise of small businesses we gradually pass to around forty in the 1980s, a number similar to that of cooperatives.[33] From the 1950s to the 1970s, the private industrials increase their productivity under the double impulse of cooperatives that maintain high milk prices and the increased salaries of the wage workers. Investments are then directed largely to reducing the amount of labor with the mechanization of the various stages of production, and a reduction of up to two-thirds of workers, with a single employee able to produce up to 30 tons of Pecorino romano per year.

The alternative to the control of the industry by private individuals is the establishment of cooperatives. The first was set up in Bortigali in 1907 by an enlightened doctor who organized, in particular, cattle ranchers.[34] Cooperation then developed after the war between the small and medium-scale sheep farmers under the impulse of the Sardinian movement, which originated among the veterans of the Brigade of Sassari and made the cooperative system a key driver of economic and political aggregation.[35] The activity is blocked by industrialists, who leverage triumphant fascism to regain control of the situation. After the Second World War, thanks to the efforts of regional laws that favor the birth of cooperatives, they are placed back on the agenda.[36] Despite the high number of associated groups, only around forty of them develop over time with positive activity.[37] Between

32 Casu G.G., "La produzione della carne in Sardegna e prospettive per le industrie ad essa connesse", in *Atti del convegno di studi per l'industrializzazione della Sardegna*, *op. cit.*, III, pp. 27, 34–5; Costa N., "L'industrializzazione dei prodotti zootecnici in Sardegna", in *Atti del convegno di studi per l'industrializzazione della Sardegna*, *op. cit.*, III, pp. 47–8.

33 Camba R. *et al.*, "Un'industria assistita", in Cannari L., Chiri S. (eds.), *Lo sviluppo economico della Sardegna*, *op. cit.*, pp. 76–80.

34 Bussa I., Longu P., Piras T., *Dalla Latteria Sociale Cooperativa di Bortigali alla Lacesa (Latteria Centro Sardegna) 1907–2007*, Cagliari, LITHOSgrafiche, 2007, pp. 15–23.

35 Sechi S., *Dopoguerra e fascismo in Sardegna*, Torino, Fondazione Luigi Einaudi, 1969; Pisano A., "Associazionismo e cooperazione tra le due guerre (1918–1940)", in Sotgiu G. (ed.), *Storia della cooperazione in Sardegna*, Cagliari, CUEC, 1991.

36 Gentili A., *Il problema della pastorizia e la sua soluzione cooperativa*, Roma, ed. "La rivista della cooperazione", 1954.

37 Porcheddu D., "Le cooperative casearie sarde come strumento di ingegneria sociale", in *Quaderni Bolotanesi*, 2004, 30, pp. 51–78; Ruju S., "I caseifici cooperativi nella Sardegna del Novecento", in Mattone A., Simbula P.F. (eds.), *La pastorizia mediterranea*, *op. cit.*, pp. 1,003–010.

the 1960s and 1990s cooperatives get satisfactory results.[38] Initially they appear much too large in relation to the production obtained. Over time, they acquire a rate of growth that leads them from the 1960s to the 1980s to a better use of their facilities: 30% to 60%, with an average productivity per cooperative between the two periods of from 100 to 340 tons per annum and a potential individual average of from 250 to 700 tons (i.e. 4 million liters per cooperative).[39] Despite the under capacity still present and the spreading of investments, the development of cooperatives at this stage enables the goal claimed for decades by shepherds to be achieved: the alignment of milk prices with those of the cheese. Conversely, the marketing sector remains firmly in the hands of manufacturers, who also buy as wholesalers part of the production of the cooperatives. In terms of transformation, the cooperation erodes part of their benefits; but at the marketing stage, private entrepreneurs still have control of the situation.

5. The forage revolution

Since the forage revolution of 1975, the volume of milk supplied by farmers increases considerably, from approximately 25,000 tons in the 1970s to more than 55,000 in the mid-1990s.[40] The two major private companies and the two major cooperatives turn an average in 1995 of around 30 million liters a year each; all the others turn an average of between 3 and 10 million. The sheep cheeses are in direct competition with those of cow's milk. In addition to Pecorino romano, which continues to be the most produced type (30,000 tons in 1994, of which 20,000 are exported to the USA, and in the following decade an annual average of 50% of the total, with fluctuations of between 65% and 47%).[41] Other table cheeses, with lighter salting and lower conservation, increasingly enter the marketplace, in particular the so-called semi-cooked cheese, which, referring to the traditional production of the island, since the early 1990s is identified as the Sardinian Flower Pecorino "Fiore sardo", produced in molds weighing two to three kilos. In addition there are the raw and short-shelf-life cheeses, the so-called soft cheeses, similar

38 Porcheddu D., "L'industria casearia sarda attraverso un confronto tra imprese capitalistiche e cooperative di trasformazione", in *Quaderni Bolotanesi*, 2001, 27, pp. 291–304.

39 Idda L. *et al.*, "La cooperazione nel settore lattiero-caseario. Indagine sui caseifici sociali della Sardegna", in *Bollettino degli interessi sardi*, 1984, 13.

40 Sabattini G., "Caratteristiche strutturali e prospettive di sviluppo delle attività di allevamento e delle attività lattiero-casearie in Sardegna", in Savona P. (ed.), *Il settore lattiero-caseario in Sardegna. Aspetti e prospettive*, Milano, Angeli, 1983; Camba R. *et al.*, "Un'industria assistita", in Cannari L., Chiri S. (eds.), *Lo sviluppo economico*, *op. cit.*, pp. 78–9.

41 Laore. Agenzia regionale per lo sviluppo in agricoltura, *Le produzioni del comparto ovicaprino in Sardegna. Dati sulle produzioni – settembre 2011.*

to the "caciotta" types produced everywhere in Italy.[42] All these other cheeses, and mainly these soft table cheeses, reach 26,000 tons in 1994, of which around 18,000 tons are for export and only 8,000 are regularly consumed domestically on the island.[43] Towards the end of the century, thanks to the good economic conjuncture, production and export more than doubled compared to during the previous two decades. This significant increase and diversification of the product does not, however, positively affect the prices of raw materials and finished product. In fact, we witness from the 1970s to the 1990s a gradual decline in the price of milk paid to farmers and of all types of cheese marketed.[44] The trend in prices of Sardinian dairy products in the second half of the twentieth century does not differ substantially from trends for those of the rest of Italy. An early period of relative stability due to the free market, despite the periodic fluctuations of agricultural products, is followed, between the 1970s and the mid-1980s, by an upward trend in prices in the first phase of Green Europe and the beginning of the European Monetary System. The next period, after initial stability, is a period of price reduction coinciding with the imposition of milk quotas on European countries as a remedy for over-production, up to the devaluation of the lira and the exit of Italy from the European monetary system.[45] An important role in this event is played by the gradual increase in the consumption of dairy products in a country like Italy which, on account of its structural deficit, uses for most of its production milk imported from abroad, undergoing all the repercussions of "price trends and monetary policies".[46] The industrial production of Sardinia, while not depending on imported milk abroad, follows the course of the prices determined by milk and bovine cheese with which it competes, because of its wrong choices. In fact, it doesn't value the local peculiarity, or neither does it value the sheep's milk over the cow's milk, but on the other hand it tries to compensate for the reduction in prices with the further growth of the cattle park and an increase in milk production per head (130/150 liters per annum for sheep).[47] This increase is achieved with the new and intensified production of forage on pasture land and with the increasing use of animal feed. The result is an increase of the production of cheese continuously less typified as competing dangerously with many similar products on the domestic and international market. Some regions of Europe with traditional shepherds behave very differently. They limit and

42 Savona P. (ed.), *Il settore lattiero-caseario in Sardegna. Aspetti e prospettive*, Milano, Angeli, 1982.

43 Laore. Agenzia regionale per lo sviluppo in agricoltura, *Le produzioni del comparto ovicaprino in Sardegna*, *op. cit.*

44 Paoli J.C., *Patrons et bergers sardes*, *op. cit.*, pp. 296–8.

45 Furesi R., Gutierrez L., "Alcune valutazioni in tema di Politica Agricola Comunitaria, Accordi Gatt e riflessi sul sistema agricolo regionale", in *Osservatorio economico e finanziario della Sardegna. Rapporto 1995*, Sassari, Banco di Sardegna, 1995.

46 Perone Pacifico C. *et al.*, "Il settore lattiero-caseario in Italia", in *Analisi economiche dei prodotti agricoli – I.N.E.A.*, 1985, 5; ISMEA, *Filiera latte*, Roma, Ismea, 1995.

47 Paoli J.C., *Patrons et bergers sardes*, *op. cit.*, pp. 302–3.

control their productions, clearly identifying them, and keep prices very high.[48] While farmers continue to intervene, under the existing criteria, in the first part of the supply chain, manufacturers and retailers accept the limitations of production and achieve the highest level of valuation.

6. The current opposing dynamics

Many of the approximately 20,000 pastoral companies in Sardinia in the 1990s are considered by the most careful observers to be at risk of deep crisis and disappearance, especially in the poorest mountain areas.[49] Only those survive who manage to grow and modernize. This prophecy comes true in the first decade of the new century, especially in the second half. The trend of milk production is characterized between 2002 and 2007 by two three-year cycles that see their peak in the third year of each cycle (year 2004, 336,319.7 tons; 332,031.2 tons in 2007). The last three-year period beginning in 2007 is the opposite of during the year 2009 and drops the absolute minimum for the decade, with 293,102.7 tons. The next two years show a strong recovery in production.[50] The regional livestock sees new growth of sheep and in 2010 reaches 3,274,518 (though Istat sets the number at 3,414,194, of a national total of 7,900,016), and in 2011 3,314,466 heads.[51] The estimated production of sheep cheeses estimated from the milk production (six liters of milk per kg) obviously indicates the maximum production in 2004 (56,053.3 tons, the lowest in 2009 with 48,850.5 tons and a significant recovery in 2010 and 2011 respectively, with 54,575.3 and 55,244.4 tons). The production of Sardinian DOP cheese in 2010 consists of 25,752.5 tons of Pecorino romano, 558.8 tons of sweet Pecorino sardo, 1,376.7 tons of mature Pecorino and 458.1 tons of Fiore sardo.[52]

At the same time, we are witnessing the collapse of the milk price (around 60 cents per liter) and a crisis of over-production of Pecorino romano, which left much of the unsold cheese at a price of just 4 euros in August 2010.[53] This situation causes difficulties, pushing many companies to closure. A crisis characterized by its long duration has few precedents in the history of this important sector. It is determined to a large extent, especially

48 Ricard D., *Les montagnes fromagères en France*, Clermont-Ferrand, CERAMAC, 1995, p. 495; Flamant J.C., Labouesse F., *L'innovation technique agricole et son contrôle dans le développement de la filière Roquefort (1950–1985)*, Montpellier, Ensam-Inra, 1991.

49 Furesi R., "La filiera lattiero casearia", in Idda L. (ed.), *Agroalimentare in Sardegna. Struttura, competitività e decisioni imprenditoriali*, Sassari, TAS (s.a.), pp. 34–5.

50 Laore. Agenzia regionale per lo sviluppo in agricoltura, *Le produzioni del comparto ovicaprino*, *op. cit.*, p. 11.

51 *Ibid.*, p. 7.

52 *Ibid.*, p. 12.

53 Sassu A., "Formazione e innovazione: le cause della crisi del Pecorino romano", in Mattone A., Simbula P.F. (eds.), *La pastorizia mediterranea*, *op. cit.*, pp. 1,077–104.

in the final phase, by a decrease in demand from the US market due to competition from local bovine and Pecorino cheeses from other countries, in particular Spain and Greece. This confirms the importance of the USA for the export of Pecorino romano, which in 2004 and 2010 alone is directed at minimum prevalence to other markets (51% to 49%).[54] If we look at the situation over a twenty-year period we have an even more complete view. From 1990 to 2010, the agricultural and livestock fell by over 48%, from 19,555 pastoral companies to 12,669, a drop of nearly 7,000 companies, representing 35% of the total. But the collapse occurs mainly in the years 2000–10, with a decrease of 43.4%. However, once again, as seen in the past, in the countryside breeders seem to respond better to the crisis. In the first decade of the millennium they go from an initial ratio of 23.8% to 33.79% compared with farmers in 2010. In the two previous decades, the year of maximum production of Pecorino romano is 1995 with 38,431 tons, followed by 2004 with 38,184. The years between 1993 and 1998 see a very high production of Pecorino sardo, which reaches its maximum again in 1995 with 25,000 tons. The dates of 2015, and of 2014, a year of significant recovery, show that the farms represent exactly one-third of all companies operating in the agricultural world. They produce 44% of the value of a regional agricultural production of around 1.6 billion euros (24.89% in 2008). In this period, around 12,000 agro-pastoral companies are counted, which nevertheless have a greater extension of land and a greater number of animals than in the past: from an average of 121 heads in 1990 to 239 in 2010.[55] With transhumance finished, shepherds leave the mountain, left to the breeding of cattle, in favor of the plains. The statistics highlight the ability of technological and quality innovation (investments of 600 million euros originating from the community and a purchase of 6,000 milking machines, but also an improvement in the hygiene conditions of the stalls, refrigeration of milk, and introduction of a measure of animal welfare certified by agronomists and veterinarians).[56] Lastly, there is the decrease in the registry age of breeders, though probably these include numerous micro-farmers, typically young people who survive the economic crisis and unemployment, devoting a minimum of resources to breeding and/or agriculture.

Confirming the cyclical trend of the dairy industry on the island, the last two years show signs of an in some ways unexpected recovery. After reaching the minimum compensation (5.50 euros per kg in 2010 and 5.68 in January 2013) the Sardinian Pecorino cheese initiates, from that point onwards, an upward race (6.93 euros in December 2013, and 8.35 in September 2014

---

54 These figures and the ones that follow can be found in Laore. Agenzia regionale per lo sviluppo in agricoltura, *Le produzioni del comparto ovicaprino in Sardegna*, *op. cit.*

55 Contu M., "Sparito un pastore su tre cresce il numero di pecore", in *La Nuova Sardegna*, 15 luglio 2015, p. 15.

56 Venusti M., "Il ruolo dell'assistenza tecnica regionale nella filiera lattiero-casearia sarda", in Mattone A., Simbula P.F. (eds.), *La pastorizia mediterranea*, *op. cit.*, pp. 1,105–10.

to 9 euros),[57] which continues in 2015 to reach a maximum of around ten euros per kg. In the same period, sheep milk, which in Sardinia reaches lows of 0.63 euros per liter in 2011 and remains stable at 0.7 euros in 2012 and 2013, increases to 0.85 euros in 2014, to reach and exceed 1.10 euros with a maximum of 1.4 euros per liter in 2015.[58]

If we conduct a comparison with the price trends of Parmesan and Grana Padano, aged respectively 12 and 15 months, we can see that Pecorino romano, with over five months of ageing in the summer of 2014, reaches and then exceeds the value of the first two, which started from a higher peak, respectively 9 and 8 euros per kg at the end of 2013 and in early 2014 began a phase of considerable decrease.[59] Another fact to note is that in the period 2010–15, against a greater demand for Pecorino romano, the cuts in production from 27,000 to 22,000 tons, due in large part to the outbreak of blue tongue affecting cattle, induce an increase in price, with sales increasing from 150 to 200 million euros per annum.[60]

It is a situation that ends up satisfying for the first time (although to varying degrees) both producers of raw materials and industrials (including private and cooperative enterprises: around sixty to seventy companies, five or six of which control much of the overall market). It is therefore necessary to properly clarify the reasons for the growth in a time when, as we have seen, the main competitors appear in production crisis. It is also necessary to identify a strategy that allows the dairy production in Sardinia to remain stable at profitable levels and so continue to generate future numeric rewards.

As regards the first point, the good times appear certain, as in the past; fluctuations in the dollar-euro exchange in this period favor the dollar and make it easier to export to the USA, which has become the main buyer of Pecorino romano. This same "driving" price achieved by this cheese, higher than that of Padana cheese of a similar age, is not a new fact, and has been witnessed in the past and seen in other periods of the twentieth century. According to some experts, reducing the amount of salt in the cheese would increase appreciation on the foreign market by transforming the product, primarily to a grating table cheese.

In this situation, some increase production of Pecorino romano (which alone constitutes over 60% of the total – approximately 50,000 tons – of the various types of cheese produced in Sardinia in 2015) to exploit the

57 Pinna P.G., "Tetti alla produzione e contratti-tipo sul Pecorino romano", in *La Nuova Sardegna*, 14 febbraio 2015, p. 15; *Id.*, "Gli operatori: ora una cabina di regia", in *La Nuova Sardegna*, 23 luglio 2015, p. 15.

58 Laore. Agenzia regionale per lo sviluppo in agricoltura, *Osservatorio filiera ovi-caprina. Serie storica 1995–2014*; Piredda P.L., "La rivincita del latte: oltre 1 euro al litro", in *La Nuova Sardegna*, 26 maggio 2015, p. 31.

59 Pinna P.G., "Tetti alla produzione e contratti-tipo", *op. cit.*

60 "Coldiretti: crescita sbilanciata, devono restituirci 50 milioni", in *La Nuova Sardegna*, 5 febbraio 2015, p. 3.

favorable circumstances, but also in a search for new markets:[61] the end of the quota system and the consequent liberalization of production imposed by the European Union do indeed create a fear of saturation of existing markets, with inevitable negative consequences for profits. The prevailing view, however, seems to be to curtail the quantity to produce and sell in a balanced way to maintain profitable prices. Pecorino romano uses on average over 50% of sheep's milk produced in Sardinia, which in 2015 amounted to 310 million liters. It therefore seems preferable to leave the shares of "romano" unchanged and increase the production of the even more profitable Pecorino and Fiore sardo and of the other soft and semi-cooked table cheeses which see a continuous increase in demand. Divergences are also seen in the price of milk: the private industrials would like to reduce it. They argue that the price now far exceeds historical highs and consider it unsustainable in the long run.[62] Farmers' representatives, on the other hand, argue that there continues to be an excessive gap between the price of milk in Sardinia and that of cheese. A further rebalance is needed, recalling that the Sardinian industrials in the past justified the low price they paid by the supply and demand laws of the market which they now refuse to apply. (In the market squares of the peninsula, where up until today around 10% of the production of Sardinia is directed, the price was considerably higher; in 2011 Tuscany and Latium respectively saw a 0.89 euro and 0.81 euro per liter price, compared to 0.65 euros and 0.63 euros in Sardinia).[63] In reality, both industrialists and breeders argue that the price should be market-driven; but while the former refers to the milk market, the latter refers to the Pecorino market.[64] With a more equitable distribution of the profits, the milk producers and the cooperatives would benefit greatly from a capitalization process that would greatly reduce the use of bank credits (currently the pastoral corporations' debt is around 160 million euro), with clear benefits for the companies.

The contrast between farmers and industry is visible (even if it doesn't touch the dramatic level reached in years of deep crisis), and the decision of the European Commission supported by "Assolatte" to allow the production of cheeses using milk dust has further exacerbated the situation.[65] This directive, of course, encounters outright rejection in a region that ensures strict controls on milk production to guarantee high quality and the absence of antibiotics.

61 Tola T.G., "Troppo latte, il Pecorino rischia grosso", in *La Nuova Sardegna*, 10 marzo 2015, p. 12.

62 "Andrea Pinna: sacrifici parziali per fare il bene di tutti", in *La Nuova Sardegna*, 5 febbraio 2015, p. 3.

63 Laore. Agenzia regionale per lo sviluppo in agricoltura, *Le produzioni del comparto ovicaprino, op. cit.*, p. 25; Laore. Agenzia regionale per lo sviluppo in agricoltura, *Osservatorio filiera ovi-caprina, op. cit.*

64 Pulina G., "Prezzo del latte di pecora, lo può stabilire un algoritmo", in *La Nuova Sardegna*, 17 novembre 2015, p. 17.

65 Fiori L., "Battaglia contro l'uso del latte in polvere"; *Id.*, "Traditi dalle industrie del latte", in *La Nuova Sardegna*, 30 giugno 2015, p. 15.

We conclude with some brief comments on other productions present in Sardinia. Nationwide statistics generally do not distinguish between sheep and goats, but report an aggregate under the heading "ovicaprids". According to available figures, however, Sardinia also holds the record in the goat sector. In the almost thirty years between 1982 and 2010 there is a reduction of 35% of the herds, but the heads increase by around 7% and the average farm size increases from 55 to 91 goats. Between 2010 and 2011 the number of goats increases from 288,088 to 294,430. Today the island harbors around 28% of the animals present in Italy and produces 46% of the milk.[66] Finally, we must report the annual production of 200 million liters of cow's milk. Over 90% is processed by the cooperative of Arborea, a rural town in Oristano, which has around 250 members and as many employees. Founded in the 1950s in an area reclaimed in the period between the two World Wars,[67] it currently figures among the top sixteen Italian companies by turnover (in 2013 it declared revenues of 131 million euro) and fourth for the production of drinking milk (fresh and UHT), with 191 million liters produced.[68] It is a reality that also looks with growing interest to the sheep and goat sector, and that is characterized by strong growth policies both in the retail market and towards the marketing of its products to Asia and Africa.

7. Conclusion

The euphoria over the growth of the Sardinian dairy sector in the years 2014–15, with a record increase in Pecorino exports and the good price achieved on the international market, is short-lived. A significant decrease in business, accompanied by a significant drop in prices, confirms the fluctuating trend of this product in 2016. From peaking at 10 euros/kg and more than 1 euro/l respectively, prices have dropped to under 6 euros/kg and under 70 euro cents/l. Once again, as noted in the past, the controversy is particularly strong among farmers, producers of raw material, industrial processors and marketers.

But above all, mutual accusations are being thrown around as to who is responsible for the increase in excessive supply driven by the favorable demand in 2015. Production increased from around 280 million to over 350 million liters of milk, an increase in the production of cheese of almost 10,000 tons. The consequence is a drop in prices when the market loses momentum. On the other hand, the end of the milk quota system decided by

66 Contu M., "L'isola è la terra delle capre. Boom di allevamenti e latte", in *La Nuova Sardegna*, 19 ottobre 2015, p. 14.

67 Lullia P., "Arborea", in Brigaglia M., Tola S. (eds.), *Dizionario Storico-Geografico dei comuni della Sardegna*, Sassari, Carlo Delfino editore, 2006, 1, pp. 69–72.

68 Selloni S., "La 3A punta verso il mercato toscano", in *La Nuova Sardegna*, 26 febbraio 2015, p. 11.

the EU has inevitably led to increased supply, with negative consequences, so that the same EU now finances the voluntary reduction of milk production to stop the falling prices.

Looking back at Sardinia, the identification of a large shipment of Pecorino from Romania directed to the island's biggest company reignites tensions. Breeders say that the industrialists, who verbally complain about the increase in supply and the excessive accumulation of unsold surplus cheese, are introducing non-controlled quantities of Pecorino from abroad and then selling it as Sardinian Pecorino, a classical counterfeit. Industrialists say, conversely, that the release of this type of grating cheese is certified with its true origin and does not compete with the typical product of the island. The reality is that beyond these skirmishes the international market, but also the Italian national market, is increasingly flooded with cheap imitation products that end up objectively entering into competition with Italian products. And most of the imports come from national companies which have settled abroad to take advantage of the low overhead costs, especially those of the local workforce, and the many tax breaks. In many ways, again, we see the same strategies as those already used by the Italian industrials nearly a century ago and repeatedly condemned by economists at the time.

As we can see, the positions once again remain distant and revive ancient controversy. Not even on production figures can an agreement be found. It is not known how much milk is actually transformed and how much Pecorino cheese is produced. The intervention of an impartial body to settle disputes and monitor the progress of production and prices is invoked. The interprofessional body of the dairy sector, promoted by the Region of Sardinia to strengthen supply chain strategies, still faces difficulties in taking off. This is because, as affirmed by the Regional Agricultural Councilor, industrialists delay data-sharing on production, claiming problems related to privacy, and as of today (September 2016) they have not joined the proposed organization. Finally, a further project supporting the Sardinian dairy sector was devised by the region, with a series of innovative financial instruments: the launch of, in particular, the "Pecorino Bond", which consists of ten agri-food companies that will emit mini-bonds at 36 months to support the working capital giving as partial collateral the shelved Pecorino. The SFIRS, the Regional Financial holding, will also support emissions using a number of innovative financial instruments that will act as a multiplier of available resources. Despite these and other attempts, Sardinia, the first global exporter of Pecorino cheese, is unable to fully exploit its wealth, because of old and new errors.

The Tuscan Ovine Milk Industry, 1950–2000

A Quantitative Analysis

Renato Ghezzi

University Magna Graecia, Catanzaro

1. Introduction

Since the time of the Etruscan civilization, sheep cheese has always had primary importance in Tuscan food culture. An untold amount of historical proof shows how the most notable characteristics of the product and its production have changed throughout the century. For example, Pliny the Elder cited, in his *Naturalis Historia, Il Lunense*, "Magnitudine cospicuum", or "Remarkable magnitude", produced in the Lunigiana and exported to Roman markets.[1] In the medieval period, several sources included *Cacio Marzolino*, with its typical oval shape, as the most famous Tuscan cheese. In 1475, via the *De honesta voluptate et valetudine* thesis, humanist Bartolomeo Platina stated that this cheese, like Parmigiano, was the absolute best in the Italian peninsula.[2] Francesco Molinelli, in his memoir entitled *Sopra i formaggi di Toscana* (Above the Tuscan Cheeses) presented at the Academy of Georgofili, after having written that Tuscany "doesn't allow any nation to surpass it in quantity and delicacy of its fresh and dried cheeses", described and analyzed the production process of *Marzolino di Lucardo*, of fresh *Terrinca* and *Cappella Cacio*, the *Raviggioli* of Cortona, Pisa and Livorno, Chianti *Marzolino*, and *Cacio* from Crete.[3] In his 1832 *Shepherd's Manual*, Ignazio Malenotti, a member of the Linneana cooperation of Paris, used for the first time the expression "Tuscan sheep cheese" to specify a product that "you use in almost all parts of Tuscany" and, aside from all the various sites of production, this had a common working method over all the region and presented specific market characteristics compared to those of other Italian cheeses.[4] Throughout the nineteenth century, Tuscan sheep cheeses consolidated their popularity abroad as well, obtaining much recognition:

1 Gaio Plinio Secondo, *Storia Naturale*, Vol. II, book 11, para. 241, Torino, Einaudi, 1997.

2 Carnevale Schianca E. (ed.), *De honesta voluptate et valetudine. Un trattato sui piaceri della tavola e la buona salute*, Firenze, Leo S. Olschki, 2015.

3 Molinelli F., *Sopra i formaggi di Toscana*, in "Atti della Reale Accademia dei Georgofili", Vol. II, Firenze, A.G. Pagani, 1795, pp. 153-166.

4 Malenotti I., *Il manuale del pecoraio*, Pisa, Tipografia Pacini e figli, 1832, p. 192.

this was awarded in Paris in 1868, in Portici in 1877, and in Genoa in 1879.[5] The importance of the ovine dairy industry for the region continued to grow until the first decade of the twentieth century, to then meet a prolonged period of difficulty tied to a recession in ovine farming.

The aim of this chapter is to detect the root causes that determine the crises in Tuscan ovine farming in the 1950s. It highlights, therefore, the structural mutations undergone by the Tuscan dairy farms during the subsequent consistent recovery of livestock. In the first part of the chapter we analyze in particular the effects on ovine farming, from the land reform to the subsequent massive migration of shepherds coming from Sardinia. The second part focuses on ovine milk production. On the basis of a quantitative analysis, it has been established that productivity increased strongly starting from the 1970s. During the same period an important cooperative movement began to be established in Tuscany. In the last part of the chapter we highlight the dairy cooperatives' role in the production of Tuscan sheep cheese and in the process that has determined the acquisition of the PDO (Protected Denomination of Origin) trade mark. This document values, furthermore, the effects of the PDO trade mark on regional ovine dairy production.

2. The livestock

At the dawn of the twentieth century, herds constituted an element of the Tuscan agricultural landscape. The amount began to progressively reduce during the 1930s due to the politics of autarchy, which encouraged grain plantations, penalizing pastoralism. Thus, if in 1908 the head count reached 1,126,000, then in 1938 it decreased to around 1,000,000.[6]

The problem grew worse during the Second World War. In 1952, there were no more than 741,000 sheep present in Tuscany. In terms of amount of herds, Tuscany was fifth in Italy after Sardinia, Latium, Abruzzo and Puglia. The province with the highest head count was by far Grosseto, which held 31% of regional sheep assets. Following were Florence at 18%, Arezzo at 17%, Siena at 11%, Pistoia at 3.6%, Massa at 3%, Livorno at 2.8% and Lucca at 2.2%.[7]

In 1961, the year of the first general agricultural census, Tuscan sheep were reduced almost by half: between 1952 and 1961 327,262 disappeared.[8]

---

5 De Bellis A., *Il cacio Pecorino tra storia e tradizioni*, Montepulciano, Editori Il Grifo, 1982.

6 Consorzio per la tutela del Pecorino toscano DOP (ed.), *In nome del Pecorino toscano. Storia di sapori e antiche tradizioni*, Verona, Oriongraph, 2006, p. 29.

7 Tornar F., *Toscana agricola*, Pisa, Pacini, 1971, pp. 164–70.

8 Istat (Istituto Nazionale di Statistica), *1° Censimento generale dell'agricoltura*, 15 aprile 1961, IV, *Bestiame*, Roma, 1976, pp. 200-203.

The region then followed a national trend. In a country that was in the midst of an economic boom and quickly industrialized, the Italian herds registered a decline of around 30%[9] due to the complexly caused decrease of the people in charge of the primary sector, while the contributions of the primary sector reduced the gross national product by half.

During those years, there were large imports of food from abroad in the hopes that there would be a trade-off between this and the increase in industrial product exports.[10] Already in 1960, the negative trade balance of agricultural products due, mainly, to import of livestock products,[11] began to raise concerns.

These alarms went unanswered, and mass migration away from the farmland, which seemed advantageous at first, went out of control because of persistent job vacancies. In fact, 1950–60 saw 2 million workers leave the Italian plantations, followed by another million and a half in the next fifteen years, without any type of intervention to guide the farming industry towards new conditions arising from the decline of manual labor.[12]

In Tuscany, the decline of sheep livestock was, however, much more clear-cut. Breaking down the data on a provincial level, one can see that the regional data are conditioned by the strong declines registered in Florence and Grosseto (-90,000 heads).

In this last case, the fall of pastoral businesses was emphasized by the effects of the land reform in the Maremma. Before the reform more than 50% of the Maremma productive surface area was occupied by arable land. The pastures and forests occupied around 13% to 29% of the land.[13] In this agricultural economy, sheep livestock had an important role. While in the province of Florence most herds were stationary, in Grosseto 232,770 heads were counted for in 1952, and 130,500 of these were nomadic. If the stationary livestock made up most of the industry and agricultural supplement, then as a rule the nomadic livestock was characterized as an independent business renting fields and pastures, and sometimes even a mountain.

With sandstone-based terrain, long winters and slopes, the Apennine mountains did not provide ideal agricultural conditions. However, their forests and pastures always allowed for livestock. In Tuscany, annual rainfall varies from 500 to 3,000 mm depending on the altitude, and the climate

9 *Ibid.*, pp. 256–74; Inea (Istituto Nazionale di Economia Agraria), *Annuario dell'agricoltura italiana*, 1955, 9, p. 95.

10 Fabiani G., *L'agricoltura italiana tra sviluppo e crisi*, Bologna, Il Mulino, 1986, p. 171.

11 These items passed from a percentage import of the 24% import total during 1951–3 to 45.1% in the period 1957–9. *Ibid.*, p. 173.

12 *Ibid.*, pp. 174–5.

13 Camera di Commercio, Industria e Artigianato di Grosseto (ed.), *La riforma fondiaria in Maremma, 1951–1954*, Roma, A. Stederini, 1955, pp. 28–48.

is characterized as warm summers and particularly dry plains. The most important effect of the climate on the livestock is determined by the seasonal pastoralism, with certain seasons better than others. The Apennine high plains offered good pastures in the second part of May, after the snows melted, up until the start of October. In the Maremma plains, grasses dried under the hot sun of June, July and August, then grew vigorously in the winter and spring. The most efficient means for livestock was to use both pastures and plains during the lush grass seasons. Nomadic pastoralism was used to solve other problems, such as the lack of water in the plains.

During the summer, sheep need around five liters of water per day. This can become difficult when one has to frequently use well pumps. Moreover, a long period of time in dry pastures produces a lack of vitamin A, the only vitamin vital for sheep. As Cianferoni wrote in 1969, "nomadic sheep normally produce double the amount of milk than that of stationary sheep".[14] In Tuscany, the flow of nomadic pastoralism originated from the entire Apennine range, from both the Tuscan side and from Emilia, Romaga, the Lunigiana, Garfagnana, Pistoia, Mugello, the Casentino, Val Tiberina and the Marchigiano. Its objective was to cover the Tuscan coastal plains, especially those including the river mouths of the Arno and Chiarone. In this area, pastoralism mixed in with the agricultural economy with a broad and seasonal characteristic, such as grain production.[15]

14 Cianferoni R., *La pastorizia nel Lazio e nell'Abruzzo*, Roma, Inea, 1969, p. 121.

15 Calzolai L., Marcaccini P., "La transumanza appenninica in età moderna e contemporanea", in Calzolai L., Kovalevich M. (eds.), *Allevamento, mercato e transumanza sull'Appennino*, Atti del Convegno Ponte Presale 29 settembre 1999, Sestino-Badia Tebalda, CREAAP, 2000, pp. 33–54; Caselli G., "La 'Via tyrrenica' e altri percorsi pastorali della Toscana collinare", in *Campagne maremmane tra '800 e '900*, Atti del Convegno di Studi *Agricoltura e società nella Maremma tra '800 e '900*, Firenze, Centro 2P, 1983, pp. 287–320.

Figure 1. Nomadic pastoralism in Tuscany

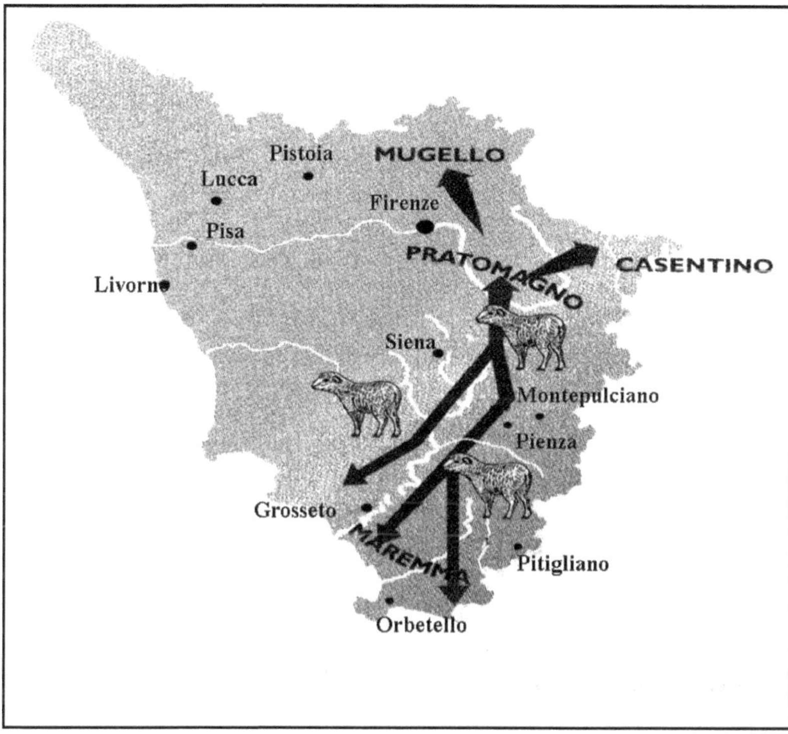

Source: Consorzio per la tutela del Pecorino toscano DOP (ed.), *In nome del Pecorino toscano, op. cit.*, p. 34.

This system had problems with agricultural reform. In favor of designating farm land, this reform caused a regression in both summer pastures and in the growth and diversity of cultivation. The breaking up of the terrain into ten to fifteen acre properties also caused a great recession for domestic herds, which were first seen as some of the largest sharecrop farms, reaching an average of sixty to seventy acres of land, made up in part by grazing pastures.[16]

Nonetheless, in the second half of the 1960s, a new and expansive management began for the Tuscan sheep hoarding business. This extended into the 1970s, when the region came to own 8% of the nation's sheep.

The second general agricultural census reveals that in 1970 the entire ovine division was estimated at around 490,000 heads.[17] Later studies by Istat present additional information, which made the estimate rise to 553,000

16 Camera di Commercio, Industria, Agricoltura e Artigianato di Grosseto (ed.), *La riforma fondiaria in Maremma, op. cit.*, pp. 31–2.

17 Istat, *2° Censimento Generale dell'agricoltura*, 25 ottobre 1970, IV, *Bestiame*, Roma, 1974, pp. 427–8.

heads on 31 December 1974. On 31 December 1976, the ovine net worth surpassed 600,000 units, with an increase of 11%. The tendency to gradual expansion stabilized in the following two-year period: in the year 1977 head counts reached 630,000, and in 1978 673,000.[18]

The recovery established itself upon new fundamentals. On a territorial level it had a redistribution of herds that benefited the western and southern provinces. The amount of livestock in the provinces of Pistoia remained more or less stable, while Lucca, Florence and Arezzo registered a rather steep decline: on a whole, between 1961 and 1978, around 25,000 ovines disappeared from these provinces, where, by contrast, the Grosseto livestock once again grew from 70,000 heads between 1961 and 1978, and the Sienese and Pisan countryside also grew in mass.

In 1961, the Siena province had 57,187 heads, which made up around 13% of Tuscan herds. The 1970 census registered 94,506 head units (19% of the region), reaching 150,000 (21.6% of the region) in 1978. In Pisa, registered head counts went from 34,000 in 1961 (8% of the region) to 52,000 in 1970, reaching 134,000 heads in 1978 (17% of Tuscany's).

The new altimetry zoning distribution for businesses emerged drastically changed. In 1961, 58% of ovines lived among the internal hills, 32% in the mountains and only 9% in the plains. In 1978, the herds from the mountainous regions were reduced to 14% of the region, where instead those in the planes and coastal hills grew significantly (32% total).[19]

In the 1960s and 1970s, an important contribution to the Tuscan ovine culture's potential was generated by Sardinian immigration. Since the 1950s, Tuscany has made up a strong part of the rural depopulation, mainly due to the crisis in share cropping as a business. From 1954 to 1973, those involved in agriculture decreased by 75% from 530,000 to 132,000, and the cultivating terrain suffered a reduction of around 25,500 acres.[20] There is a link between the migration of the share croppers, the abandoned farm fields, the drop in land values, the absence of land market competition especially in the mid- and high hills, and the Sardinian migration. The immigration phenomenon was most intense in the areas that had more share cropping and where the small farming properties were less widespread, most of all in Siena, Pisa, Livorno and Grosseto. Instead, the immigration phenomenon was less notable in the provinces where small farms were greater, such as in Pistoia, Massa and Lucca.[21] The Sardinian shepherds joined in with their savings to fill in a gap for the less productive regional lands: they asked to rent pastures, and they agreed upon agistment contracts with the landowners. In the early

18 Istat, *Annuario statistico della zootecnia, pesca e caccia*, 1975, 16, p. 27; *La pastorizia in Toscana: aspetti, problemi, proposte*, Preparatory document at the Regional Conference of Siena, 28–9 March 1981, Tuscan Region, 1981, p. 6.

19 *La pastorizia in Toscana: aspetti, problemi, proposte*, *op. cit.*, p. 6.

20 Tornar F., *Toscana agricola*, *op. cit.*, pp. 62–5.

21 Meloni B., "Pastori sardi nelle campagne toscane", in *Meridiana*, 1995, 26, pp. 167–202.

1960s, the first arrivals registered with the Sienese and Grosseto provinces, where they found abandoned properties and buildings to be restored.

They joined in as shepherds when they purchased the land and became farmers. They gave themselves a head of operations, they promised valuable work in recovery and reconstruction of the old colonial farm houses, and they contributed to the conservation of the fertile land and handling of the territory.

It is difficult to calculate this phenomenon precisely. However, in 1978, there were 667 Sardinian members in the provincial welfare of Arezzo, Grosseto, Florence, Pisa and Siena, who made up 86% of Tuscan ovines, and they raised around 163,000 sheep, around 67% of the total.[22]

According to a study by research groups undertaken by Pier Giorgio Solinas, in the province of Siena in 1986 there were 1,256 new Sardinian shepherds, 340 families owning 16,000 acres of land, and around 100,000 livestock, with an average head count of around 300 per business.[23]

The Sardinian shepherds arrived with their herds, introducing a Sardinian race to the region whose spread proved an even more expansive influence on the area from Mugello to Valdarno, and to the internal areas of the Livorno, Pisa and Grosseto provinces.

Up until the start of 1960s, the predominant breeds in Tuscany were the massese, found throughout the region and known for their double predisposition to produce milk and meat; the Appeninnica, mostly found in the provinces of Pistoia and Florence, who even though they were not great producers of milk and meat were valued for their great ability to adapt to different environments; the Comisana, a dairy sheep mostly found in Grosseto; the Sopravvisana, typical of the province of Pisa, belonging to the Merino group and therefore good producers of wool; and the Garfagnina, raised primarily for meat production.[24]

In 1972, the Sardinian breed had already become the most popular in the region: the census shows that there were 178,319 heads and that they made up for 33% of Tuscan counts.[25] In 1978, its population grew even more: the Sardinian breed grew to more than 200,000 heads, around 42% of Tuscan herds. The massese continued to prevail only in Northern Italy, especially in

22 Dini M., *Il contributo dei coltivatori e dei pastori sardi nell'economia agricola toscana*, Firenze, Confederazione Nazionale Coltivatori Diretti, 1980, pp. 15–16.

23 Solinas P.G., *Pastori sardi in provincia di Siena*, 3 Vols., Siena, Etno-Anthropology Laboratory, Department of philosophy and social sciences, 1990.

24 Ministero dell'agricoltura e delle foreste, *Indagine nazionale su alcuni aspetti degli allevamenti e delle produzioni ovine*, Roma, Associazione nazionale della pastorizia, 1972, pp. 104–9.

25 Castor (Caseifici Toscani riuniti) *et al.* (ed.), "Il riconoscimento della denominazione d'origine del Pecorino toscano", in *Conferenza regionale di Siena 28-29 marzo 1981*, Regione Toscana, 1981, pp. 5–6.

the Lunigiana, the Versilia, Lucchesia and Pistoiese. The Appenninica mainly stayed in Mugello and the Casentino.[26]

In this period, not only did the breeds change, but the dimensions and the organizations of the businesses were transformed. In the 1950s, sheep livestock, representing a vital sector of the regional zoo technical division, functioned mainly as a traditional pastoralism relegated to secondary and notably poor terrains. According to the findings of Istat and the first general agricultural census, in 1961 ovine businesses in Tuscany numbered 16,789, though their sizes were very moderate, with each herd at around 24 heads. The most popular type of business was that made up of 1–5 heads, whereas at the opposite end there were only around sixteen businesses with over 500 heads, concentrated exclusively in Grosseto, with twelve examples, and in Siena, with four.

To begin with, the sudden change in quality occurred with the purchase in part by the farmers of adequate land areas offering pastures destined to be used for herds. A good part of the hill country and plateau, abandoned or misused, established in the early 1970s a stable foundation for ovine livestock. In 1978, the land properties interested around 63% of the ovine businesses, with a strong hold in the Grosseto area (85% of the businesses).[27] Credit and tax benefits played an important role in the development of the farming properties. From 1965 to 1980, such breaks interested 3,340 pastoral businesses in a total area of 61,532 acres. With a land area of around 34,009 acres, 803 of these properties benefited from low rate mortgages for a total amount of 18 billion lire, equal to a third of the total mortgages given for the development of farming properties in Tuscany.[28] Instead, there were 2,537 properties with 27,123 acres which only gained from tax benefits.[29] As a whole, 47% of land purchased with credit and tax benefits were to be used for shepherds originating from Sardinia.[30]

As regards the types of businesses, in 1978 farming businesses constituted 80% of the businesses and over 70% of the ovine assets. The economic businesses, with salaried employees and shareholders, made up 7% of total businesses and 18% of ovine assets. They were mainly found in Siena, Grosseto, Florence and Arezzo. The sharecropping type of businesses made up less than 6%, and agistment contracts were completely gone. Herds became more numerous than in the past. In 1970, each business had on

26 *La pastorizia in Toscana: aspetti, problemi, proposte*, *op. cit.*, pp. 6–14.

27 Rent showed instead still prevalent in Livorno, Pisa and Florence.

28 Of these, 328, with a surface area of 21,590 acres, belonged to shepherds originating from Sardinia. *La pastorizia in Toscana: aspetti, problemi, proposte*, *op. cit.*, p. 8.

29 210, with a surface area around 7,699 acres (47% of purchased land with financial benefits), belonged to Sardinian shepherds, *ibid.*

30 Dini M., *Il contributo dei coltivatori e dei pastori sardi*, *op. cit.*, pp. 21–2.

average 54 heads,[31] then in 1975 63 heads,[32] with a value almost triple that of 1961.[33]

During the 1980s, the percentage of Tuscan ovines at national level remained largely unchanged. However, Tuscan herds continued to grow in unlimited value: they grew from 673,000 heads in 1978 to 712,500 in 1981, 831,300 in 1985 and 864,100 in 1987 (the highest value registered for the second half of the twentieth century), to then lower in 1989 to 817,700.[34] Tuscany had by then surpassed Puglia and Abruzzo, and after Sardinia, Latium and Sicily was the fourth region in Italy for ovine assets.[35]

During the same years, the focus on the business process continued. In 1975 there were 8,476 Tuscan ovine businesses, which had on average 63 heads. The 1982 census revealed 8,101, which later dropped to 7,556 in 1990. Each business, however, had herds made up of an average of 95 heads.[36]

On a territorial level, the Grosseto countryside grew significantly. The fourth general agricultural census identified 31,137 businesses in the province, which had herds made up of an average of 111 heads. Therefore, not only did Grosseto have the majority of businesses, but they were also the largest of the region. In 1990, 54% of all businesses and 59% of ovines were mainly in six counties only: Scansano, Manciano, Sorano, Grosseto, Magliano and Roccalbegna. The largest herds were in Scansano and Manciano, which alone had 25% of the sheep and 24% of the businesses in the province. Thus, livestock prevailed in the central-southern part of the province.[37]

With the strong growth of Grosseto, the reduction of head counts in the rest of the region is evident. Herds from the province of Pisa suffered large cutbacks (from 134,000 heads in 1978 to 63,089 in 1990) and Livorno (-12,400 heads between 1978 and 1990), as well as those in Siena, with -7,300 heads, Pistoia (-3,100 heads) and Arezzo (-3,000 heads). At this point, Tuscan production associated itself more with the Maremma.[38]

---

31 Istat, *2° Censimento Generale dell'agricoltura, op. cit.*, pp. 427–8.

32 Istat, *Annuario statistico della zootecnia, pesca e caccia*, 1976, 17, p. 27.

33 Of the 8,476 businesses counted in 1975, 678 (8%) belonged to Sardinian immigrant shepherds. This pertained mainly to large-scale businesses: on average they had 197 heads. Dini M., *Il contributo dei coltivatori e dei pastori sardi, op. cit.*, p. 16.

34 Istat, *Annuario Statistico della zootecnia, pesca e caccia*, 1982, 23, p. 27; Istat, *Statistiche dell'agricoltura, zootecnia e mezzi di produzione*, 1985, 33, p. 288; Istat, *Statistiche dell'agricoltura, zootecnia e mezzi di produzione*, 1987, 35, p. 340.

35 In 1987, Sardinia alone possessed 34% of the entire national ovine assets.

36 Istat, *Censimento agricoltura 2010, Serie storiche*, <http://daticensimentoagricoltura.istat.it/index.aspx?queryid=13>.

37 *Ibid.*

38 Righini A. "Il ruolo del settore lattierocaseario ovino nell'economia agricola toscana", in Tuscan Region, Department of Economic, *Atti della Conferenza Regionale dell'Agricoltura della Toscana* 1996, Firenze, Angeli, 2007, pp. 143–7.

Figure 2. Sheep population in the various provinces of Tuscany between 1970 and 2000 (%)

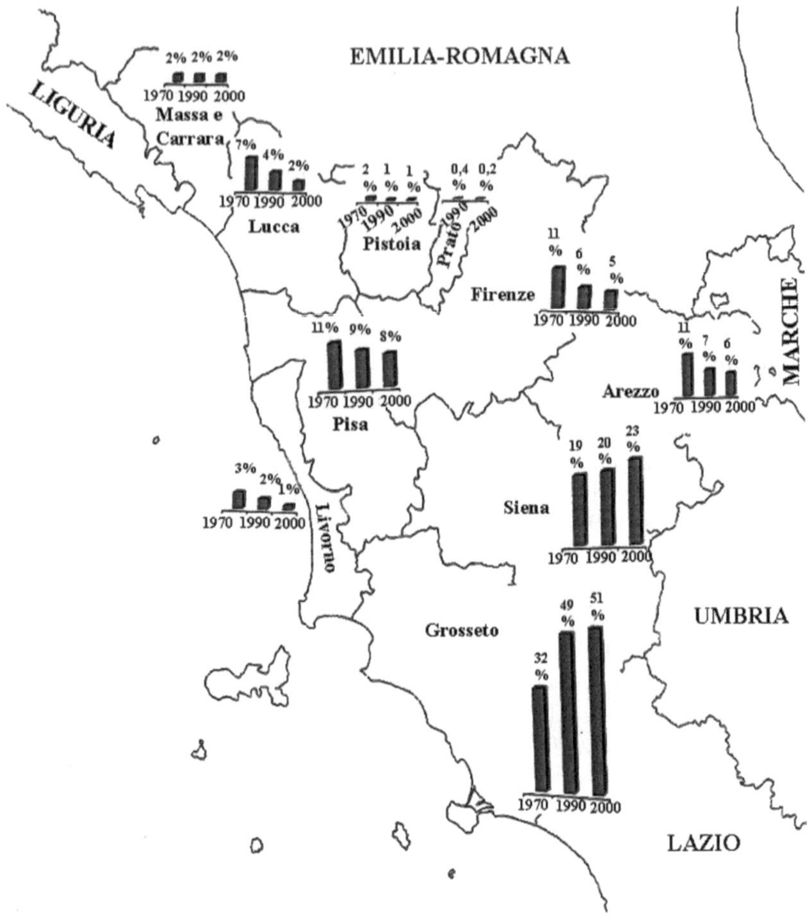

Sources: Istat, *2° Censimento Generale dell'agricoltura*, *op. cit.*, pp. 427–8; Istat, *Censimento agricoltura 2010*, *op. cit.*, <http://dati-censimentoagricoltura.istat.it/index.aspx?queryid=1333#>.

The expansion process, which ended at this time and which had characterized the Tuscan ovine culture in a continuing way, was abruptly interrupted during the last ten years of the century. The 1990s, and especially the years 1995–2000, represented a difficult period for the sector. According to the fifth general agricultural census, there were only 471,064 ovines left in Tuscany by the year 2000. In just ten years, 161,674 heads disappeared and Tuscany returned to a lower rank than in the 1980s. Overall, the province that registered the largest cutbacks was Grosseto, at –67,320 heads. However, percentage-wise, Grosseto suffered a minor reduction of herds compared with the average for the region. In fact, in the year 2000 its position in the sector even showed greater growth than in the previous ten years.[39]

In a way, the crisis was the cause of the difficulties that many farmers dealt with in order to adjust to the county laws No. 46–7/92, which, in order to guarantee better sanitary conditions, imposed a structural modification for livestock: the creation of new stalls and equipped milking rooms.[40] As the fifth agricultural census shows, Tuscan businesses were more commonly family run, with few capital endowments, and therefore rarely capable of confronting high costs.[41] The adjustment to new quality standards proved even more difficult during the decade, characterized by a substantial lull in the ovine milk prices. As Fig. 4 shows, the average prices of ovine milk in Italy dropped in 1990–2, followed by a brief comeback, and then another reduction between 1996 and 2000.[42]

Figure 3. Average prices of ovine milk in Italy (€/l)

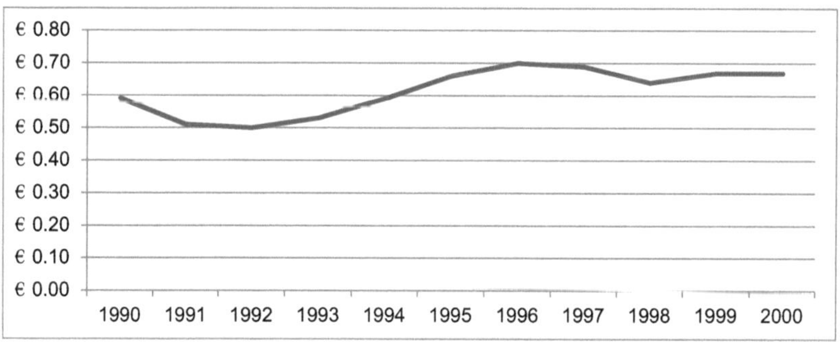

Sources: Istat, *Statistiche dell'agricoltura, zootecnia e mezzi di produzione*, anno 1990, 38, p. 41; Istat, *Statistiche dell'agricoltura, zootecnia e mezzi di produzione*, anno 1991, 39, p. 41; Istat, *Statistiche dell'agricoltura, zootecnia e mezzi di produzione*, anno 1993, 41, p. 44; Istat, *Statistiche dell'agricoltura*, anno 1994, 39, p. 40; Istat, *Statistiche dell'agricoltura*, anno 1995, 43, p. 31; <http://www.ismea.it>.

---

39 The province of Grosseto lost around 19% of its ovines, compared to a regional average loss of –22%. Istat, *Censimento agricoltura 2010, op. cit.*

40 Community regulations required respect for the specific level measures in all the phases of dairy farming, including the entire process from the raw materials to the final product. They were recognized in Italy by the DPR 54/97. Rossi A., "L'evoluzione della filiera lattierocasearia tra vincoli e opportunità", in Pacciani A. *et al.* (ed.), *Strutture e dinamiche nel sistema agroindustriale toscano*, Florence, agroindustrial observatory for Tuscany, 1998, pp. 158–9.

41 The 5th agricultural census identified 4,635 businesses with ovines in Tuscany; of these only 92 were run by salaried workers. Istat, 5 ° *Censimento generale dell'agricoltura, La zootecnia in Italia, Caratteristiche tipologiche delle aziende agricole: fascicolo regionale Toscana*, Roma, 2003.

42 An influential factor of the dairy price adjustment is represented by Sardinian milk for its large quantity of production, whose price in 1996 dropped drastically following the Pecorino romano and fiore sardo market crisis. Rossi A., "L'evoluzione della filiera lattiero-casearia tra vincoli e opportunità", *op. cit.*, pp. 180–1.

For the remainder, in the final decade national data indicate a livestock reduction in practically all Italian regions: between 1990 and 2000 around 1,895,000 ovines disappeared (22% of the total), and 57,400 businesses (39%). This demonstrates the fact that the farmers crisis, which was gripped by growing costs alongside lowered earnings, did not pertain only to Tuscany.

Tuscan ovine livestock emerged from the crisis downsized for sure, and deeply transformed as well. In 2000, only 4,409 businesses were in operation in Tuscany, 2,767 fewer than in 1990. However, each had a herd made up of on average 125 heads, 26 more than in the previous decade. On a national level, at that time only Sardinia had an average-scale enterprise, with a larger head count of 194.

Figure 4. Average ovine count per business in Tuscany and in Italy

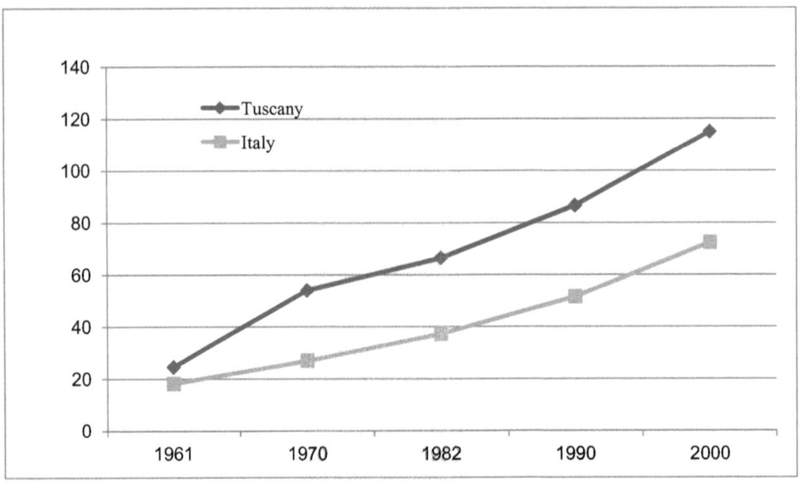

Sources: Istat, *1° Censimento Generale dell'agricoltura, op. cit.*, pp. 200–10; Istat, *2° Censimento Generale dell'agricoltura, op. cit.*, pp. 424–32; Istat, *Statistiche della zootecnia, pesca e caccia*, 1982, 23, p. 27; Istat, Serie storiche, *Numero di aziende con allevamenti, numero di capi delle aziende con allevamenti per forma di conduzione, tipo allevamento*, <http://dati-censimentoagricoltura.istat.it/index.aspx?queryid=1333#>.

3. The milk production

During the second half of the twentieth century, Italian ovine milk was characterized by the prevalence of milk from Sardinia, strong and growing. At the beginning of the 1950s, herds from the island provided around 26% of the national ovine milk;[43] in the 1970s their contribution reached around

[43] The data refer to 1952. Istat, *Annuario di statistica agraria*, 1954, pp. 90–1.

38%, to rise to 40% in the 1980s, with a continuing rise to 57% by the 1990s.[44]

Contrastingly, Tuscan production started with a marginal position. In 1952, Istat reveals, there were around 382,800 q of milk gathered by farmers, only 3.4% of the national production.[45] Fourteen years later, volumes produced were reduced to 219,000 q as a result of the Tuscan ovine farming crisis of the first half of the 1960s.[46] Not only in these years did production maintain a very low level, but the relation between milk and the number of heads was also among the less favorable in Italy. In 1952, Tuscany produced 0.53 q per head, while on a national level the average was 0.7 q: as the most productive region, Sicily obtained 1.02 q of milk per head.[47]

To blame for the low production in Tuscany is not only low farming efficiency, but also the fact that the region's ovine farming was not directed mainly towards the production of milk. Annual statistics from the 1950s show that Tuscany actually was a large producer of wool. For example, in 1952 the region produced 10,708 q of wool, 7% of the national total, compared to the 8,552 q produced in Sicily, which even gained a majority of ovine assets (over 3,600 heads).

Over the next fourteen years, the Tuscan farm structure did not show any significant changes. It was only with a settling of the Sardinian shepherds, with the drastic introduction of the Sardinian race and the farming method directing business towards milk and volume production, that growth began: 328,000 q in 1972; 346,000 in 1974; 432,000 in 1977; 494,900 in 1979; 530,115 in 1981; 516,800 in 1984; 540,700 in 1985; and 547,400 in 1987.[48]

On a national level, Tuscan production went from 7% in 1972 to 11% in 1987.

---

44 Ministero dell'agricoltura e delle foreste, *Indagine nazionale su alcuni aspetti degli allevamenti e delle produzioni ovine, op. cit.*, 1972, pp. 103–9; Istat, *Statistiche dell'agricoltura, zootecnia e mezzi di produzione*, 1985, 15, p. 290; Istat, *Statistiche dell'agricoltura, zootecnia e mezzi di produzione*, 1987, 35, p. 346; Istat, *Statistiche dell'agricoltura, zootecnia e mezzi di produzione*, 1989, 37, p. 336; Istat, *Statistiche dell'agricoltura, zootecnia e mezzi di produzione*, 1990, 38, p. 330; Istat, *Statistiche dell'agricoltura, zootecnia e mezzi di produzione*, 1991, 39, p. 316; Istat, *Statistiche dell'agricoltura, zootecnia e mezzi di produzione*, 1993, p. 316; Istat, *Statistiche dell'agricoltura*, 1998, p. 281; Istat, *Statistiche dell'agricoltura*, 1999, p. 292; Istat, *Statistiche dell'agricoltura*, 2000, p. 307; Istat, *Statistiche dell'agricoltura, zootecnia e mezzi di produzione*, 1987, 35, p. 346.

45 Istat, *Annuario di statistica agraria*, 1954, p. 90.

46 Istat, *Annuario di statistica agraria*, 1968, 15, tomo I, p. 239.

47 Inea (National Institute for Farming Economy), *Annuario dell'agricoltura italiana*, 1955, 9, p. 106; Istat, *Annuario di statistica agraria*, 1954, pp. 90–2.

48 Ministero dell'agricoltura e delle foreste, *Indagine nazionale su alcuni aspetti degli allevamenti e delle produzioni ovine, op. cit.*, pp. 103–9; *La pastorizia in Toscana: aspetti, problemi, proposte, op. cit.*, p. 12; Istat, *Statistiche dell'agricoltura, zootecnia e mezzi di produzione*, 1985, 15, p. 290; Istat, *Statistiche dell'agricoltura, zootecnia e mezzi di produzione*, 1987, 35, p. 346.

Figure 5. Tuscan ovine milk production, 1950-2000 (q)

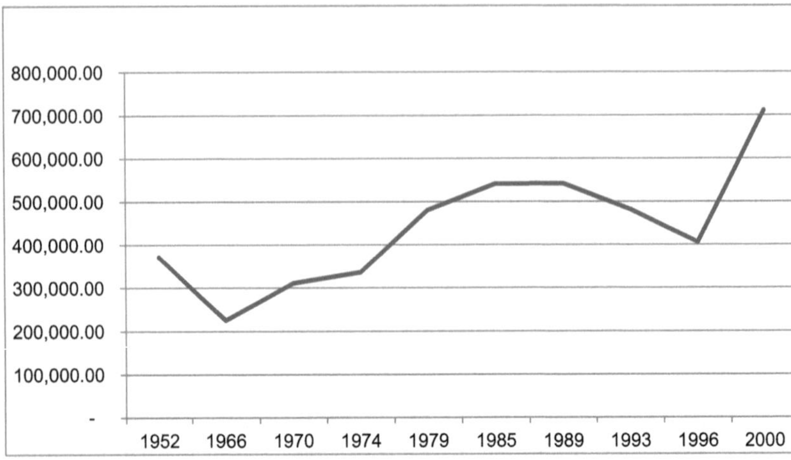

Sources: Istat, *Annuario di statistica agraria*, 1954 p. 90; Istat, *Annuario di statistica agraria*, 1968, 15, tomo I, p. 239; *Conferenza agraria di zona giugno 1973-maggio 1974*, pp. 117–56; *Indagine nazionale su alcuni aspetti degli allevamenti e delle produzioni ovine*, Roma, Associazione nazionale della pastorizia, 1972, pp. 103–9; *La pastorizia in Toscana: aspetti, problemi, proposte, op. cit.*, p. 12; Istat, *Statistiche dell'agricoltura, zootecnia e mezzi di produzione*, 1985, 15, p. 290; Istat, *Statistiche dell'agricoltura, zootecnia e mezzi di produzione*, 1987, 35, p. 346; Istat, *Statistiche dell'agricoltura, zootecnia e mezzi di produzione*, 1989, 37, p. 336; Istat, *Statistiche dell'agricoltura, zootecnia e mezzi di produzione*, 1990, 38, p. 330; Istat, *Statistiche dell'agricoltura, zootecnia e mezzi di produzione*, 1991, 39, p. 316; Istat, *Statistiche dell'agricoltura, zootecnia e mezzi di produzione*, 1993, p. 316; Istat, *Statistiche dell'agricoltura*, 1998, p. 281; Istat, *Statistiche dell'agricoltura*, 1999, p. 292; Istat, *Statistiche dell'agricoltura*, 2000, p. 307.

In the 1990s, as an effect of the above-mentioned events, growth stabilized: in 1993, production stopped at 495,000 q, and in 1996 it went down to 406,000 q.[49]

The business method of concentration, and reasoning imposed by the crisis, stimulated, nevertheless, strong productivity growth. Therefore, in 2000, farmers were able to produce 732,031 q of milk, having only 554,011 heads available with an average of 1.32 q per head. This was the highest value registered for the entire fifty years under examination.[50]

[49] Istat, *Statistiche dell'agricoltura, zootecnia e mezzi di produzione*, 1993, 41, p. 316; Istat, *Statistiche dell'agricoltura*, 1996, 44, p. 281.

[50] Istat, *Statistiche dell'agricoltura*, 2000, 48, p. 307.

Figure 6. Ovine milk production in Tuscany and Italy: 100 kilograms per head

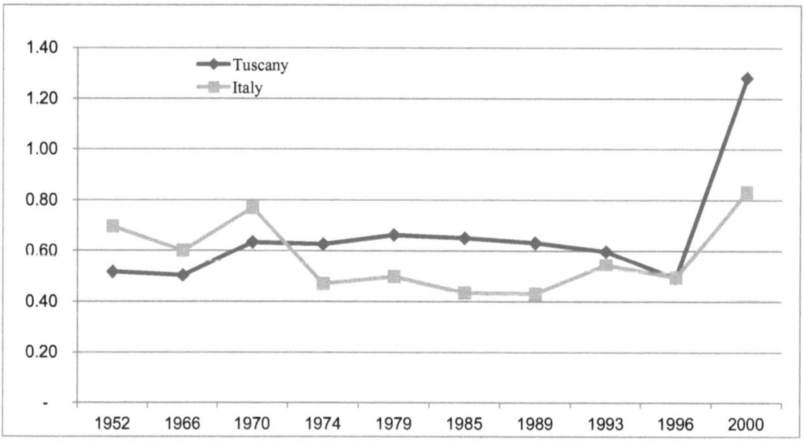

Sources: Istat, *Annuario di statistica agraria*, 1954, p. 90; Istat, *Annuario di statistica agraria*, 1968, 15, tomo I, p. 239; Istat, *Annuario di statistica agraria*, 1970, 17, pp. 166–7; Istat, *Statistiche dell'agricoltura*, 1996, p. 192; Istat, *Statistiche dell'agricoltura*, 2000, p. 307; Istat, *Annuario statistico della zootecnia, pesca e caccia*, 1975, 16, pp. 71–3; Istat, *Statistiche dell'agricoltura, zootecnia e mezzi di produzione*, 1985, 33 p. 292; Istat, *Statistiche dell'agricoltura, zootecnia e mezzi di produzione*, 1987, 35, p. 290; Istat, *Statistiche dell'agricoltura, zootecnia e mezzi di produzione*, 1989, 37, p. 336; Istat, *Statistiche dell'agricoltura, zootecnia e mezzi di produzione*, 1990, 38, pp. 331–2; Istat, *Statistiche dell'agricoltura, zootecnia e mezzi di produzione*, 1991, 39, pp. 317–18; Istat, *Statistiche dell'agricoltura, zootecnia e mezzi di produzione*, 1993, 41, p. 328; Banche dati <http://web.inea.it:8080/>.

At a provincial level, there was always a strong prevalence of ovine milk production in Grosseto, not surprising in view of how much it previously showed. Already in 1952, the province produced 28% of Tuscan milk[51] and in the following years its strength grew in a directly proportional way to its herds, without productivity being surpassed by the regional average. It reached: 40% in 1970;[52] 41% 1980;[53] and 44% 1993.[54] In 2000, it also produced 51% of the regional ovine milk, as it obtained 51% of the heads in Tuscany.[55]

The second half of the twentieth century was also characterized by the continuing rise of Siena. In 1952, the province produced barely 36,310 q of milk, not only a modest amount compared to the 108,000 q gathered the

---

51 Istat, *Annuario di statistica agraria*, 1954, p. 90.

52 *Conferenza agraria di zona giugno 1973-maggio 1974*, pp. 117–56.

53 Castor *et al.* (ed.), "Il riconoscimento della denominazione d'origine del Pecorino toscano", *op. cit.*, p. 9.

54 Tosi L., *La produzione e la commercializzazione del Pecorino toscano tra tutela della tipicità e orientamento al mercato*, Firenze, INEA, 1999, p. 13.

55 Caldelli M., Giannone M., "Allevamento ovino in Toscana e razza sarda", in *Rivista di Agraria.org*, 2008, 57, <http://www.rivistadiagraria.org/articoli/anno-2008/allevamento-ovino-in-toscana-e-razza-sarda>.

same year in Grosseto, but also less than the production of Florence, Lucca and Arezzo.[56]

Change occurred in the second half of the 1960s. In 1970, Sienese countryside production almost tripled, reaching 21% of regional production. In the following years, with this big leap, other advances were recorded, even if they were less significant (in 1980 Siena produced 22% of Tuscan ovine milk, in 1993 23%, and in 2000 24%), and Siena established itself in second place among Tuscan provinces.[57]

With the rise of the southern provinces, northern provinces, contrastingly, began to decline. The rank of Massa, Lucca and Pistoia, which in 1952 had produced 22% of Tuscan ovine milk, rose negligibly as the century went on. They arrived at the point that in 2000 they contributed only 4% of regional production. This occurred not only because of the continuing decline of their herds but also because the small-scale businesses of the province remained tied to the traditional farming method, with triple aptitude and signs of productivity inferior to the Tuscan average.[58]

4. Dairy production

In 1952, the estimated production of Tuscan sheep cheese for commercial products was equal to around 9,000 q, around 2% of the national total, in a market which was dominated by Roman sheep cheese (80,000 q products in 1952) and Sardinian sheep cheese (60,000 q).[59] The transformation was effected by the same producers of milk, or in small dairy farms. In Tuscany in 1955, 83 were counted. Of these, only 14 were included in the 6–50 registered category; 28 registered 3–5 employees, while around 41, or 49%, were managed by only one or two dairy farmers.[60]

In 1963, Tuscan production dropped to the lowest low of the century. It is estimated that just 8,100 q of Tuscan sheep cheese wee allowed onto the market.[61] This crisis period, nonetheless, did not see the dairy business

56 Table 10 in the Appendix.

57 In 2000, the province of Grosseto and Siena both produced 77% of Tuscan ovine milk.

58 In 2000, farmers from Massa, Lucca and Pistoia obtained around 100 kg of milk per head; in the same year farmers from Grosseto and Tuscany obtained 189.

59 Inea, *Annuario dell'agricoltura italiana*, 1955, 9, p. 95; Istat, *Annuario di statistica agraria*, 1954, p. 90.

60 On a whole, in the facilities with a 6–50 employee category, there were 147 employed individuals, 99 people worked in the facilities with 3–5 in charge, 41 were employed in the small farm houses with up to 2 in charge. By way of comparison, it is considered that during the same year in Latium there were 1,542 employees in dairy production, 1,075 in Sardinia. Inea, *Annuario dell'agricoltura italiana*, 1955, 9, p. 108.

61 During the same year there were 82,000 q of Roman sheep cheese and 53,000 q of Sardinian sheep cheese. Inea, *Annuario dell'agricoltura italiana*, 1966, 20, p. 326.

remain immobile. During the 1960s, there began to settle in the area an important cooperative business which brought about important innovations, introducing rules and regulations to both improve the cheese quality and make the production process more sanitary.

In 1961, a cooperative made up of twenty-one members formed the Dairy Cooperative of Manciano. In 1963, the Dairy Cooperative of Sorano joined in; a year later, the Val D'Orcia Dairy Farm was founded in Radicofani, and in 1970 the Pratomagno Zootechnical Cooperative was created. In 1975, there were fourteen Tuscan cooperatives working in the ovine business, and in 1980 there were twenty.[62]

Most of the cooperatives were joined by two groups. Nine of them, which represented 2,000 producers, were joined in Castor, with a branch in Siena. Four cooperatives for processing and one for collection for every 33,000 farmers collectively were in collaboration in Lugalat, with a branch in Lucca. Both the unions and many of the cooperatives began and prospered in their businesses thanks to the support of Ente Maremma.[63]

Between 1977 and 1979 the ovine milk quantities bestowed an increase of 30% on the cooperatives.[64] Most Tuscan ovine milk production continued, nonetheless, to be processed in private dairy farms.[65]

Not all the milk was used to produce ovine cheeses:. A good percentage was used for ovine-vaccined mixed cheeses, among which were the Toscanello and Tuscan Caciotta, whose production was found throughout the region and which today still makes up a good part of production as the most important of Tuscan cheeses.[66]

In order to place a limit on specifications to bring together the production of the small multiple dairy farms that were operating in culturally and geographically different areas so as to stand out from each other (like Roman and Sardinian sheep cheese), the Castor promised to provide a group of technical workers from the Department of Agriculture and Forestry, with an agreement to hold revision by the Ministry of Agriculture and ask that Tuscan sheep cheese be recognized as a registered name.[67] The procedure worked and was approved in 1981. It concluded in February 1987 with the

---

62 Table 11 in the Appendix.

63 In 1980, Ente Maremma was member of ten cooperatives. It created three facilities for milk processing and seven others managed the development through licensing guarantees. *Indagine conoscitiva sugli aspetti lattierocaseari della pastorizia. Incontro del 6 dicembre 1975*, Sorano, 1975, pp. 37–51.

64 *La pastorizia in Toscana: aspetti, problemi, proposte*, *op. cit.*, p. 16.

65 Table 12 in the Appendix.

66 Tosi L., *La produzione e la commercializzazione del Pecorino toscano tra tutela della tipicità e orientamento al mercato*, *op. cit.*, p. 15.

67 Castor *et al.* (ed.), "Il riconoscimento della denominazione d'origine del Pecorino toscano", *op. cit.*, pp. 2–3.

publication in the *Gazzetta Ufficiale* that the approval recognized Tuscan sheep cheese as a traditional product.[68]

In 1992, regulation CEE 2081 replaced the previous national procedure for protection of traditional food products with new and valid universal regulations at a community level. In the case of the traditional cheeses, the registered name slogan was replaced by European slogans: protected registered name, protected denomination of origin (PDO).

Tuscan producers promoted another procedure to obtain a new European brand, and in June of 1996 Tuscan sheep cheese finally became a recognized PDO product. It entered as part of a selective group of 30 Italian dairy products and of 117 community cheeses that could use this brand.[69] In the meantime, the best dairy farm cooperatives formed the Union for the Protection of Tuscan Sheep Cheese. Between 1985 and 1987 the Union limited itself to developing business focused on commercializing. In 1988, however, it acquired an institutional role attributed to the function of checking and protecting the registered name entitled by the Ministry of Agricultural Resources. All dairy farm producers could be part of this who were located in the designated area for Tuscan sheep cheese, or even in the entire territory of Tuscany and in some counties of northern Latium (Acquapendente, Onano, San Lorenzo Nuovo, Grotte di Castro, Gradali, Valentano, Farnese, Ischia di Castro, Montefiascone, Bolsena and Capodimonte), or from Umbria (Castiglione del Lago and Allerona), which would have liked to produce a registered name cheese.

---

68 In 1954, when the Ministry of Agriculture legislated the registered name of Italian cheeses, the Tuscan ovine milk sector was going through a period of crisis. Tuscan sheep cheese, which was produced in small distributed quantities on the local market, was not recognized as a traditional product. Tosi L., *La produzione e la commercializzazione del Pecorino toscano tra tutela della tipicità e orientamento al mercato*, *op. cit.*, p. 24.

69 The protected registered name "PDO" is a brand that began to be recognized for foods that had characteristics which depended on the geographical area in which they are produced. The geographical area included both environmental factors (the climate and surroundings) and the way in which the production allowed for creating the original product outside its native area. Cf. Consorzio per la tutela del Pecorino toscano DOP (ed.), *In nome del Pecorino toscano*, p. 131.

Figure 7. The zone of production of PDO Tuscan Sheep cheese

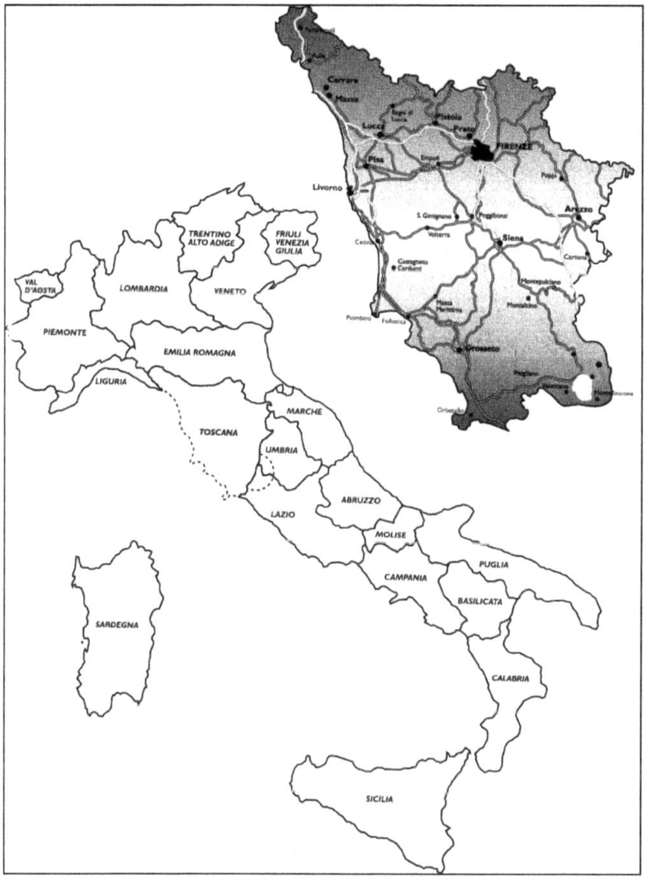

Source: Consorzio per la tutela del Pecorino toscano DOP (ed.), *In nome del Pecorino toscano*, *op. cit.*, p. 55.

The production of sheep cheese, however, required specific production methods and was costly (demanding also requirements for partner businesses to finance their assets). For this reason, only large-scale dairy farms, who were hoping to join and expand their markets, followed the rules. In 1996, the year in which Tuscan sheep cheese was recognized as a PDO registered brand, the members of the Consorzio Union of Protection numbered nineteen (twelve private dairy farms and seven cooperatives). Of these, eleven were located in the province of Grosseto, four in the province of Siena, two in the province of Florence, one in the province of Arezzo, and one outside the region in Viterbo.

Pisa, Livorno, Prato, Pistoia and Massa did not participate directly in the production of Tuscan sheep cheese. They limited themselves to merging a part of their milk production with the processing facilities in Grosseto, Siena and Florence.[70]

In 1989, the year in which the Union of Protection reached full operation, they produced 6,000 q of sheep cheese. This was followed by a slight fallback in 1990, and then production increased tenfold in three years (from 4,140 q in 1990 to 40,500 q in 1993) – a tremendous feat. This growth continued until 1996, when it produced 42,000 q of Tuscan sheep cheese.[71] The second half of the 1990s, however, saw an overall shift in trends, coinciding with the difficulties experienced in that time by Tuscan farmers, and in 2000 production froze at 16,640 q.[72]

On a national level, in a sector continuously dominated by Sardinian sheep cheese and most of all Roman sheep cheese,[73] Tuscan sheep cheese continued to occupy a modest amount of the national production despite its growth which earned it second place in Italian PDO ovine cheese:[74] from 1.10% in 1989 to 6.23% in 1996, to then drop again to 2.72% in 2000 (Fig. 8).

It is surely more difficult to evaluate the effect of Tuscan sheep cheese on the regional ovine dairy production as a whole because it is extremely variable and complex. Mixed cheeses like Caciotta or Raveggiolo were popular throughout the whole region. Aside from them, there are many types of sheep cheeses with restricted markets but very popular on a local level, and whose production is quantifiably difficult.

It is, however, significant that of the 541,832 q of ovine milk processed in Tuscany in 2000 only 99,840 q, or 18.42% of the total, were processed into PDO cheeses. For the remaining partnership businesses of the Union of Protection, they used around 50% of their milk to produce other ovine or mixed cheeses during the mid-1990s.[75] At the end of the twentieth century, however, in relation to the notable productive capacity of the regional ovine

70 This shows that contributions from these provinces to the regional production of ovine milk were limited.

71 Tosi L., *La produzione e la commercializzazione del Pecorino toscano tra tutela della tipicità e orientamento al mercato*, *op. cit.*, p. 48.

72 Irpet (Istituto Regionale Programmazione Economica della Toscana), *9° Rapporto Economie e politiche agricole in Toscana*, 2007, 9, p. 81.

73 In 1989, 273,000 q of Roman sheep cheese were produced, 37% of Italian sheep cheese production, 353,486 q in 1996 (47% of Italian production), and in 2000 336,501 q (48% of the national total). Inea, *Annuario dell'agricoltura italiana*, 1991, 45, p. 391; Inea, *Annuario dell'agricoltura italiana*, 1997, 51, p. 390; Laore Sardegna (Agenzia Regionale per lo Sviluppo dell'Agricoltura), *Le produzioni del comparto ovicaprino in Sardegna. Dati sulle produzioni settembre 2011*, pp. 12-13, <http://www.sardegnaagricoltura.it/documenti/14_43_20120203134032.pdf>.

74 Again in 1994, Sardinian sheep cheese was the second Italian ovine cheese with a production level of 240,000 q. In 2000 its production level dropped to 3,600 q and that of Tuscan sheep cheese rose to second place, at a great distance from Roman sheep cheese.

75 Tosi L., *La produzione e la commercializzazione del Pecorino toscano tra tutela della tipicità e orientamento al mercato*, *op. cit.*, p. 42.

division, the dairy industry for Tuscan sheep cheese still had great potential for development.

Figure 8. Production of main sheep cheeses in Italy, 1952–2000

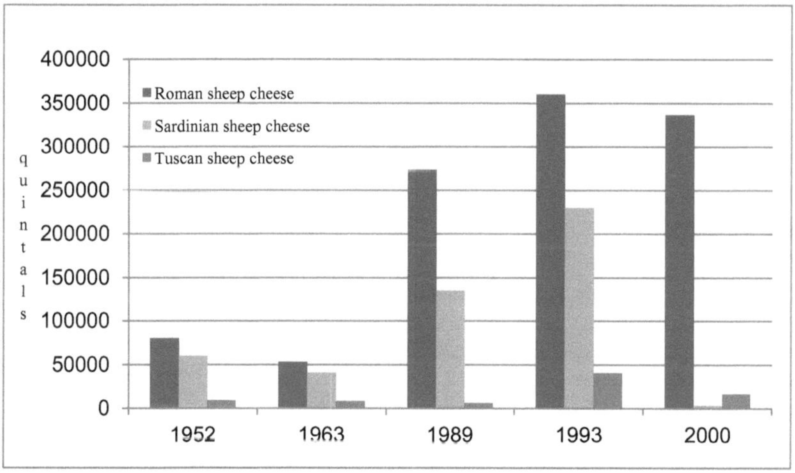

Sources: Inea (Istituto Nazionale di Economia Agraria), *Annuario dell'agricoltura italiana*, 1955, 9, p. 106; Inea, *Annuario dell'agricoltura italiana*, 1966, 20, p. 320; Inea, *Annuario dell'agricoltura italiana*, 1991, 45, p. 391; Inea, *Annuario dell'agricoltura italiana*, 1997, 51, p. 390; Irpet (Istituto Regionale Programmazione Economica Toscana), *9° Rapporto Economie e politiche agricole in Toscana*, 2007, 9, p. 81; Laore Sardegna (Agenzia Regionale per lo Sviluppo dell'Agricoltura), *Le produzioni del comparto ovi-caprino in Sardegna. Dati sulle produzioni settembre 2011*, pp. 12–13, <http://www.sardegnaagricoltura.it/documenti/14_43_20120203134032.pdf>.

5. Conclusion

In the second half of the twentieth century Tuscan ovine farming underwent drastic changes. The transhumance stopped, farmers bought enough farming land to house and pasture feeding for animals, they introduced the Sardinian race in the region, and business was decisively directed towards milk production. Even the average number of single herds grew significantly compared to in the past. On a territorial level, the strength of the Siena, and most of all the Grosseto, provinces grew decisively. During the same years, with the sponsorship of Ente Maremma, an important line of cooperatives began and consolidated itself. These associations, particularly active in the field of ovine milk modification, proposed improving and unifying cheese factory processing techniques for the value of Tuscan sheep cheese and creating a consistent quality product, clearly noticeable on the Italian and international markets. From 1977 to 1979, the amounts of ovine milk given to the cooperatives grew by 30%. In 1985, the best cheese factories of Tuscany founded the Union for the legal protection of Tuscan sheep cheese. In 1987, Tuscan sheep cheese was recognized as a registered named product, and in 1996 it obtained the PDO trade mark. During the 1990s, despite these undeniable successes, Tuscan sheep cheese was unable to make itself the leading product in the Tuscan cheese industry. To avoid high derivative costs from regulation surveillance, many businesses preferred to focus their production on other ovine and blended cheeses. With a restricted market, though very popular on a local level as other niche products, these were often able to obtain superior selling prices to those of Tuscan PDO sheep cheese.

Appendix

Table 1. Population of Sheep in Tuscany, 1952

Province	Sheep allocated	Pastoralism	Total
Massa	12,747	12,239	24,986
Lucca	9,600	7,200	16,800
Pistoia	19,850	7,270	27,120
Firenze	131,508	6,850	138,358
Livorno	9,784	11,022	20,806
Pisa	36,400	30,814	67,214
Arezzo	126,361	0	126,361
Siena	86,500	0	86,500
Grosseto	102,270	130,500	232,770
Tuscany	535,020	205,895	740,915

Source: Tornar F., *Toscana agricola, op. cit.*, pp. 164-170.

Table 2. Population of sheep in Tuscany, 1952–1961

Province	1952	1961	Difference 1961-1952
Massa	24,986	8,852	-16,134
Lucca	16,800	38,263	+21,463
Pistoia	27,120	11,254	-15,866
Firenze	138,358	49,780	-88,578
Livorno	20,806	8,599	-12,207
Pisa	67,214	34,340	-32,874
Arezzo	126,361	63,334	-63,027
Siena	86,500	57,187	-29,313
Grosseto	232,770	142,044	-90,726
Tuscany	740,915	413,653	-327,262

Sources: Tornar F., *Toscana agricola, op. cit.*, pp. 164–70; Istat, *1° Censimento generale dell'agricoltura, op. cit.*, pp. 200–3.

Table 3. Population of sheep in Tuscany, 1961-1978

Province	1961	1970	1978	Difference 1978-1961
Massa	8,852	8,725	12,000	3,148
Lucca	38,263	32,833	25,000	-13,263
Pistoia	11,254	12,083	11,000	-254
Firenze	49,780	53,677	46,000	-3,780
Livorno	8,599	16,699	28,000	19401
Pisa	34,340	52,573	134,000	99,660
Arezzo	63,334	58,392	55,000	-8,334
Siena	57,187	94,506	150,000	92,813
Grosseto	142,044	160,739	212,000	69,956
Tuscany	413,653	490,227	673,000	259,347

Source: Istat, *1° Censimento generale dell'agricoltura, op. cit.*, pp. 200–3; Istat, *2° Censimento Generale dell'agricoltura, op. cit.*, pp. 427-428; *La pastorizia in Toscana: aspetti, problemi, proposte, op. cit.*, p. 6.

Table 4. The Sardinian sheep in Tuscany, 1978

Province	Heads	% of the ovine total
Arezzo	7,890	14.34
Firenze	28,925	62.88
Grosseto	35,060	16.53
Livorno	7,840	28.00
Lucca	93	0.37
Massa Carrara	0	0
Pisa	106,750	79.66
Pistoia	0	0
Siena	94,860	63.24
Total	281,418	41.81

Source: Castor (Caseifici Toscani riuniti) *et al.* (ed.), "Il riconoscimento della denominazione d'origine del Pecorino toscano", *op. cit.*, pp. 4–9.

Table 5. Development of pastoral farming property in Tuscany, 1965–1980

	With Tax and Credit Benefits			With Tax Benefits			Total		
	families	area		families	area		families	area	
Province	No.	Acre	%	No.	Acre	%	No.	Acre	%
Arezzo	137	1,095	5.60	479	5,165	19.04	616	7,070	11.57
Firenze	53	1,822	5.30	52	911	3.36	105	2,733	4.47
Grosseto	335	12,235	36.00	1,725	14,847	54.74	2,060	27,082	44.30
Livorno	13	772	2.30	22	115	0.42	35	887	1.45
Lucca	4	46	0.14	41	131	0.48	45	177	0.29
Massa	2	21	0.06	1	13	0.05	3	34	0.06
Pisa	80	3,349	9.85	25	841	3.10	105	4,190	6.85
Pistoia	2	16	0.05	83	200	0.74	85	216	0.35
Siena	177	13,843	40.70	109	4,900	18.07	286	18,743	30.66
Total	803	33,199	100.00	2,537	27,123	100.00	3,340	61,132	100.00

Source: *La pastorizia in Toscana: aspetti, problemi, proposte, op. cit.*, pp. 8–9.

Table 6. Ovine businesses in the provincial counties of Grosseto, 1982–2000

Year	1982		1990		2000	
Counties	**Heads**	**Businesses**	**Heads**	**Businesses**	**Heads**	**Businesses**
Arcidosso	6,915	102	8,398	102	7,781	65
Campagnatico	13,094	182	20,084	150	17,722	89
Capalbio	10,061	83	14,181	99	18,044	60
Castel del Piano	1,794	52	1,645	36	1,251	28
Castell'Azzara	2,951	53	3,549	54	1,274	16
Castiglione della Pescaia	2,858	41	3,244	37	1,975	22
Cinigiano	7,369	135	10,321	137	8,814	71
Civitella Paganico	6,478	69	9,185	78	6,505	40
Follonica	626	4	810	5	5	1
Gavorrano	5,587	50	8,833	79	4,994	36
Grosseto	21,406	233	27,440	246	24,353	147
Isola del Giglio	0	0	0	0	5	1
Magliano in Toscana	26,605	265	40,683	243	26,010	153
Manciano	32,380	412	46,957	370	40,865	245
Massa Marittima	6,062	85	6,130	73	5,040	57
Monte Argentario	9	1	9	1	0	0
Monterotondo Marittimo	3,568	32	6,387	33	6,374	22
Montieri	2,410	41	2,139	36	1,268	31
Orbetello	5,483	64	10,566	76	6,972	41
Pitigliano	5,255	96	10,048	107	7,685	67
Roccalbegna	15,249	224	17,371	180	13,416	109
Roccastrada	8,452	83	8,469	100	8,167	38
Santa Fiora	2,800	40	3,182	29	3,144	26
Scansano	37,179	436	51,821	399	39,564	201
Scarlino	1,516	19	1,375	26	1,433	14
Seggiano	1,219	35	1,320	26	1,077	17
Semproniano	12,268	224	13,176	151	8,898	83
Sorano	21,109	355	22,117	264	19,484	164
Province Total	260,703	3,416	349,440	3,137	282,120	1,844

Source: Istat, *Censimento agricoltura 2010, op. cit.*, <http://dati-censimentoagricoltura.istat.it/index.aspx?queryid=1333#>.

Table 7. Ovine businesses in Tuscany

Province	1990			2000			difference 2000–1990		Difference 2000–1990 %	
	Businesses	Ovine	Ovine/ Businesses	Businesses	Ovine	Ovine/ Businesses	Businesses	Ovine	Businesses	Ovine
Massa	459	12,203	26	319	9,483	29	-140	-2,720	-30.50	-22.29
Lucca	491	26,853	54	234	14,255	61	-257	-12,598	-52.34	-46.91
Pistoia	184	7,905	43	98	5,294	54	-86	-2,611	-46.73	-33.03
Firenze	550	42,507	77	382	27,424	72	-168	-15,083	-30.54	-35.48
Livorno	163	15,555	95	100	7,874	78	-63	-7,681	-38.65	-49.38
Pisa	517	63,089	122	322	44,061	137	-195	-19,028	-37.72	-30.16
Arezzo	783	52,140	66	535	34,323	64	-248	-17,817	-31.67	-34.17
Siena	830	142,723	171	533	128,081	240	-297	-14,642	-35.78	-10.26
Grosseto	3,137	349,440	111	1,844	282,120	153	-1,293	-67,320	-41.22	-19.26
Prato	62	3,270	52	42	1,096	26	-20	-2,174	-32.26	-66.48
Tuscany	7,176	715,685	99	4,409	554,011	125	-2,767	-161,674	-38.56	-22.59

Source: Istat, *Censimento agricoltura 2010, op. cit.*, <http://dati-censimentoagricoltura.istat.it/index.aspx?queryid=1333#>.

Table 8. Businesses with ovine livestock in Italy

Region	1990		2000		difference 1990–2000			
	Businesses	Ovine	Businesses	Ovine	Businesses	Ovine	Businesses %	Ovine %
Piemonte	3,010	87,461	1,960	87,384	-1,050	-77	-34.88	-0.09
Valle d'Aosta	212	3,811	130	2,101	-82	-1,710	-38.67	-44.87
Liguria	1,779	22,284	925	17,172	-854	-5,112	-48.00	-22.94
Lombardia	3,559	94,350	2,564	90,425	-995	-3,925	-27.95	-4.16
Trentino Alto Adige	2,226	45,837	2,355	59,940	129	14,103	+5.79	+30.76
Veneto	932	35,786	859	30,371	-73	-5,415	-7.83	-15.13
Friuli-Venezia Giulia	297	5,395	201	6,201	-96	806	-32.32	+14.93
Emilia Romagna	2,642	97,716	1,765	83,466	877	-14,250	-33.19	-14.58
Toscana	7,176	715,685	4,409	554,011	-2,767	-161,674	-38.55	-22.59
Umbria	5,710	205,590	3,450	148,866	-2,260	-56,724	-39.58	-27.59
Marche	7,921	217,892	3,489	157,664	-4,432	-60,228	-55.95	-27.64
Lazio	18,209	874,130	10,635	630,510	-7,574	-243,620	-41.59	-27.87
Abruzzo	18,205	453,528	8,871	279,504	-9,334	-174,024	-51.27	-38.37
Molise	6,192	135,445	3,701	112,647	-2,491	-22,798	-40.23	-16.83
Campania	13,737	237,924	7,987	225,834	-5,750	-12,090	-41.85	-5,08
Puglia	4,352	349,112	2,309	217,515	-2,043	-131,597	-46.94	-37.69
Basilicata	10,605	354,426	7,759	334,713	-2,846	-19,713	-26.83	-5.56
Calabria	9,263	328,469	5,398	236,216	-3,865	-92,253	-41.72	-28.08
Sicilia	10,756	1,290,491	5,979	706,835	-4,777	-583,656	-44.41	-45.23
Sardegna	19,766	3,129,687	14,405	2,808,450	-5,361	-321,237	-27.12	-10.26
North West	8,560	207,906	5,579	197,082	-2,981	-10,824	-34.82	-5.20
North East	6,097	184,734	5,180	179,978	-917	-4,756	-15.04	-2.57
Central	39,016	2,013,297	21,983	1,491,051	-17,033	-522,246	-43.65	-25.93
South	62,354	1,858,904	36,025	1,406,429	-26,329	-452,475	-42.22	-24.34
Islands	30,522	4,420,178	20,384	3,515,285	-10,138	-904,893	-33.21	-20.47
Italy	146,549	8,685,019	89,151	6,789,825	-57,398	-1,895,194	-39.16	-21.82

Source: Istat, *Censimento agricoltura 2010, op. cit.*, <http://dati-censimentoagricoltura.istat.it/index.aspx?queryid=1333#>.

Table 9. Ovine milk production in Italy, 1972 and 1987

Region	1972		1987	
	Total milk (q)	Total milk in %	Total milk (q)	Total milk in %
Sardegna	1,626,900	37.19	1,983,600	40.45
Sicilia	476,700	10.89	452,250	9.22
Lazio	333,800	7.63	663,000	13.52
Toscana	328,000	7.49	547,400	11.16
Puglia	326,300	7.45	258,900	5.28
Campania	246,000	5.62	129,600	2.64
Abruzzo	235,200	5.37	251,900	5.13
Calabria	183,600	4.19	142,000	2.89
Basilicata	177,400	4.05	71,500	1.45
Marche	120600	2.75	110,350	2.25
Molise	73,700	1.68	59,000	1.20
Umbria	62,600	1.43	104,200	2.12
Piemonte	56,900	1.30	30,850	0.62
Emilia Romagna	48,500	1.10	68,700	1.40
Lombardia	29,700	0.67	6,050	0.12
Liguria	17,400	0.39	9,050	0.18
Veneto	11,600	0.26	12,550	0.25
Trentino Alto Adige	9,300	0.21	0	0.00
Friuli Venezia Giulia	5,100	0.11	1,750	0.03
Valle d'Aosta	4,700	0.10	450	0.01
Italy	4,374,000	100.00	4,903,100	100.00

Sources: *La pastorizia in Toscana: aspetti, problemi, proposte, op. cit.*, 1981, p. 24; Istat, *Statistiche dell'agricoltura, zootecnia e mezzi di produzione*, 1987, 35, *op. cit.*, p. 346.

Table 10. Ovine milk production in Tuscany

Year	1952				1970				2000			
Province	Total milk q	Total milk %	Milk trans-formed q	Milk trans-formed %	Total milk q	Total milk %	Milk trans-formed q	Milk trans-formed %	Total milk q	Total milk %	Milk trans-formed q	Milk trans-formed %
Massa Carrara	18,390	4.80	11,860	64.49	16,732	3.72	12,468	74.51	4,130.00	0.64	3,297.12	79.83
Lucca	52,010	13.58	49,370	94.92	25,471	5.67	18,733	73.54	17,087.70	2.63	14,591.79	85.39
Pistoia	12,600	3.29	12,240	97.14	7,000	1.56	5,090	72.71	7,344.00	1.13	6,941.80	94.52
Firenze	60,670	15.87	53,290	87.69	36,949	8.23	23,510	63.62	30,294.01	4.66	27,852.27	91.93
Livorno	16,420	4.28	9,720	59.19	32,442	7.23	25,775	79.44	9,516.05	1.46	9,128.02	95.92
Pisa	34,350	8.97	30,400	88.50	25,471	5.67	18,733	73.54	57,652.17	8.87	49,687.21	86.18
Arezzo	43,970	11.48	36,680	83.42	28,361	6.36	21,036	74.17	23,450.13	3.61	19,158.02	81.69
Siena	36,310	9.48	31,620	87.08	95,590	21.31	80,059	83.75	161,655.79	24.88	134,414.87	83.14
Grosseto	108,020	28.21	91,760	84.94	180,609	40.25	112,767	62.43	337,408.04	51.94	275,990.92	81.79
Prato	–	–	–	–	–	–	–	–	1,143.20	0.18	1,036.78	90.69
Tuscany	382,840	100.00	326,94	85.39	448,625	100.00	318,171	70.92	649,681.09	100.00	542,098.80	83.39

Sources: Istat, *Annuario di statistica agraria*, 1954, *op. cit.*, p. 90; *Indagine nazionale su alcuni aspetti degli allevamenti e delle produzioni ovine*, *op. cit.*, pp. 103–9; Caldelli M., Giannone M., "Allevamento ovino in Toscana e razza sarda", *op. cit.*

Table 11. Cooperatives in the Tuscan ovine business, 1980

Cooperative	County	Province	Annual amount of ovine milk (q)
Sestinese	Sestino	Arezzo	380
Zootecnica di Pratomagno	Talla	Arezzo	10,800
Dal produttore al consumatore	Palazzuolo sul Serio	Firenze	–
Il Forteto	Barberino Mugello	Firenze	628
Le Torri	Calenzano	Firenze	130
Pastori Mugellani	Barberino Mugello	Firenze	3,600
C. A. F. Comuni amiatini	Castell'Azzara	Grosseto	100
Caseificio di Manciano	Manciano	Grosseto	20,000
Caseificio di Sorano	Sorano	Grosseto	41,000
Montebamboli	Massa Marittima	Grosseto	–
Pastori Garfagnana	Pieve Foscana	Lucca	–
Apualat	Fivizzano	Massa	1,300
Armentizia media e alta Val di Cecina	Montecatini	Pisa	6,500
Avola	Pisa	Pisa	–
Agrizoo Casolese	Casole d'Elsa	Siena	–
Caseificio Val d'Elsa	Casole d'Elsa	Siena	8,100
Caseificio Val d'Orcia	Radicofani	Siena	9,100
G. Boscagli	Radicondoli	Siena	–
Il Poggio	Castiglion d'Orcia	Siena	–
Rinascita	Piancastagnaio	Siena	–
Total			101,638

Source: *La pastorizia in Toscana: aspetti, problemi, proposte*, *op. cit.*, p. 17.

Table 12. Ovine milk to be sent to various branches for processing, 1980 (value percentages)

Province	Personally Owned Dairy Farms	Cooperatives	Private Dairy Farms
Massa-Carrara	100.00	0.00	0.00
Lucca	60.00	15.00	25.00
Pistoia	87.50	0.00	12.50
Firenze	42.10	30.70	27.20
Livorno	7.10	3.60	89.30
Pisa	20.80	34.20	45.00
Arezzo	64.70	23.20	12.10
Siena	5.90	42.60	51.50
Grosseto	14.90	49.00	36.10
Total	44.77	22.03	33.18

Source: Castor *et al.* (ed.), "Il riconoscimento della denominazione d'origine del Pecorino toscano", *op. cit.*, pp. 10–11.

From Milan to Novara

Production and Sale of Gorgonzola Cheese in the Twentieth Century*

Vanessa Pollastro

Università Cattolica del Sacro Cuore, Milan

1. Growth in production and export volumes of Gorgonzola from the end of the nineteenth century through the first four decades of the twentieth century

Traditionally, the Stracchino-type cheese known as Gorgonzola was produced in a number of provinces across an area encompassing Eastern Piedmont and Western and Central Lombardy. Like other types of Stracchino, it is made from full-cream cow's milk using a low temperature treatment process. It is most recognizable by the veins of mold within its interior that give it its characteristic appearance and distinctive taste.[1]

* I would like to thank the Consortium for the protection of Gorgonzola cheese [il Consorzio per la tutela del formaggio Gorgonzola] and Gianpiero Mellone of the Latteria Sociale di Cameri (Novara) for their help in providing me with information and documentation; special thanks are due too to Claudio Besana for his invaluable support and advice.

1 The manufacture of Gorgonzola in two phases was a common practice in the 19th century and involved the combination of milk obtained from two successive milking sessions: curds made from the milk of the previous evening were placed in molds together with fresh milk produced in the morning; as the cheese aged, holes were pierced through the moulds, causing mold spores to grow inside the cheese giving it its characteristic appearance. This manufacturing procedure gave the product a markedly strong and piquant taste. Throughout the 20th century a single-phase process began to be widely used: full-cream milk from one milking session was pasteurized before adding starter bacteria and mold spores to encourage the mold to grow inside the cheese; it was then heated during the curdling phase along with the remold of the whey, after which it was placed in moulds and left to ripen; the actual ageing process was shorter than that for the two-phase system. A crucial element in this manufacturing process was the move from powdered to liquid spores.
Published volumes and studies on Gorgonzola include: Cattaneo L., *Sul modo pratico di fabbrica il formaggio grasso detto Stracchino di Gorgonzola*, Milan, Molina, 1840; Massara F., *Dello Stracchino di Gorgonzola*, Milano, 1866; Muoni D., *Melzo e Gorgonzola e loro dintorni: studi storici con documenti e note*, Milan, Tipografia di F. Gareffi, 1866; Bochicchio N., *I diversi tipi di formaggio*, Milan, Hoepli, 1902; Cornalba G., *La tecnica di fabbricazione del Gorgonzola*, Casale Monferrato, Cassone, 1909; Fascetti G., *Caseificio*, Milan, Hoepli,

The cheese itself is named after a small village which was, probably as far back as the medieval period, the first staging area on the way down from the mountain pasture lands for migratory livestock herds, many of which were owned by so-called "bergamino" families; at the end of September these family groups came down from the valleys of Lecco and Bergamo, heading for the plains irrigated by the Rivers Adda and Sesia where they would winter their herds, passing through Gorgonzola or other villages in the territory crossed by the Martesana.[2]

The arrival of the bergamino herders in farmsteads on the plains coincided with a specific phase in the yearly milk production cycle. In the mid-nineteenth century many farmhouses suspended their butter and Grana cheese production from early autumn until the first chills of winter to concentrate on that of Stracchino cheeses, which were made both with and without mold,[3] products that they found easy to sell, since they were consumed by all classes including the poorest members of society. Another

1910; Besana C., *Caseificio*, Torino, UTET, 1916; Ministry of Agriculture and Forestry of the Kingdom of Italy, with the collaboration of the General Fascist Confederation of Italian Industry (ed.), *L'industria del latte e dei suoi derivati in Italia*, 1931; Minguzzi L., *Caseificio*, Milan, Antonio Vallardi, 1933; Savini E., *Il "Gorgonzola" e gli altri formaggi erborinati*, Rome, Società anonima Arte della stampa, 1936; Schiavi C., *Il Gorgonzola e gli altri formaggi erborinati*, Novara, Poligrafica Moderna s.p.a., s.d.; Delforno G., *Il formaggio Gorgonzola*, Rome, Tipografia Atena, 1971; Besana C., *Tra monte e piano. Allevamento transumante ed attività casearie nell'area lecchese tra Ottocento e Novecento*, in Fontana G.L., Leonardi A., Trezzi L. (eds.), *Mobilità imprenditoriale e del lavoro nelle Alpi in età moderna e contemporanea*, Milan, Cuesp, 1998, pp. 55–74; Barisonzo R., Giordani M. (eds.), *Il Gorgonzola. Ieri e oggi*, Novara, Consorzio per la tutela del formaggio Gorgonzola, Istituto Geografico De Agostini, 2000; Battilani P., Bigatti G. (eds.), *Oro bianco. Il settore lattiero-caseario in Val Padana tra Otto e Novecento*, Milan, Giona, 2002; Besana C., *Tra agricoltura e industria: il settore caseario nella Lombardia dell'Ottocento*, Milan, Vita e Pensiero, 2012; Corti M., *La civiltà dei bergamini. Un'eredità misconosciuta. La tribù lombarda dei malghesi tra la montagna e la pianura dal quattordicesimo al ventesimo secolo*, Sant'Omobono Imagna, Centro Studi Valle Imagna, 2014.

2 As Fedele Massara has noted, the populous province of Milan was, in the 19th century, "the first port of call for those who, travelling down from the Alps, made their way through the lowlands of Lombardy, and one might say that it was also the last stage of the journey down towards the vast, fertile plains known as the Valley of the Po" (Massara F., *Dello Stracchino di Gorgonzola*, *op. cit.*, p. 1). The term "bergamino" was coined by authors in the 19th century to indicate livestock owners, some of them relatively wealthy, who lived for most of the year in towns spread across the Bergamasque Pre-alps, and came down each year to the plains in search of pasture lands and hay for their herds of milk-producing livestock. For more on these protagonists of rural life in the mountains and plains of Lombardy and Eastern Piedmont, and on the trails followed by their migrating herds, see Corti M., *La civiltà dei bergamini. Un'eredità misconosciuta. La tribù lombarda dei malghesi tra la montagna e la pianura dal quattordicesimo al ventesimo secolo*, *op. cit.*

3 Along with Gorgonzola, other full-cream, cold processed cheeses were produced, like Quartirolo and Taleggio (Schiavi C., *Il Gorgonzola e gli altri formaggi erborinati*, *op. cit.*, p. 16; Delforno G., *Il formaggio Gorgonzola*, *op. cit.*, p. 49).

benefit of Stracchino cheeses was that, unlike Grana, they did not require an extended ageing process.[4]

Despite constituting a highly profitable use of milk, as confirmed by contemporary experts like Carlo Besana,[5] Stracchino cheeses, and Gorgonzola in particular, could not be produced during the hotter months of the year due to the lack of refrigeration facilities in the farmhouses where the cheeses were produced. Things began to change in the 1870s and 1880s, when a growing demand for Stracchino cheeses on the domestic market and the new rail network that allowed perishable goods to be distributed across a wider area led to an increase in the volume of production of these cheeses, with manufacturing extended to the entire year and the development of a new ripening method to counter the product's rapid deterioration.

In the first half of the nineteenth century there was a widespread belief that Gorgonzola was produced only in the town that gave it its name and in surrounding areas, though we know from contemporary sources that it was also produced in large quantities in the areas of Abbiategrasso (towards Novara) and Codogno (towards Piacenza).[6]

---

4 Massara F., *Dello Stracchino di Gorgonzola, op. cit.*, pp. 1–11. In this period Gorgonzola was ripened for 5–6 months and in some cases for as much as a year; this gave it a piquant flavor, similar to that of the French cheese Roquefort. The modern variety of single-phase Gorgonzola undergoes a far shorter ripening process, as little as 50 days, and has a soft, creamy taste and a color close to white.

5 Carlo Besana (1849–1929), with a degree in chemistry and physics from the University of Pavia, worked for the Royal technical institute [Regio istituto tecnico] in Milan, and in 1880 become director of the experimental dairy station in Lodi [La Stazione sperimentale di caseificio di Lodi], later known as the Istituto superiore di caseificio. He published numerous studies and articles, and participated in a number of conferences as well as the compilation of technical manuals. He founded new academic courses at the Lodi station, helping to introduce general improvements in its working practices (Gaudiano A. (ed.), *Besana, Carlo*, in *Dizionario biografico degli italiani*, online).

6 Gorgonzola production in areas beyond the town of Gorgonzola had already been noted by Luigi Cattaneo in 1840. See also the investigation by Karl Czoerning published in Faccini L. (ed.), *Agricoltura e condizioni di vita dei lavoratori agricoli lombardi (1835–1839)*, Milan, 1986.

Figure 1. The zone of production of Gorgonzola cheese

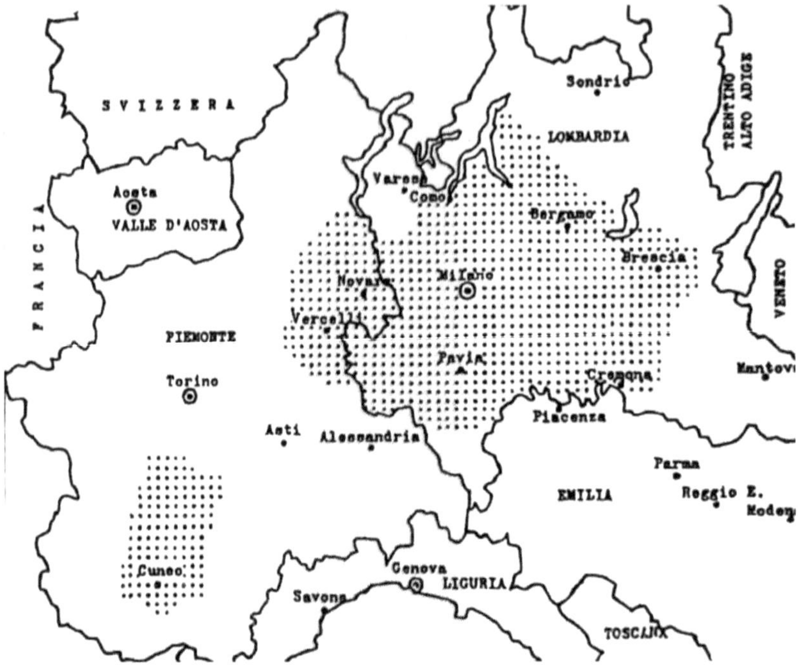

Source: "Produzione e commercio del formaggio Gorgonzola", in *Mondo Latte*, August 1970, 8, p. 579.

After the unification of Italy, production of Gorgonzola saw a significant increase in the province of Milan, as well as in Lomellina and the province of Novara. To the west of the River Ticino production grew in response to changes in the agricultural landscape of the area after irrigation systems became widely available, and due to the decision of numerous bergamino families to abandon migratory herding and establish themselves permanently in local farmsteads where they combined livestock farming with dairy production.[7]

Gorgonzola's growing success benefited further from the presence of the large Milanese market that represented a profitable consumer base as well as a trading point for vast quantities of dairy products. This is borne out by sources that mention the presence in Milan, during the second half of the nineteenth century, of numerous retailers, including the Guscetti and Trovati families, who also specialized in the export of cheeses to international markets.[8] Gorgonzola's position within the Italian dairy industry acquired

7 Corti M., *La civiltà dei bergamini. Un'eredità misconosciuta. La tribù lombarda dei malghesi tra la montagna e la pianura dal quattordicesimo al ventesimo secolo, op. cit.*, pp. 241–59.

8 See Besana C., *Tra agricoltura e industria. Il settore caseario nella Lombardia dell'Ottocento, op. cit.*

further prominence when several large-scale ripening facilities were opened in the Valsassina Valley at the foot of the Lecco Alps.[9] The first of these facilities were opened in 1860 in Ballabio by Mattia Locatelli, whose son and successor, Giovanni, opened outlets in London and Buenos Aires at the beginning of the twentieth century with Giovanni carrying their business as far as New York.[10] In a recent publication by Sylvie Vabre on the activities surrounding the production and sale of Roquefort, a contemporary French source is quoted as recording that Gorgonzola was originally ripened in the caves of the Valsassina by a businessman from Lodi, Antonio Zazzera.[11]

It is plausible that the ripening of Gorgonzola in natural caves was the idea of a business partner of the Polenghi brothers from Lodi who was faced with the problem of finding a cheese that was easy to make and could be produced throughout the year in order to satisfy the growing market in Europe, which was especially strong in Great Britain. It may also be that the process of ripening cheese in grottoes was first carried out by merchants from Lecco, like the Locatelli family, who traded with the bergamini families that spent the summer in the alpine pasture lands and whose products included soft cheeses. What matters is that, from the last two decades of the nineteenth century, Gorgonzola began to be produced in ever greater quantities and was ripened in the natural grottoes of the Valsassina or in structures built up against the rock face in the same area, with the same methods as those used to ripen the already famous French cheese.[12] Gorgonzola production, then, began to take place in spring and summer too, especially in Lomellina and in the vicinity of Novara. Thanks to the new rail network, Gorgonzola could be delivered in Lecco fifteen to twenty days after it had been packaged. It was transported to the natural or artificial ripening facilities that were becoming ever more numerous, at altitudes between 500 and 900 meters above sea

---

9 The Valsassina is a valley in the province of Como made up of five districts stretching from Bellano to Lecco. In the 19th century cattle rearing and milk processing activities were managed by businessmen involved in the transhumance industry who earned money from the soft cheeses that they produced.

10 Delforno G., *Il formaggio Gorgonzola*, *op. cit.*, p. 13. For the commercial activities of the Locatelli family, see their entry edited by Amilcare Mantegazza in the *Dizionario biografico degli italiani*.

11 The French viewed Zazzera's apparent imitation of Roquefort's production model with a sense of apprehension due to the success that Gorgonzola had achieved in European markets, since their cheese was bound to suffer in that it involved more expensive production costs because it was made with sheep's milk. See Vabre S., *Le sacre du roquefort. L'émergence d'une industrie agroalimentaire. Fin XVIIIe siècle-1925*, Rennes-Tour, Presses universitaires de Rennes-Presses universitaires François Rabelais de Tours, 2015, pp. 291–2.

12 Savini E., *Il "Gorgonzola" e gli altri formaggi erborinati*, *op. cit.*, pp. 36–7. The rock faces in question were dolomitic and contained numerous fissures. The first ripening facilities consisted of caves dug out of the mountainside, and were followed by purpose-built edifices containing various levels. After 1880 the storehouses began to be built underground. See Besana C., *Tra monte e piano. Allevamento transumante ed attività casearie nell'area lecchese tra Ottocento e Novecento*, *op. cit.*, pp. 55–74.

level.[13] It was not long before the caves and the first storehouses were joined by ripening facilities equipped with artificial refrigeration able to store up to fifty or sixty thousand cheeses.[14]

As noted above, the cheese produced in those times was very different from that of today and was made in two phases, with a cheese paste obtained by cold-curdling the milk produced by the cows in the evening that was mixed with a second cheese paste obtained from the milk produced the following morning.[15] With regard to estimated production levels,[16] it seems that in 1866, 2,500 quintals of cheese were produced in the environs of Gorgonzola. In 1902 the overall production of this cheese was recorded as 20,000 quintals, a figure which rose to 100,000 by 1909, 125,000 in 1910 and 250,000 by 1926.[17] In 1936, 130–150,000 quintals were produced in the province of Novara (which counted 275 dairies, of which 20 could be described as industrial dairies), 100,000 in the province of Milan, 70–80,000 in the province of Pavia (where production was concentrated in the Lomellina area and in some towns in the south of the province) and 14–20,000 in the province of Vercelli (where production had begun to grow after 1890, and benefited from the presence of modern ripening and storage facilities that were built in the 1920s in Cigliano and Santhià, and of a refrigeration room in Vercelli); in Valsassina, 800,000 cheeses were stored each year in the same period (64–80,000 quintals), equivalent to one fifth of the entire national production of Gorgonzola.[18]

The survey carried out in 1937 demonstrates the highly fragmented nature of Gorgonzola production, divided as it was among a large number of different manufacturers.[19] As of 25 May 1937, 917 manufacturing plants[20]

---

13 Delforno G., *Il formaggio Gorgonzola*, p. 37.

14 Savini E., *Il "Gorgonzola" e gli altri formaggi erborinati*, p. 38.
The years following the First World War saw the growth of the refrigeration industry and the appearance of a number of artificially cooled storehouses which were also built in the plains, close to the areas where the cheese was produced: they were built below the level of the road with one, two or three levels, and were fitted with radiators that cooled the air. Many of these storehouses also appeared in Turin, and later in Novara. The use of artificial storage facilities was to prove fundamental in the period between the two wars because their refrigeration equipment was able to reproduce the constant temperatures typical of natural caves (source: *L'industria del latte e dei suoi derivati in Italia*, *op. cit.*, p. 24).

15 Minguzzi L., *Caseificio*, *op. cit.*, p. 97.

16 On the unreliability of production statistics for the agricultural industry in the late 19th century see Romani M., *L'agricoltura in Lombardia dal periodo delle riforme al 1859. Struttura, organizzazione sociale e tecnica*, Milan, 1957.

17 Delforno G., *Il formaggio Gorgonzola*, *op. cit.*, p. 115; Savini, E., *Il "Gorgonzola" e gli altri formaggi erborinati*, *op. cit.*, pp. 71–5. Delforno used data collected by Carlo Besana, Nicola Bochicchio, Gaetano Cornalba, Giuseppe Fascetti.

18 Savini E., *Il "Gorgonzola" e gli altri formaggi erborinati*, *op. cit.*, pp. 71–5.

19 Istituto centrale di statistica del Regno d'Italia (ed.), *Censimento industriale 1937-XV. Industria della lavorazione del latte e dei prodotti derivati. Censimento del 25 maggio 1937-XV*, Rome, Tip. Failli, 1937.

20 In 1937 the term "esercizio industriale" (translated here as "manufacturing plant") was used to describe a space or collection of spaces used exclusively for the processing of milk

owned by 862 companies were involved in the production of Gorgonzola;[21] 229 of these plants were located in Piedmont and 684 in Lombardy.[22] Most manufacturers (320) achieved an annual output of between 100 and 200 quintals of Gorgonzola, whereas only five plants produced more than 1,000 q of Gorgonzola in a year.[23] With regard to overall production, the 1937 figures show an interruption in the trend of the previous years, with sales figures for Gorgonzola of approximately 183,500 q.[24] By the mid-1930s almost all Gorgonzola was sold fresh; around 165,000 q passed through 116 ripening and storage facilities belonging to 110 different companies.[25] Thirty-seven of these storehouses were located in the province of Novara, 23 in the province of Milan, 11 in the province of Pavia, and 10 in the province of Como.[26]

Increased production levels were accompanied by efforts to improve the quality of manufacturing techniques which, in the late nineteenth century, were still substantially based on the empirical knowledge of the bergamini and the cheese-makers working in traditional farmsteads. Manufacturing techniques were the subject of studies carried out by Arnaldo Maggiora in 1891, which looked at the composition of "stravecchio" matured cheeses by analyzing samples that had exceeded their maximum ripening period.[27] At the beginning of the twentieth century, studies were carried out on the ripening and storage of Gorgonzola in the Valsassina at the Experimental Dairy Institute in Lodi by its director Carlo Besana along with his assistant Gaetano Cornalba and Giuseppe Fascetti, who would go on to replace Besana as director of the Institute.[28] A crisis in the Gorgonzola export market took place in the years 1906–7, when the use of barium sulphate to coat the cheese was discontinued,[29] forcing the researchers at the Lodi station

derivatives. The production of Parmigiano Reggiano was even more fragmented; in May 1937 the survey recorded 2,818 plants for the production of Parmigiano-Reggiano in Emilia Romagna and the province of Mantua (*Censimento industriale 1937-XV, op. cit.*, p. 109, tab. 67).

21 800 of the plants surveyed produced only Gorgonzola, while the other 117 also made other cheeses (*Censimento industriale 1937-XV, op. cit.*, p. 112, tab. 69).

22 181 plants were recorded in the province of Novara, 272 in the province of Milan and 381 in the province of Pavia. The figures quoted were published in *Censimento industriale 1937-XV, op. cit.*, p. 112, tab. 69.

23 In particular, two Stracchino plants, in the province of Pavia, produced between 1,000 and 2,000 q of Gorgonzola and another three produced more than 2,000 q in the province of Novara.

24 *Censimento industriale 1937-XV, op. cit.*, p. 102, tab. 61. Since the average weight of a cheese wheel was around 7 kg, this would have amounted to around 2.6 million wheels.

25 *Censimento industriale 1937-XV, op. cit.*, p. 150, tab. 91.

26 Around 23,000 q of Gorgonzola, according to official surveys, were ripened in the ten facilities present in the province of Como, all of them in the Valsassina area.

27 Savini E., *Il "Gorgonzola" e gli altri formaggi erborinati, op. cit.*, p. 50.

28 *L'Istituto sperimentale lattiero caseario di Lodi dal 1871 al 1922. Un secolo di storia lombarda*, s.l., Ministero delle risorse agricole, alimentari e forestali, 1996, p. 161.

29 France and the USA went so far as to ban Gorgonzola imports from Italy, accusing Italian producers of using materials that could be damaging for human health (barium

to engage in an active search for new products that could be used to coat Gorgonzola wheels in order to facilitate their conservation and make it easier to transport them over long distances.[30]

The growth in production was favored by other manufacturing innovations. In particular, attempts were made to standardize manufacturing procedures in order to achieve product uniformity. In 1907, Filippo Samarani attempted a procedure whereby penicillium conidiophores were introduced into the cheese paste before curdling or during the ripening process.[31] In 1910 penicillium cultures for Gorgonzola were produced for the first time in Italy at the Lodi station.[32] The cultures took the form of a powder: a method that was already used in France for the production of Roquefort. The powder was sprinkled between layers of curd or introduced into the cavities that were created in the cheese during the ripening process.[33] After a period of inactivity during the war, experiments began once more in the 1920s, when workable penicillium species were discovered[34] thanks to studies carried out in the USA (by Thom and Raper, and by Biourge and Weidemann).[35] At the same time, M. Bondiol was using a microscope to study the action of bacteria during the Gorgonzola ripening process. A few

---

sulphate) to coat Gorgonzola wheels in order to make them less vulnerable to the attack of insects and easier to transport. The British health authorities also placed restrictions on Gorgonzola imports.

30 Schiavi C., *Il Gorgonzola e gli altri formaggi erborinati, op. cit.*, pp. 25–6; *L'Istituto sperimentale lattiero caseario di Lodi dal 1871 al 1922. Un secolo di storia lombarda, op. cit.*, p. 163.
Criticism came from the USA and Great Britain. The Ministry for Agriculture appointed a commission whose members included professors Gorini, Carlo Besana and Giuseppe Fascetti. The coating used for Gorgonzola had always been made from barytes, powdered brick, and "Rossetto di Germania" mixed with animal fats; barytes were seen as economically advantageous but were opposed by other countries due to their effect on the weight of the cheese (Savini E., *Il "Gorgonzola" e gli altri formaggi erborinati, op. cit.*, p. 32). The commission concluded that barium sulphate was harmless, but it advised using other substances like talcum or white clay, powdered brick, "Vicenza Earth", kaolin or innocuous plastic materials. The efforts to satisfy the requests of foreign buyers were a result of the prominence that Gorgonzola had assumed on the export market; the use of barium had also been criticized by the Parisian Union Syndacaledes Marchands en gros des fromages at a time when Roquefort, a direct rival of Gorgonzola, was twice as expensive as the Italian cheese. At the beginning of the 20th century "the fluctuating quality" of the cheese was one of the main obstacles to increasing Gorgonzola's position in the export market (Colli A. (ed.), "Galbani, Egidio", in *Dizionario biografico degli italiani*, online).

31 Schiavi C., *Il Gorgonzola e gli altri formaggi erborinati, op. cit.*, pp. 22–3.
The idea of using penicillium produced especially for use in manufacturing was first proposed by Johan Olsen.

32 *L'Istituto sperimentale lattiero caseario di Lodi dal 1871 al 1922. Un secolo di storia lombarda, op. cit.*, p. 163.

33 It was only later that the easier to use liquid cultures was widely introduced.

34 Various experiments were carried out by Carbone and Savini (Delforno G., *Il formaggio Gorgonzola, op. cit.*, p. 63). *L'industria del latte e dei suoi derivati in Italia, op. cit.*, p. 45.

35 Schiavi C., *Il Gorgonzola e gli altri formaggi erborinati, op. cit.*, pp. 23, 30.

studies on penicillium for Gorgonzola were published between 1926 and 1928 by the microbiologist Carlo Arnaudi,[36] and involved the cultivation of artificial molds that had already been experimented by Roquefort producers. Arnaudi's studies also highlighted the role played by the acidity level of curds and concentrations of sodium chloride in determining the multiplication of penicillium rather than that of other molds. In the same period, Professor D. Carbone carried out a number of experiments on the use of penicillium in Gorgonzola production at various dairies in Milan,[37] demonstrating that the addition of spores to milk before coagulation was a more practical method than sprinkling them on the curds, and led to a more uniform growth of molds on blue cheeses. Other experiments were also carried out by Elia Savini.[38] These studies meant that it was possible to reduce ripening periods by a month and to achieve uniformity in the formation of blue cheeses,[39] leading to a reduction in wastage due to the lack of mold growth. It is also worth noting that, although the addition of spores did not require any new technology, it was not carried out in an organized manner or on a large scale due to the lack of technical and scientific advancement in the sector. Also, during the 1920s, the Lodi Experimental Dairy Institute was investigating the subject of salt penetration in cheese.[40]

36 *Ibid.*, p. 21.

37 *Ibid.*, p. 25.

38 Savini E., *Il "Gorgonzola" e gli altri formaggi erborinati*, *op. cit.*, p. 56.

39 Savini compared ripening the processes of cheeses without mold content to those of other cheeses produced by adding moulds or spores.

40 *L'Istituto sperimentale lattiero caseario di Lodi dal 1871 al 1922. Un secolo di storia lombarda*, *op. cit.*, p. 168.

Table 1. Total cheese exports and exports of soft cheeses (Gorgonzola, Stracchino, Fontina) in the period 1905–1938

Year	Total export figures in quintals	
	Total cheese exports	Exports of soft cheeses
1905	171,000	55,337
1906	192,000	67,700
1907	211,000	72,031
1908	198,000	66,493
1909	200,000	58,896
1910	261,000	78,863
1911	279,000	75,355
1912	306,000	89,747
1913	328,000	91,624
1914	299,000	84,777
1915	298,000	78,831
1916	178,000	40,895
1917	11,000	364
1918	4,000	18
1919	8,000	1,036
1920	13,000	3,016
1921	76,000	16,331
1922	145,000	50,662
1923	203,000	78,558
1924	336,000	117,552
1925	391,000	128,488
1926	331,000	103,602
1927	318,000	104,195
1928	365,000	101,537
1929	329,000	99,715
1930	367,000	103,817
1931	404,000	106,961
1932	301,000	85,798
1933	239,000	72,760
1934	251,000	70,405
1935	278,000	66,016
1936	194,000	26,964
1937	239,000	50,956

Source: Savini E., *Il "Gorgonzola" e gli altri formaggi erborinati, op. cit.*, p. 78; "Tavola rotonda sulla produzione ed esportazione dei prodotti lattiero-caseari", in *Mondo Latte*, April 1971, 4, p. 251; ISTAT, *L'Italia in 150 anni. Sommario di statistiche storiche 1861–2010, 2011*, Table 16.10, pp. 740–1.

Available data on cheese exports confirm the growing success of Gorgonzola on international markets. According to scholarly estimates, Italy was producing around 2.5 million quintals of cheese by the eve of the First World War, with a total export volume of 328,000 q; soft cheese exports contributed to this figure, with 91,624 q, most of which was represented by Gorgonzola.

When the war began exports initially fell, going on to collapse completely in 1917, when all exports were banned. The post-war recovery took some time, not least because unrestricted sale of this cheese abroad was still banned until the end of 1921, a situation which led the most prominent exporters of Gorgonzola to write a letter of complaint in January 1922.[41] In 1922, export figures returned to their previous levels, reaching the high point of the period in 1925 when they exceeded 128,000 q; in the mid-1920s around 60% of all Gorgonzola produced in Italy was destined for the export market. The industry encountered difficulties once again in the early 1930s when Gorgonzola exports, along with those of all Italian cheeses, suffered a drastic reduction due to the effects of the Great Depression, the devaluation of the pound against the lira, and competition from an imitation Gorgonzola product made in Denmark. Foreign sales, nonetheless, were far from being wiped out. At the end of the 1930s around 55,000 q. were still being sold abroad, with sales in Great Britain reaching 32,000 q.[42] London continued to be the principle destination for this cheese. A proportion of Gorgonzola exports also found their way to France and the USA.

The outbreak of the Second World War was, for obvious reasons, followed by a downturn in export volumes. In France in particular, Gorgonzola's absence was made up for by a similar product called Saingorlon, produced in collaboration with Italian cheese-makers and other specialists. Gorgonzola itself was produced in Pontarlier thanks to the efforts of a businessman who had connections with the Italian emigrant community there. These products were the forerunners of the Blue di Bresse cheeses that became popular from the 1950s.[43]

2. Manufacture and sale of Gorgonzola after the Second World War: industrial scale production and the introduction of protection policies

In the period immediately after the Second World War Gorgonzola production continued to be centered around Southern Lombardy and the province of Novara. In the 1950s most milk in Novara was processed on small family farmsteads, with only around twenty industrial dairies;[44] in

41 Cf. Archivio di Stato di Como, fondo Camera di commercio di Lecco, b. 182. The letter was signed by ten Italian companies, including Polenghi Lombardo and Locatelli, which also operated commercially on the London market.

42 Ferrari P., *L'industria del latte in Italia*, Piacenza, Camera di commercio di Piacenza [after 1970], p. 311.

43 Over the next few years a number of cooperative milk associations were formed in the area of Ain, and overall annual production grew to 20,000 tons, similar to that of Roquefort. Furthermore, it was sold in relatively small packages (Schiavi C., *Il Gorgonzola e gli altri formaggi erborinati*, *op. cit.*, p. 110–11).

44 "I caratteri economici della provincia di Novara", in *Sintesi economica*, No. 10, October 1950.

Lombardy, on the other hand, the numerous small businesses were joined by several large companies, including Galbani and Invernizzi, which were involved in the production of many different types of cheese. The market as a whole was characterized by the presence of a variety of actors: producers of fresh Gorgonzola, cheese ripeners and wholesalers; in most cases, each actor was responsible for only one part of the manufacturing process.

What turned out to be fundamental for the survival of small and medium sized producers was the creation of custom-built storehouses for the ripening and conservation of cheeses. These appeared in Mantua, Cremona, Novara, Villa Poma (MN), Pegognaga (MN), Lodi and Pavia, thanks to investments from the largest bank in Lombardy, the Cassa di risparmio delle provincie lombarde (Cariplo). They were equipped with refrigeration and conservation facilities and possessed weighing machines, cranes and internal transport systems. The storehouse in Novara,[45] built between 1955 and 1957, had 54,000 m^2 of cold storage and 70,000 m^2 of storage space, able to store 300,000 Gorgonzola wheels. With the introduction of reduced ripening periods, the storehouse was able to deal with a turnover of one million cheeses.[46] With its special refrigeration cellars, freezing tunnels and processing halls, the Novara storehouse boasted its own station connected to the state rail network, customs offices, and an internal bank where wholesale prices were agreed with other parties. Thanks to this investment, a large number of small and medium sized producers were able to count on an efficiently operated processing center for their products and have access to financial resources, since the internal bank granted loans with the stored cheeses as collateral.[47]

With regard to production volumes, the entire milk and dairy sector experienced a period of slow growth in the 1950s followed by a phase of stagnation in the 1960s.[48] Gorgonzola production remained stable throughout

---

45 The Novara Agricultural Storehouses Society [Società Magazzini Agricoli Novaresi] was set up in 1953 with a starting capital of 660 million lire: 224,400,000 lire were underwritten by the Provincial Agrarian Consortium of Novara and the remaining sum by the Agrarian Credit branch of Cariplo bank; the latter, in 1967, voted to purchase the shares possessed by the Consortium, becoming full owners of the society (Cassa di risparmio delle province lombarde, *25 anni di iniziative economiche, sociali e culturali della Cariplo*, supplement to *Cà de Sass*, No. 60, Milan 1978, pp. 24–5).

46 According to another publication (Cassa di risparmio delle province lombarde, *25 anni di iniziative economiche, sociali e culturali della Cariplo*, *op. cit.*, pp. 21–32), the Novara storehouse had a capacity of 500,000 Gorgonzola wheels and 100,000 wheels of Fontina. In the course of a single year around 2,400,000 Gorgonzola wheels were ripened at the storehouse.

47 Cariplo, *25 anni di iniziative economiche, sociali e culturali della Cariplo*, supplement to *Cà de Sass*, No. 60, *op. cit.*, p. 26. Cariplo granted low-interest loans to businessmen up to a maximum of 70% of the value of the goods held in the storehouse; the businessman could receive money for marketing and sales before his product was actually ready to sell (Porteri A., "Il credito agrario", in *La Cassa di Risparmio delle Provincie lombarde nel cinquantennio 1923–1972*, Vol. II, Milan, Giuffrè, 1973, pp. 1,967–8).

48 Felice E., "Il settore lattiero-caseario in Italia dal dopoguerra al Duemila", in Bertagnoni G. (ed.), *Una storia di qualità. Il Gruppo Granarolo fra valori etici e logiche di mercato*, Bologna, il Mulino, p. 42.

the 1950s but fell over the following decade by up to 30–40%. According to Ferrari, it was during the 1960s that meat prices increased, encouraging cattle farmers to increase the number of animals destined for the slaughterhouse rather than for the production of milk. This led to a drastic reduction in the amount of milk available to the cheese-making industry, a shortfall which imported milk was not able to make up for. The result was a reduction in the amount of cheese produced.[49] At the same time, the birth of the European Community led to a substantial increase in imports from Denmark, France and Germany of cheeses that were generally cheaper than their Italian counterparts.[50] Tab. 2 shows how Gorgonzola production fell during the 1960s.

Table 2. Gorgonzola production compared to the overall production of Italian cheeses (1938–1971)

Years	Gorgonzola: volume of production (quintals)	Italian cheeses: volume of production (quintals)	Gorgonzola as a percentage of total Italian cheese production (quintals)
1909	100,000		
1923	200,000		
1926	250,000		
1949	310,000	2,470,000	12.55
1950	320,000	2,700,000	11.85
1951	330,000	2,900,000	11.37
1952	380,000	3,105,000	12.24
1953	350,000	3,270,000	10.70
1954	350,000	3,365,000	10.40
1955	350,000	3,482,000	10.05
1956	360,000	3,450,000	10.43
1957	320,000	3,360,000	9.52
1958	330,000	3,473,000	9.50
1959	330,000	3,940,000	8.37
1960	330,000	4,380,000	7.53
1961	330,000	4,513,000	7.31
1962	330,000	4,516,000	7.30
1963	180,000	3,745,000	4.80
1964	207,000	3,855,000	5.37
1965	174,000	4,228,000	4.11
1966	250,000	4,531,000	5.52
1967	250,000	4,524,000	5.52
1968	220,000	3,880,000	5.67
1969	235,000	3,900,000	6.02
1970	242,000	3,950,000	6.13
1971	245,000	4,007,000	6.11

Sources: "Produzione e commercio del formaggio gorgonzola", in *Mondo Latte*, March 1973, 3, pp. 159, 161.

49 Ferrari P., *L'industria del latte in Italia*, *op. cit.*, pp. 272–3.

50 Delforno G., *Il formaggio Gorgonzola*, *op. cit.*, p. 118.

In 1973, a convention was organized in Novara on the Gorgonzola industry, during which a number of problems emerged. With regard to the export sector, Gorgonzola represented 13.5% in 1938 (27% Grana, 22% Italico and soft cheeses, 21% Pecorino cheeses, 16.5% Provolone) and only 6% in 1971 (36% Grana, 33% Italico and soft cheeses, 5% Pecorino cheeses, 20% Provolone).[51] The difficulties it was encountering on foreign markets were believed to be connected to the price of milk (which was particularly expensive in Italy) and to official listed prices of cheese.[52]

All the same, Gorgonzola's space for manoeuvre had become more restricted on the domestic as well as on the foreign market due to a problem with the product itself, not only its pricing. Consumer tastes were changing: product uniformity was becoming more widespread and brought with it the tendency towards a blander flavor which worked to the benefit of mild cheeses, a tendency which became even more pronounced from the 1970s.[53] Changes in consumer demand and the birth of the Common European Market, therefore, caused problems, but they were also a stimulus for change.

The dramatic increase in imports of European cheese encouraged Italian companies to modernize their production techniques and increase the size of their companies.[54] New machinery was introduced along with the use of clean working environments (by eliminating mold on the walls, for example), making it possible to improve the quality of the product and at the same time increase uniformity. Some phases were automated, like the creation of holes in the cheese with metal needles. At the same time European legislation introduced increased quality controls on milk production, leading to a rise in the overall quality of raw materials and the standardization of dairy produce.[55]

Changes in consumer tastes provoked an intensification of the relationship between science and industry. After the Second World War a number of studies were carried out on the properties of molds, whose authors included the American scholar Charles Thom of the National Center for Agricultural Utilization Research in Illinois.[56] Certain penicillium variants were isolated

---

51 "Produzione e commercio del formaggio Gorgonzola", in *Mondo latte*, March 1973, 3, pp. 159–63.
Immediately after the war the Gorgonzola industry experienced a series of export-related difficulties caused by the presence of numerous foreign-made Italian style cheeses. One of the markets that was lost in this period was that of the UK, which had been conquered by a Gorgonzola rival called Danish Blue ("Esportazione dei prodotti caseari italiani", in *Mondo latte*, October 1947, p. 14).

52 "Produzione e commercio del formaggio Gorgonzola", *op. cit.*

53 Felice E., "Il settore lattiero-caseario in Italia dal dopoguerra al Duemila", in Bertagnoni G. (ed.), *Una storia di qualità. Il Gruppo Granarolo fra valori etici e logiche di mercato, op. cit.*, p. 38.

54 *Ibid.*, pp. 46–51. The European Community set aside Feoga funds for the enlargement of manufacturing plants used by cooperatives and consortia.

55 *Ibid.*, p. 51.

56 Schiavi C., *Il Gorgonzola e gli altri formaggi erborinati, op. cit.*, p. 28. The publication is *A Manual of the Penicillia*, 1949.

and developed in the laboratory for industrial use:[57] the creation of cultures by institutes and laboratories made it possible to select various strains (each strain produced different effects) and compounds. In particular, the Centro sperimentale del latte in Milan introduced the use of penicillium in liquid form. The selected molds were added straight to the milk. A new method became widespread, whereby selected milk enzymes (selected bacteria cultures) prepared in the dairies were added to the milk during the heating process before introducing the spores; milk enzymes became a necessary ingredient following the introduction of the pasteurization process to eliminate bacteria in the milk. Some companies began to collaborate with research institutes as early as the 1950s, and by the 1960s the new techniques were available on an industrial scale[58] despite the fact that they were not yet fully perfected. These two scientific innovations opened the way for mass production and the uniform application of technological parameters. The advantages were obvious: fully rationalized manufacturing processes, elimination of waste products and poorly developed molds, uniformity in organoleptic characteristics (consistency in mold growth), controlled ripening, and shorter ageing times leading to reduced storage costs.[59] The introduction of enzymes and spores led to technological improvements, with advances in milk treatment and laboratory testing. Improved manufacturing techniques and milk quality, together with the selection of cultures, were all factors that contributed to the creation of the final product. These advances led to the development of a new type of Gorgonzola, with a delicate and soft taste, less piquant than the traditional variety. The 1970s and 1980s saw the production of a cheese that found more favor with consumers; over the following decades its flavor was modified further to produce the sweeter taste that was most appreciated by the public.

At the aforementioned convention of 1973, it emerged that many felt the need for enhanced recognition of milk production and for the granting of guarantee of origin status to dairy products,[60] and recommended a collective approach to the promotion of traditional products on the international market in collaboration with the State. Among their aims was that of reconquering the British market, "to redirect the ever more manifest tendency of the British consumer to prefer imitation Gorgonzola products

---

57 *Ibid.*, p. 30.

58 *Cent'anni di latteria (1914–2014). Una storia di imprenditoria e coraggio*, Cameri (Novara), Latteria di Cameri, December 2014, p. 29; Delforno G., *Il formaggio Gorgonzola*, *op. cit.*, p. 81.

59 "L'impiego di penicilli selezionati nella fabbricazione del Gorgonzola", in *Mondo Latte*, February 1948, pp. 3–6.

60 "Produzione e commercio del formaggio Gorgonzola", *op. cit.* Ever since 1934, producers had asked that Italy adhere to international agreements protecting designation of origin status.

made in Denmark".[61] As Table 3 shows, the UK had, in the past, represented a substantial proportion of overall exports.

Table 3. Final destination of Gorgonzola exports (1938–1967)

Years	1938	1953	1964	1965	1966	1967
	q.	q.	q.	q.	q.	q.
France	10,912	2,006	5,520	14,160	15,550	16,630
Switzerland	5,974	11,263	8,500	8,160	7,940	8,000
UK	32,158	6,578	5,880	5,680	5,640	5,700
West Germany	–	164	1,630	1,690	2,060	2,390
Sweden	–	260	690	860	880	990
Belgium and Luxembourg	–	80	520	630	630	660
USA	2,885	561	490	670	580	480
Austria	–	1,500	500	410	420	350
Holland	–	20	160	220	180	160
Others	3,603	984	840	830	940	870
Total	55,532	23,416	24,730	33,310	34,820	36,230
Total EC	–	2,270	7,830	16,700	18,420	19,840

Source: Ferrari P., *L'industria del latte in Italia*, Camera di commercio di Piacenza [197–], *op. cit.*, p. 311.

By 1951 some countries had already come together at Stresa to establish a common policy for the protection of designation of origin status and of cheese denominations. Italy, Austria, Belgium, France, Holland and Switzerland all signed an agreement to protect the traditional characteristics of cheeses, not least in order to provide consumers with quality guarantees. This was a first example of international cooperation towards the recognition of designation of origin status for agricultural products. The agreement was introduced into Italian law with the 1955 legislation "on the protection of designation of origin and typicalness of cheeses"; with this legislation, Gorgonzola, too, obtained legal recognition for the first time.[62]

Consortia for the protection of other cheeses like Parmigiano-Reggiano and Grana Padano were established shortly afterwards; Gorgonzola, on the other hand, had to wait until the end of the 1960s, while in the meantime it continued to experience problems on international markets and was faced with increasing competition from foreign products on the domestic market. In 1968 two voluntary consortia were established: one founded in Novara at the Chamber of Commerce was formed by small producers and cheese ripeners from the province of Novara, and another founded in Milan at the

[61] "Esportazione dei prodotti lattiero caseari (relazione presentata dal dott. Antonio Masutti direttore dell'Associazione Italiana Lattiero Casearia al Convegno economico svoltosi nel settembre c.a. in occasione della Fiera del Latte di Lodi)", in *Mondo latte*, November 1950, 11, pp. 611–16.

[62] Schiavi C., *Il Gorgonzola e gli altri formaggi erborinati*, *op. cit.*, p. 137.

headquarters of the Italian Milk and Dairy Association was composed mainly of large companies. In May 1970 the two consortia came together in one single body, though not without a certain amount of friction between the two different categories. The first statute was compiled by 35 producers and cheese ripeners from both the industrial and the agricultural sectors.[63] The aim of the consortium, which had its headquarters in Novara, was to provide protection for the production and retailing of Gorgonzola, defend its designation of origin status, increase sales, improve the facilities available for its manufacture and for improvements in quality, propose standardized packaging, oversee its production and sales networks, and collaborate with government authorities in the protection and safeguarding of the cheese's PDO (protected designation of origin).[64] In 1972, a label indicating its origin was approved, consisting of a metallic stencil displaying the consortium's logo and the identification number of the dairy. The following years saw the introduction of stamps impressed onto fresh cheese along with the application of a countermark on the product's aluminum packaging: each producer and cheese ripener had to pay a fee for the use of the metallic stencils which was calculated according to the quantity of cheeses produced, and each retailer had to pay a fee based on the amount of embossed tinfoil packaging that they purchased. This system was designed to limit the fraudulent marketing of blue cheeses under the name "Gorgonzola". The activities of the consortium also included checking on the provenance of Gorgonzola, verifying which milk was used (whether it was from the same territory or not), and quality control checks on the products sold by its members. In order to make sure the checks were effective, the consortium employed a specially designated inspector. Marketing was at the heart of the responsibilities of the consortium. A substantial proportion of its financial resources were destined for its advertising fund; towards the end of the 1970s the first advertising campaigns were launched on television, on the radio and in newspapers: these included advertising slots in television programs, newsreels, newspaper and magazine supplements, and poster campaigns; publicity material was also produced, like labels, adhesive transfers, leaflets, stickers, recipe cards and promotional penknives. The consortium also made contact with other similar trade associations with the aim of organizing joint advertising campaigns to promote traditional Italian products in other European countries; at the same time it promoted the participation of its members in trade fairs and exhibitions in Italy and abroad.

Despite all these efforts, the market remained unstable during the 1980s, too: milk was produced in excess and was more expensive than that of other European countries, while the cheese industry was still under threat from the flood of more neutral-tasting imports in superior packaging. In 1981, the

63 See: Barisonzo R., Giordani M., *Il Gorgonzola. Ieri e oggi*, *op. cit.*

64 Founding acts of the Gorgonzola Consortium [Consorzio Gorgonzola], Novara, 14 May 1970, in *Consorzio per la tutela del formaggio Gorgonzola*, Meeting minutes, Novara, 1970–9, available from the Consortium headquarters, Novara.

Ministry for Agriculture and Forestry asked the consortia to set out annual programs for the control of production, so as to avoid creating a surplus. In the case of Gorgonzola, each producer was designated a specific number of cheeses to produce in relation to an overall maximum of 3 million cheeses for the year 1984, a figure which rose over the following years. These programs operated successfully throughout the 1980s.

With regard to the product itself, there were already plans by the end of the 1970s to package Gorgonzola together with mascarpone and to sell it in the form of slices or triangles, while by the end of the 1980s it was clear that the "intensely flavored", piquant variety of Gorgonzola had lost its position on the market. The consortium also rode the wave of the increasingly popular "*Made in Italy*" brand. In 1986 a group of Italian food products were presented in New York at the Bloomingdales department store, and on other occasions Gorgonzola and wine tastings were organized for members of the foreign press, while gastronomical shows and presentations of Gorgonzola-based recipes were precursors to the Slow Food movement that would begin at the end of the decade with the purpose of promoting traditional Italian foods. In the 1990s, Gorgonzola was presented along with other products at the Italian Institute for Foreign Commerce offices in Montreal, Toronto and Düsseldorf, in order to highlight the high standards and nutritional quality of the product. This was followed soon afterwards by sponsorship of sporting events and television programs. Despite the continuing presence of imitation cheese,[65] production levels of Gorgonzola grew thanks to a larger domestic market and to early signs of growth in the export sector (Tabs 3 and 4); the latter was made possible above all thanks to the mass production techniques that had replaced traditional manufacturing methods and were providing far more competitive advantages. The year 1992 saw Europe-wide recognition of protected designation of origin and protected geographical indication products, which helped Italian and European products to penetrate non-European markets. Meanwhile Gorgonzola itself benefited from an increase in scientific research, with the consortium co-financing a study on milk bacteria aimed at controlling or eliminating pathogenic bacteria. The study was carried out by the Milk and Dairy Institute in Lodi which, a few years earlier in 1988–9, had investigated the possibility of pre-packaging the cheese "into portions in a controlled atmosphere, defining the most appropriate procedures and types of plastic wrapping and the optimum composition for gases usable in the manufacturing process" of Taleggio and Gorgonzola.[66]

"Mild" Gorgonzola achieved popularity on foreign markets thanks to its soft, creamy texture which distinguished it from all its immediate

65 These included the Austrian Oesterzola and the German Cambozola. The consortium took legal action against these products' allusion to the name of Gorgonzola and imitation of its properties.

66 *L'Istituto sperimentale lattiero caseario di Lodi dal 1871 al 1922. Un secolo di storia lombarda*, *op. cit.*, p. 189.

competitors, which were closer to the piquant variety. Tab. 4 shows this growth in production and export volumes from the 1990s.

Table 4. Gorgonzola production and export volumes (1976–2013)

	Production				Export		
Year	Cheese wheels produced n.	Difference compared to previous year: Total value	Difference compared to previous year: %	Quantity q	Quantity q	Growth compared to previous year %	exports as % of total production
1976	2,156,039			258,725	44,312		17.13
1977	2,263,153	107,114	5	271,578	45,392	2.44	16.71
1978	2,457,097	193,944	8.6	294,851	46,601	2.66	15.80
1979	2,474,893	17,796	0.7	296,987	63,942	37.21	21.53
1980	2,563,156	88,263	3.6	307,578	62,890	-1.65	20.45
1981	3,015,331	452,175	17.6	361,840	65,461	4.09	18.09
1982	3,152,935	137,604	4.6	378,042	65,951	0.75	17.44
1983	2,968,544	-184,391	-5.8	356,225	66,898	1.44	18.78
1984	3,150,356	181,812	6.1	378,042	70,127	4.83	18.55
1985	3,172,020	21,664	0.7	380,642	78,787	12.35	20.70
1986	3,298,866	126,846	4	395,863	78,842	0.07	19.92
1987	3,292,873	-5,993	-0.2	395,144	85,179	8.04	21.56
1988	3,282,702	-10,171	-0.3	393,925	85,433	0.30	21.69
1989	3,417,309	134,607	4.1	410,077	76,787	-10.12	18.73
1990	3,211,806	-205,503	-6	385,416	80,833	5.27	20.97
1991	3,157,604	-54,202	-1.7	378,912	85,503	5.78	22.57
1992	3,223,800	66,196	2.1	386,856	88,944	4.02	22.99
1993	3,304,613	80,813	2.5	396,553	105,616	18.74	26.63
1994	3,360,414	55,801	1.7	403,249	96,370	-8.75	23.90
1995	3,408,782	48,368	1.4	409,054	100,420	4.20	24.55
1996	3,532,796	124,014	3.6	423,935	108,250	7.80	25.53
1997	3,441,908	-90,888	-2.6	413,028	110,440	2.02	26.74
1998	3,616,146	174,238	5.1	433,937	102,936	-6.79	23.72
1999	3,646,725	30,579	0.8	437,607	106,020	3.00	24.23
2000	3,844,592	197,867	5.4	461,350	114,070	7.59	24.73
2001	3,872,339	27,747	0.7	464,680	116,850	2.44	25.15
2002	3,842,653	-29,686	-0.8	461,118	122,810	5.10	26.63
2003	3,878,297	35,644	0.9	465,395	131,870	7.38	28.34
2004	3,968,547	90,250	2.3	476,225	136,640	3.62	28.69
2005	4,040,061	71,514	1.8	484,807	141,370	3.46	29.16
2006	4,011,211	-28,850	-0.7	481,345	144,470	2.19	30.01
2007	4,071,610	60,399	1.5	488,593	147,060	1.79	30.10
2008	4,060,142	-11,468	-0.3	487,217	131,410	-10.64	27.00
2009	3,970,390	-89,752	-2.2	476,447	129,850	-1.19	27.25
2010	4,052,028	81,638	2.1	486,243	136,060	4.78	27.98
2011	4,194,458	142,430	3.5	503,335	146,680	7.80	29.14
2012	4,156,966	-37,492	-0.9	498,836	152,540	3.99	30.58
2013	4,175,610	18,644	0.4	501,080	160,320	5.10	31.99

Source: data provided by Consortium for the protection of Gorgonzola cheese.

Along with this simultaneous focus on quality and uniformity, changes were also taking place on the supply side. From the 1990s the three separate functions carried out by fresh cheese producers, cheese ripeners and retailers began to merge: many producers started to ripen their own cheeses in order to have full control over the production process, while wholesalers gradually lost their importance due to the introduction of large distribution chains and the falling number of small retailers. In this context the Cariplo storehouse also began to lose its relevance, with a reduction in the number of its associated producers, although they did manage to diversify their position within the market. Three different manufacturing procedures emerged in the dairies: the first and most traditional was the introduction of heating vats with a capacity of six quintals and ripening on wood with static cooling, the second, semi-industrial, set of procedures involved multi-purpose heating vats with a capacity of 30/50 quintals and the mechanization of several manufacturing phases, and finally, industrial manufacture on a large scale introduced curdling tanks and manufacturing processes that involved minimum human intervention. These different manufacturing procedures led to different types of product due to the effects produced by different ripening procedures, and were therefore each aimed at different markets: the more traditional product was generally intended for over-the-counter sales where the cheese was sliced before purchase, whereas the most industrial sector produced a ready-packaged product and, in even greater volumes, fed the export market. At the same time, advances in cattle feed improved the quality of milk, inspiring Gorgonzola producers to introduce a system of incentives to reward high quality milk.

Despite losing sales to a series of dietary and nutritional fads and fashions, production and export volumes of Gorgonzola grew throughout the 1990s and the cheese consolidated its position on the market. The recognition by the European Community of designation of origin status gave rise to new studies of the product's innate characteristics and provided study centers and laboratories, with an incentive to continue to develop innovations in the Gorgonzola industry. The application of PDO status required the creation of an authority qualified to grant certificates of conformity to companies, implying a need to redefine the structure of the consortium. Changes in the market also led to a process of natural selection among producers and cheese ripeners, and the shrinking of the agricultural component of the sector. With regard to production oversight, all quotas were abandoned in the 1990s in order to improve sales trends and comply with anti-trust legislation.

Conclusions

Throughout the nineteenth and twentieth centuries, a number of factors contributed to the success of the manufacture of Gorgonzola cheese. Some of them belonged to the features of this product, such as its shortened ageing process; some concerned the technological changes which took place in that period. Beginning in the 1870s-1880s, new ripening methods (natural grottoes and artificial facilities), the spread of the new rail network and the irrigation systems in Northern Italian agriculture and the permanent settlement of bergamino families allowed the production levels of this cheese to increase. After the First World War, even some "scientific" attempts were considered in the search for product uniformity in order to increase production and export volumes. All these factors led to the success of Gorgonzola in international markets, especially in Great Britain, at least until the Great Depression of the 1930s. After the Second World War, things changed. Periods of slow growth (the 1950s), stagnation (the 1960s) and competition (the 1970s) were experienced. The common policy for the protection of designation of origin status and cheese denominations and the foundation of the Consortium for the protection of Gorgonzola contributed to defending this product from competitors. However, at the beginning, they were not enough to increase its exports. Two stimuli had an impact on Gorgonzola production: consumer tastes – requesting a mild product – and the birth of the Common European Market. The latter resulted in a dramatic increase in imports of European cheese, as well as legislation concerning quality control on milk. Responses from producers were: a modernization of production techniques, increases in company size, the introduction of new machinery, interaction between science and industry, and the merging of ripening and manufacturing processes within the same plant. All these elements led to industrial-scale production and product uniformity: this product was meant to feed export markets, while traditional manufacturing procedures still remained for niche markets. In addition, the Consortium's activities deeply contributed to promoting the Gorgonzola brand and financing scientific research.

Buffalo Mozzarella in Campania from Its Origin to the Twenty-First Century

Franca Pirolo

University of Catania

Introduction

The importance assumed by the dairy products market over time on both national and international level leads us to examine particularly, despite the scarcity of documentation, the importance of buffalo Mozzarella production. This takes place almost always in the same areas as buffalo farming is situated. As a matter of fact, the figure of the farmer often coincides with that of the producer and the seller. Therefore, buffalo Mozzarella production, thanks to the particular type of milk used and the close interaction with the territory, constitutes one of the most important productions of the dairy farming sector located in some areas of Campania and Southern Lazio.

This chapter aims to offer a historical and economic reconstruction of buffalo Mozzarella production and its manufacturing process. It traces its history from its origins, following its evolution from the first production processes to the manufacturing techniques improvements, then to the gradual and progressive expansion of the sales markets worldwide and then, a very important step, the establishment in 1981 of the *Consorzio Nazionale per la tutela della Mozzarella di bufala*, which in 1993, after the emanation of a protective EEC Regulation, changed its name to *Consorzio per la tutela del formaggio Mozzarella di bufala campana*. The aim of the "Consorzio" is to preserve this product from frauds and adulteration. This need induced the traditional producers to claim the creation of a trademark identified with the name "Mozzarella di bufala campana DOP", to be attributed only to the dairy products made exclusively from whole buffalo milk and manufactured by rigidly selected farms and dairies located in specifically designated areas of the Campania and Southern Lazio coastal strip.

1. From tradition to typicality: an overview

Buffalo Mozzarella is a symbol of wealthy society, and represents one of the culinary specialities that make Campania and the Mediterranean diet world-famous. However, its story originates in the strain and pain of the poor people of Southern Italy.[1]

Before talking about this fresh cheese made of buffalo milk, salt and rennet, we need to give some information about the animal from which the raw matter comes, about its origin and its first appearance in Italy. First of all, it is important to notice that both documentary and biographical sources on buffalos in Italy and the origin of Mozzarella made from their full protein milk are not sufficient and are incomplete.[2] Although we can easily quantify buffalo livestock in the Campanian provinces of Caserta and Salerno and in the Lazio provinces of Latina and Frosinone, we cannot do the same for Italian buffalo livestock and Mozzarella production as a whole.[3] since ISTAT sources, specifically the agricultural census "Statistiche dell'agricoltura, zootecnia e mezzi di produzione", and data collected by cow farming organizations such as ISMEA and ASSOLATTE and European Union figures, provide contrasting data.[4]

On the origins of the buffalo, a strong and easy-to-tame bovine that lives in boggy areas and eats low quality fodder, authors have advanced some hypotheses. It is certain that the boggy lands and the unhealthy climate

1 Scotellaro R., *L'uva puttanella. Contadini del sud*, Roma, Laterza, 1986.

2 Pirozzi N. (ed.), *Mozzarella di bufala campana. Storia, tradizioni e immagini di un formaggio nato all'ombra della Magna Grecia*, Napoli, Consorzio di tutela del formaggio Mozzarella di bufala campana, 2004; Associazione culturale Feudo Ron Alfrè, *Bellizzi, la Mozzarella, le masserie, la piana*, Battipaglia, 2011; Campanile Castaldo M., "La bufala e il suo prodotto tipico la Mozzarella", relazione presentata all'Associazione Provinciale Agricoltori di Caserta, in *Rivista di zootecnia*, XXXIII, 1960, 7–8; Faraglia N.F., *Storia dei prezzi in Napoli dal 1131 al 1860*, Napoli, G. Nobile, 1878; Guadagno G., "La Mozzarella nella storia", in *Caserta Zootecnica*, 1990, 5–6; Giardiello A. (ed.), *Prodotti agro-alimentari tipici della Campania*, Portici, Arte tipografica, 1995; Cupo C. (ed.), *Struttura e competitività dell'industria lattiero-casearia in Campania*, Roma, CNR-RAISA, 1995; Correale E., Citro A., *Allevamento del bufalo*, Bologna, Edagricole, 1995; Meloro M., "Salone della Mozzarella di bufala campana. Vetrina delle tipicità salernitane", in *Il Follaro*, I, 2006, nuova serie, 3–4; "Latte di bufala evoluzione e tendenze", in *Il mondo del latte*, 1989, 6, pp. 237–8; Correale E., "La Mozzarella di bufala campana", in *Il mondo del latte*, 1995, 6, pp. 445–7; "La Mozzarella di bufala campana: lo sapevate che …", in *Il mondo del latte*, No. 1, 2005, pp. 12–13; "La Mozzarella di bufala campana: volumi e fatturati in crescita", in *Il mondo del latte*, 2005, 1, pp. 55–6.

3 Despite the numerous contributions, the survey presents limitations. Data processing of the Agricultural census derives from Guariglia A., *Economia dell'allevamento bufalino in Italia*, Salerno, Istituto di Economia Politica, Università degli Studi, 1982 and from Cerrato (M.)'s works *La filiera della "Mozzarella di bufala campana" nell'area della Denominazione di Origine Protetta (D.O.P.)*, Salerno, Bubalus bubalis, 1999 and Cerrato M. (ed.), *Le filiere zootecniche in provincia di Salerno*, Salerno, Angri, Editricegaia, 2005.

4 There are no statistical surveys about animal products until 1949. From 1979 on, special investigations were launched to balance the data of different countries in compliance with Community directives, with two different data collections: milk industry and dairy products data. The most accurate statistic until 1933 is that contained in *L'industria del latte in Italia*, ed. Federazione Nazionale Fascista del Latte, derivati ed affini, Roma, XII, 1933.

of Salerno and of the swamps in the Pontina flatland have facilitated the settlement of these "big animals with inflamed eyes", monstrous creatures "with the appearance of a hippopotamus with wild and bloody eyes": that is how the German poet Goethe described them in his book *Viaggio in Italia* [Italian Journey].[5] Buffalo herds and the production of Mozzarella can be found in the documents of mapmakers who passed through the Paestum flatland at that time.[6] The first documents, dating to the thirteenth century, refer to manor farms of buffalos grazing on the extended fields of "Mensa Arcivescovile of Salerno", as studies by Sinno,[7] Galasso,[8] Avagliano[9] and Musi[10] reveal. Moreover, these farm animals represented the most important baronial income.[11] According to some academics, the *bubalus bubalis* or domestic Indian buffalo bred in the province of Salerno, Caserta and Roma was imported to Italy between the fourth and sixth centuries. Other academics argue that the Arabians imported it from Egypt to Campania in the eighth century,[12] while others say that it has a local origin.[13]

Since the eighteenth century, buffalo herds have represented the economic entity of the Sele flatland, even if in a rudimentary and primitive form. "Butteri" and watchmen used to work for herds owners on extended grazing lands and on wild plots along coastal areas.

Buffalo Mozzarella can be traced back to the fifteenth century when the Benedictine monks used to offer it to the pilgrims going to the abbey of San Lorenzo ad Septium in Aversa.[14] The abbey of Santa Barbara, in the province of Salerno, also had the tradition of making *mozzata* and offering

5 Goethe J.W., *Viaggio in Italia*, Milano, Arnoldo Mondadori, 1987.

6 Mottola O., "Rollin assaggiò e trovò squisita la 'provatura'", in *Agire*, 26 settembre 2004.

7 Sinno A., *Commercio e industrie nel salernitano dal 13⁰ ai primordi del 19⁰ secolo*, Salerno, Camera di Commercio, Industria e Agricoltura, 1954.

8 Galasso G., *Motivi, permanenze e sviluppi della storia regionale in Campania*, Napoli, Guida, 1973.

9 Avagliano L., "Terre e feudi della chiesa nel Mezzogiorno. I beni della Mensa Arcivescovile di Salerno nei secoli XVI e XVII", in Massafra A. (ed.), *Problemi di storia delle campagne meridionali nell'età moderna e contemporanea*, Bari, Dedalo Libri, 1981.

10 Musi A., "Il Principato Citeriore nella crisi agraria del XVII secolo", in Massafra A. (ed.), *Problemi di storia delle campagne meridionali nell'età moderna e contemporanea, op. cit.*

11 *Ibid.*, pp. 181–3.

12 Still today buffalo breeding is widespread in countries like Asia, Africa and Latin America because of their hot and humid weather. Milk, meat and leather are obtained from buffalos. In China, instead, buffalos are used as working animals.

13 Cantalupo P. (ed.), *Il bufalo nella storia e nell'economia del salernitano*, Acciaroli, Centro di Promozione Culturale per il Cilento, 1990, p. 14; also in Cantalupo P., "Il bufalo nella storia e nell'economia del salernitano", in *Annali Cilentani*, 1990, 1; Pirozzi E., "Allevamento bufalino nella Piana del Sele" (I parte), in *Il Follaro*, XXVI, 5 maggio 1971, p. 19.

14 Gambardella C., Giovannini M., Martusciello S. (eds.), *Le vie dei mercanti. Cielo dal Mediterraneo all'Oriente*, Atti del sesto forum internazionale di studi, Caserta, Capri, 5, 6, 7 June 2008, Napoli, La Scuola di Pitagora, 2009; De Franciscis G., in "Tutto sulla bufala", in *L'allevatore*, 1969, cites the *Acta Imperia Seculi XII e XIV* reporting a buffalo trading price higher than the bovines'.

it with bread to the pilgrims.[15] In the Sele flatland Ferdinando II established an agricultural colony populated by farmers, wealthy owners with rustic farmhouses, stables, stalls and *bufalare*,[16] in which was produced the *mozzata*, a buffalo Provola slightly smoked to preserve it.[17]

We have clearer knowledge about the production of Mozzarella in the eighteenth century. In the *Reggia di Carditello*, in "Terra di Lavoro", the Bourbons introduced buffalo herds, boosting the production of the refined Mozzarella of Cardito in the *Reale Industria della Pagliara delle bufale*. From "Terra di Lavoro" the production of Mozzarella expanded and reached the Sele flatland, in the Salerno area, where the Bourbons owned the *Reale Tenuta di Persano*, another swampy flatland used to breed animals. Thanks to crossbreed improvements between buffalos of Caserta and those of the Sele flatland and to more accurate techniques for milk transformation, the production of Mozzarella of the *Reale Industria della Pagliara delle Bufale* reached 30,840 kilos using 130,000 liters of buffalo milk.[18] In a short period of time the production of Mozzarella became the most important income source for the *bufalari*, *butteri* and *casari* of the Salerno and Caserta areas. Moreover, the net income resulting from the exploitation of buffalos was 20% higher than that coming from other bovine species.[19] At the end of the nineteenth century in the Citra Princedom some livestock fairs and Agrarian Exhibitions were organized to promote the business of dairy products. In 1878, livestock breeds in the province of Salerno included 368,694 animals, 5,134 buffalos, 33,110 cows, 8,016 horses, 101,970 goats and 38,878 pigs.[20] Between the seventeenth and nineteenth centuries buffalo herds were widespread in Southern Italy, even though old breeding systems, which will last at least until the following century, were still used. The profitability of the herds was due to the fact that minimal investments

15 Citro A. (ed.), *La Mozzarella di bufala campana. La corretta tecnica di filatura quale risanamento termico della Mozzarella*, Unione Regionale delle Camere di Commercio Industria Artigianato e Agricoltura della Campania, Collana di Studi e Ricerche dell'Unioncamere Campania, Salerno, 1999.

16 *Bufalare* were circular masonry constructions with a big chimney in the middle. They were the first examples of organized work because they supplied large quantities of hot water which were used to heat up the milk employed for curd and were also used for shaping cheese.

17 Cestaro A., *Il terremoto del 1857 e la fondazione della "colonia agricola" di Battipaglia*, Venosa, Osanna Venosa, 1995, p. 10.

18 From the archive registers of the *Reale Industria* kept in the Royal Palace of Caserta. Forgione A., *Made in Naples*, Milano, Magenes, 2013, p. 202; Guadagno G., *Campania antica. Storia, archeologia, vita quotidiana*, Benevento, Vereja, 2009, fig. 3, *Saggio della produzione di latte di bufala e di provole e mozzarelle in tre anni campione nel decennio 1780–1790 nella Tenuta Reale del Cardito o "Carditello"*, p. 302.

19 Pirozzi N. (ed.), *Mozzarella di bufala campana. Storia, tradizioni e immagini di un formaggio nato all'ombra della Magna Grecia*, *op. cit.*, p. 117.

20 Reale Società Economica, *Verbale della tornata ordinaria del 23 ottobre 1870*, in Il Picentino, giornale della Reale Società Economica ed organo del Comizio agrario di Salerno. 7. serie, A.13 (1870), pp. 289–94.

were required and to the presence of diffused swampy areas which could not be exploited otherwise. Thanks to the process of reclamation[21] and land reforms, these areas were gradually converted into arable fields to the detriment of the swamps.[22] With the new land arrangement, the traditional breeding system was modernized.[23]

In the twentieth century, in most of the Sele flatland areas, buffalo breeding, combined with tobacco and tomato plantations and Mozzarella production, which keeps its handcrafted connotation even today, coexisted. In fact, the technique used for the transformation of buffalo milk has not been modified during the past years. In this industry, cheese-makers' experience and operators' manual labor cannot be replaced by machines. This is why such craftsmanship, which is carried out in agricultural farms and which guarantees high quality products, is highly valued by the national and international market and does not produce leftovers.

Therefore, buffalo milk has a specific use.[24] It is processed and transformed into Mozzarella, Provature, Caciottine, Caciocavalli and Mascarpone cheese. To preserve its taste and its organoleptic characteristics it is almost never processed with milk coming from other bovine species.

Empirical and original production of milk products used to take place in manor farms. In big masonry houses, improvised breeders and cheese-makers collected fresh milk, heated it up at the right temperature in a copper *caccavo* (a receptacle), blended rennet and processed the dough on wooden rudimentary tables without respecting any hygiene rule. The cheeses were then sold in weekly local markets.[25] In the first decades of the twentieth century, thanks to the high quality of the fresh buffalo cheese production that took place in Lazio, Puglia and mostly in Campania (from Capua to the Agro Aversano, and from the Salerno flatland to Battipaglia), and thanks also to local breeders, some little cheese factories with modern equipment were established. They were used as points of sale as well.

21 Developments in agriculture and changes in farming systems in the Sele flatland were possible thanks to the reclamation started by the Bourbon Government and completed by 1930s by the Società Farina-Valsecchi, later changed to the Società Anonima Bonifica.

22 Ciampi P., Paiani M., "Il bufalo italiano la sua storia", in *Ingegneria alimentare*, 1993.

23 As regards the transformations of the Sele flatland after the reclamation see Pirozzi E., "Allevamento bufalino nella Piana del Sele" (II parte) in *Il Follaro*, Anno XXVI, 6, giugno 1971, pp. 59–60.

24 Sordi C., "L'approvvigionamento del latte nei centri urbani"; Fascetti G., "Conservazione industriale del latte (Sterilizzazione-Condensazione-Polverizzazione)", in *L'Italia Agricola*, special issue dedicated to *Il Latte e i suoi derivati*, anno 63, 12, dicembre 1926.

25 D'Alfonso C., "Il caseificio meridionale", in *L'Italia Agricola*, special issue dedicated to *Il Latte e i suoi derivati*, anno 63, 12, dicembre 1926; Cornalba C., "Sguardo alla produzione del latte in Italia", in *L'Italia Agricola*, special issue dedicated to *Il Latte e i suoi derivati*, anno 63, 12, dicembre 1926. In a document of the Federazione Nazionale Allevatori Bufali, Sottosezione di Salerno, Battipaglia, 26 settembre 1910, we can read about a merchant of Emilia affirming that local producers were "great but naive workers", lacking in commercial attitude.

If we can easily confirm the precise localization of the buffalo, we cannot do the same when evaluating the amount of the assets, the number of the companies involved and the production of milk and Mozzarella, due to the non-homogeneity of the sources and fluctuations in the data. In 1926 the total amount of animals in the South and in the Islands covered the 13.7% of the national data and buffalo herds reached 0.04%. Approximately speaking, the milk used in the dairy industry represented 1/3 of the production.[26] In the first years of the 1900s in the area of Salerno there were 20,000 buffalos, reduced to 19,362 in 1908 and to 16,000 in 1931.

In 1947, after the war and due to reclamation works, the number of buffalos was 12,000. However, these estimates are neither official nor reliable, but they effectively highlight the importance attributed to the buffalo, which was considered a synonym of underdeveloped zootechnics and unhealthy lands. For all these reasons, in the following official statistics the buffalo data were not separated from bovine animals data.[27]

It was in the 1930s that a sudden growth of the production was registered in the Sele flatland: in each local cheese factory production reached 40 quintals of fresh dairy products per day. Later, in the 1950s, automatic buffalo milking machines were introduced for the first time.[28] The modernization of breeding techniques caused the increase in the average number of items per company on both a national and a global level,[29] which meant that there was a greater quantity of milk available and a greater opportunity to meet the buffalo Mozzarella market demand. In 1961 the number of buffalos was 22,747, and it more than doubled in 1970 with 50,853 buffalos. In the following years there was a gradual increase: in 1980 87,000 buffalos, 137,900 in 1988, 158,336 in 1992 and 220,000 in 2002.[30]

The increase in the demand for Mozzarella and for the milk to produce it caused the increase in the number of buffalos in the breeding farms. After all, the seasonality of milk production and the inability to satisfy the needs of the market during the period of great demand has to be taken

---

26 Cornalba C., "Sguardo alla produzione del latte in Italia", *op. cit.*

27 Pirozzi N. (ed.), *Mozzarella di bufala campana. Storia, tradizioni e immagini di un formaggio nato all'ombra della Magna Grecia*, *op. cit.*, p. 117.

28 Thanks to technical developments, sophisticated new equipment was introduced into breeding farms along with *paddocks*, open stables covered by canopies that had artificial small lakes and troughs. Mondini E., "Tradizione e novità nei nomi delle bufale", in *Zootecnica*, Vols. V, VI, 1987; Bianchi A., Dal Sasso D., Manera C., "Piccoli caseifici per la produzione di formaggi a pasta filata", in *Genio rurale*, 1985, 7–8, pp. 35–46; Zicarelli L., "La bufala mediterranea italiana: esempio di una razza autoctona in espansione", in *Scienza e tecnica lattiero casearia*, 2001, 52(4).

29 There was an increase in animals of 12.7% per company in 1961, 26.3% in 1970 and 40.1% in 1990. At the same time there was an 178% increase in milk production and in buffalo mozzarella production, which reached 238.500 tons per annum. Cerrato M., *La filiera della "Mozzarella di bufala campana" nell'area della Denominazione di Origine Protetta (D.O.P.)*, *op. cit.*; De Franciscis G., "Tutto sulla bufala", *L'allevatore*, 1969.

30 *Ibid.*, p. 122.

into account for it forces the producers to use an inferior percentage of buffalo milk and to produce Mozzarella with a mixture of cow and buffalo milk, milk powder and casein, and not just with whole buffalo milk, which has a higher cost.[31] Hence the need to protect the product with rules that control the production of Mozzarella "Campana" with the PDO logo, by acknowledging the Community laws related to the branch, and the foundation of the "Consorzio per la tutela del formaggio Mozzarella di bufala campana".

In the culinary tradition of Campania, as we can read in the *Relazione* of 1928: "there are traditional flavors and aromas of typical products [...] and the production of fresh and dry cheeses constitute the elements forming the basis of the Mediterranean diet that represents a complete and balanced nutrition plan based on the consumption of the local food produced in the Southern Mediterranean area".[32] Buffalo breeding represented a unique feature of Salerno flatland and it was "kept in the open fields even though the introduction of breeding in stables led to great results for the production of the milk used for cheeses and milk production".[33] In the final part of the *Relazione* a comparison of the data collected by the general census of livestock in 1918 and in 1929 shows an increase of almost 2,000 buffalos.

Tab. 1. Livestock present in the province of Salerno according to species for the years 1918 and 1929

	1918	1929	
Horses	10,209	10,000	-209
Donkeys	31,897	40,000	+8.103
Cattles	49,055	52,000	+2.945
Pigs	55,119	58,000	+2.881
Sheep	197,585	250,000	+52.415
Goats	116,414	100,000	-16.414
Buffalos	6,018	8,000	+1982

Source: Consiglio Provinciale dell'Economia di Salerno, *Relazione sull'andamento economico della Provincia di Salerno nell'anno 1928*. Also in Santoro G. (ed.), *L'economia della provincia di Salerno nell'opera della Camera di Commercio 1862–1962*, Salerno, F.lli Di Giacomo, 1966, pp. 145–8.

31 As regards the problem of production seasonality and of deseasonalization techniques of buffalos' births, see Addeo F., Chianese L., Anelli G., "Riconoscimento e dosaggio del latte vaccino nel formaggio Mozzarella di Bufala", in *Rivista italiana di Scienze Alimentari*, 1982, pp. 10–38; Valletrisco M., "La produzione lattiero-casearia e la struttura delle aziende nella penisola sorrentina", in *Industrie Alimentari*, 1984, febbraio, pp. 92–8.

32 Consiglio Provinciale dell'Economia di Salerno, *Relazione sull'andamento economico della Provincia di Salerno nell'anno 1928*.

33 Consiglio Provinciale dell'Economia di Salerno, *Relazione sull'andamento economico della Provincia di Salerno nell'anno 1928*, p. 96. The main breeding farms were: Governativo di Persano, Comm. Alfani, Fratelli Farina, Fratelli Moscati, Fratelli Morese, Fratelli Iemma, Fratelli Pastore, M. Pinto.

But from the following year to 1936, animal population consistency began to decrease. In particular, buffalos decreased from 6,290 to 5,198 in 1936. The Consiglio Provinciale dell'Economia (Provincial Council of the Economy) decided to intervene by launching a support policy for animal husbandry with subsidies for breeders and initiatives aimed at promoting improvement of the breeds. In particular, the improvement of fodders and the introduction of innovative machines and equipment into stables were very important.[34] In the *Relazione*, pig and bovine herds are said to "have a certain importance in the province of Salerno [...] The aim of these companies is to produce milk for cheese and dairy products".[35]

Most of the food industries in the Battipaglia flatland, Eboli and Capaccio were little farms specializing in processing buffalo milk into dairy products, such as Provolas and Mozzarellas, which were sold in the Salerno area and also in Naples, Rome, Calabria and Sicilia. In the province of Salerno, the processing of milk into cheese and dairy products was "one of the most important branches of big farms' production"[36] as well as the production of selected Caciocavalli and Provoloni made with cow's milk, along the Amalfi coast, in the Agro Nocerino and in the valley of Diano. But these last products had to compete with packed cheese coming from Northern Italy. Actually, this was a problem that fresh products made with buffalo milk did not have to face thanks to the milk produced by local buffalos which were a local speciality.

A significant boost to the dairy industry came from the Società Agricola Industriale Salernitana (SAIS), which established a cheese factory in Battipaglia with the purpose of collecting and processing the milk produced by other companies, and by the Società Industria e Commercio Latticini e Affini that produced 12 quintals of Mozzarella per day between 1925 and 1926. This result would not be achieved again in the following years.[37]

In 1930, the Consiglio Provinciale dell'Economia di Salerno wrote:

> The dairy industry evolved in normal conditions on the basis of the cost of raw materials and the price of the finished product. This strategy allowed keeping a more constant proportion than the one of the former years. The price of milk and dairy products seems to be regular; in particular the average price of buffalo dairy products was 175 lire for each quintal of processed milk, which corresponds to an average of 9 lire per kilo of Mozzarella, which represents the most important product made from milk.[38]

---

34 Santoro G. (ed.), *L'economia della provincia di Salerno nell'opera della Camera di Commercio 1862–1962*, Salerno, F.lli Di Giacomo, 1966, p. 205.

35 Consiglio Provinciale dell'Economia di Salerno, *Relazione sull'andamento economico della Provincia di Salerno nell'anno 1929*, p. 123.

36 *Ibid.*, p. 156.

37 Consiglio Provinciale dell'Economia di Salerno, *Relazione sull'andamento economico della Provincia di Salerno nell'anno 1928*.

38 Consiglio Provinciale dell'Economia di Salerno, *Relazione sull'andamento economico della Provincia di Salerno nell'anno 1930*, p. 113.

The dairy industry grew increasingly important thanks to its special buffalo milk products called "Provolas and Mozzarella", whose good taste and fine quality were strongly demanded by the market.

In 1931 in the Province of Salerno alone, there were 50 dairy establishments, mainly concentrated in the areas of the Amalfi Coast, Valle di Teggiano and Bassa Valle del Sele. In the first two areas Provoloni, Caciocavalli, butter and Provolas made mostly of cow's milk were produced. In Bassa Valle del Sele instead, the well-known and very tasty Provolas and Mozzarellas were made almost exclusively with buffalo's milk. Although the dairy industry in Salerno had considerably progressed in the last years, it still had some deficiencies from both a technical and an organizational point of view.[39]

In those years, the production of buffalo Mozzarella amounted to 27,000 quintals, and it "can be found in the regions where buffalo breeding still survives although cow's milk is largely used for the production of cheese [...] The Province of Salerno is the area with the greatest production: more than half of the total production. Most of Mozzarella is produced in industrial companies that had to specify which kind of animal the processed milk came from in their product quality statement".[40] In Southern Italy, Provola was produced everywhere, especially in the Province of Salerno where 75,000 hectolitres of buffalo milk were used to make Mozzarella. According to the 1937 census, 314,558 hectolitres of milk were processed into 40,880 quintals of Mozzarella, which corresponded to 1.3% of the total production.[41]

In the 1930s a consortium of milk producers, whose aim was to organize production, was founded in Battipaglia. It also managed an important cheese factory in the city and placed the products on the market. Although the Consortium was very important, its actions were limited to specific areas. Most of the production was managed by directors who worked for more than one cheese factory, and the dairy industry was run using not very rational methods. A survey conducted in those years reveals that the total production of the dairy products in the Province of Salerno was around 100,000 quintals per annum and that, although prices were decreasing, this amount tended to increase them, "because compared to other industries, the dairy industry used to make good profits, and also because in 1931 the number of buffalo livestock was higher than in 1930 and its breeding was mostly aimed at milk production".[42]

---

39 Consiglio Provinciale dell'Economia di Salerno, *Relazione sull'andamento economico della Provincia di Salerno nell'anno 1931*.

40 Federazione Nazionale Fascista del Latte, derivati ed affini, *L'industria del Latte in Italia*, *op. cit.*, p. 101.

41 *L'industria della lavorazione del latte e dei prodotti derivati* al 25 maggio 1937, XV.

42 Consiglio Provinciale dell'Economia di Salerno, *Relazione sull'andamento economico della Provincia di Salerno nell'anno 1931*, pp. 246–8.

The increase in the cost of living and the need to stabilize prices and salaries in 1935 and in 1936 forced the Provincial Office of the Corporations to control the maximum prices to be applied in the province. Four maximum wholesale and retail sale price lists for essential goods like flour, pasta and bread, and for animal fats like butter, lard, cheese and meat, were established in Salerno.

For buffalo Mozzarella, cow Mozzarella and Provolone the following maximum wholesale and retail prices were applied respectively: 650 lire (per q.le) and 7.5 lire (per kg); 450 lire (per q.le) and 5.5 lire (per kg); 850 lire (per q.le) and 9.5 lire (per kg).[43]

2. The PDO market of Mozzarella di Bufala Campana

The production of Mozzarella takes place in the same areas as milk is produced and often the breeder is also the producer and trader of these products. Even today, cheese factories are not completely mechanized because, unlike for processing cow's milk, it is not possible to employ machines for processing buffalo's milk. Most of the production is consumed in the areas where it is made, and is sold directly by the producer, while only 20% is sold on domestic and foreign markets, in particular in American and German markets,[44] through sellers and sale representatives. The shortage of buffalo milk and the inequality between demand and supply induces the producers to create fakes and defraud buyers with an inauthentic product. The production chain for buffalo Mozzarella is short and involves breeders, cheese factories and sales channels due to the perishable nature of the product and its need to be consumed fresh. Salerno and Caserta are the areas where most of buffalo Mozzarella is produced and, in 1981, thanks to three companies in Caserta and one in Salerno, the Consorzio Nazionale per la tutela della Mozzarella di bufala was founded, which was recognized by the Ministry of Agriculture and Forests two years later. During the following ten years the law changed, in particular with regard to the territoriality of the designation of origin. Therefore, the typical product made exclusively with the buffalo milk produced in Caserta, Salerno, Napoli, Benevento, Latina, Frosinone and Roma was certified as PDO (DOP) buffalo Mozzarella (Fig. 1). Later in 1993, Regulation EEC number 1107 granted the buffalo Mozzarella the PDO logo, a guarantee of quality. The production process, the use of the logo and the product labeling and packaging protect both the producer and the consumer. In particular, the consumer is protected from any possible *free riding* behavior such as product imitation or fraud, but also from the producers' *moral hazard* behavior. In the same year, the

43 Santoro G. (ed.), *L'economia della provincia di Salerno nell'opera della Camera di Commercio 1862–1962*, Salerno, F.lli Di Giacomo, 1966, pp. 227–8.

44 Giardiello A. (ed.), *Prodotti agro-alimentari tipici della Campania*, *op. cit.*, pp. 331–3.

Consortium[45] (which will become a guarantee and supervisory body) will change its own name to Consorzio per la tutela del formaggio Mozzarella di bufala campana. In 2004, on the association's initiative, a book edited by Pirozzi, whose aim was to value this typical cheese,[46] was published.

Figure 1. The zone of production of PDO Buffalo Mozzarella

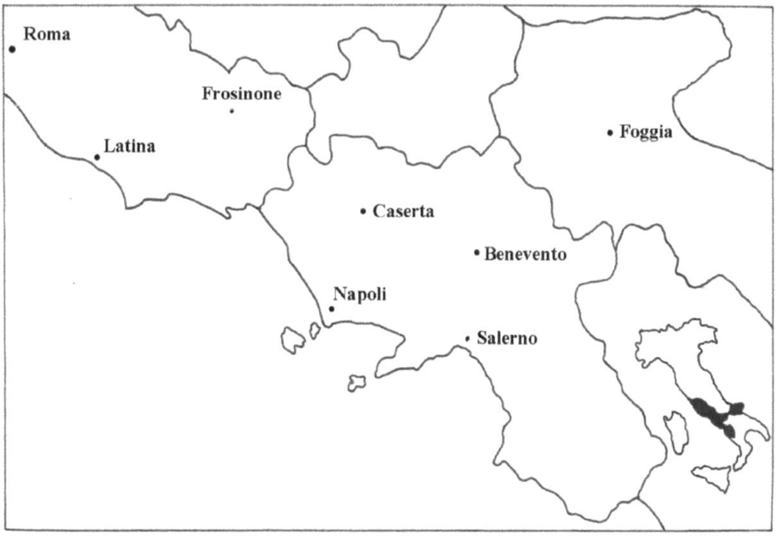

Elaborated by author.

[45] For the importance of the relation between associationism and land development see Flora A., "La filiera della Mozzarella di bufala: un laboratorio dello sviluppo locale", in *QA, La Questione agraria*, 2002, 1.

[46] Pirozzi N. (ed.), *Mozzarella di bufala campana. Storia, tradizioni e immagini di un formaggio nato all'ombra della Magna Grecia, op. cit.*

Figure 2. Water buffalos wallowing in a pool

Source: Archivio digitale del comune di Eboli (Sa), Fondo Gallotta.

At the beginning of the 1990s this industry started to develop in particular in Southern Italy. Between 1961 and 1990, there was a record increase in the numbers of zootechnical farms, especially in the province of Caserta.[47] The increase in demand and the impossibility of meeting it led to product adulteration that forced local producers to claim a branding policy in order to identify, by means of a DOC[48] logo, the fresh cheese made only with buffalo milk, and produced and processed in breeding farms and cheese factories mainly located on the coastline of Campania (Napoli, Caserta and Salerno) and in lower Lazio (Frosinone, Latina and Roma) as the original *mozzarella di bufala campana*. Moreover, strict controls on the cheese's genuineness are important for the consumer as well. Nowadays, the buffalo Mozzarella market represents a unique economic entity due to strong ties to the territory. This may open up new possibilities of inserting the product in

47 Flora A., *La filiera della Mozzarella di bufala: un laboratorio dello sviluppo locale, op. cit.*, p. 9.

48 Citro A., Correale E., "La Mozzarella di bufala campana. Nubi ed ombre sulla D.O.C.", in *Il Follaro*, XXXVII, 2, nuova serie, 1993.

a regulated telematics market, a Telematics Commodity Exchange Market whose role could support business operators in the identification of new national and foreign markets.[49]

According to the surveys conducted by the Consortium in PDO areas, the cheese factories that apply the symbol to commercialize the Mozzarella produced with buffalo milk as required by product specifications are 107 in 2014, with an almost unchanged allocation of the production in the interested provinces throughout the last five years, 2009–14. These data indicate a decrease in production in lower Lazio and in the Napoli and Foggia areas, and an increase in the provinces of Caserta and Salerno which register 58% and 33% respectively.

Graf. 1. Production of PDO Buffalo Mozzarella 1993–2009 (Tons)

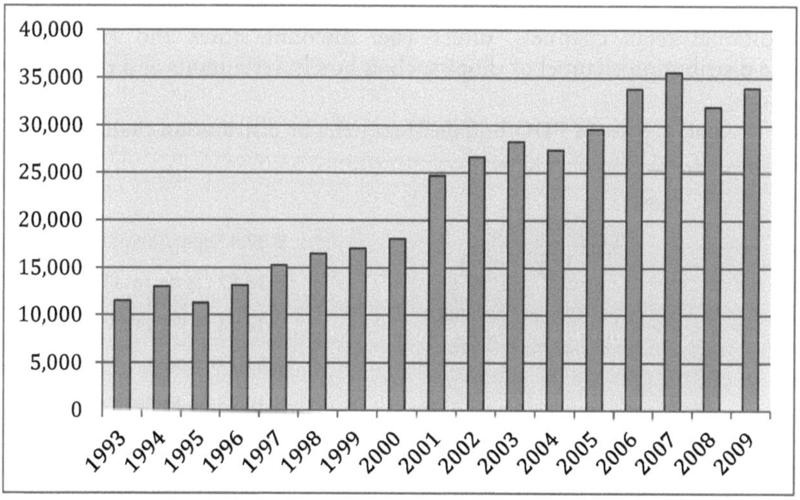

Source: Consorzio per la Tutela del Formaggio Mozzarella di Bufala campana di Caserta.

In 2014, data were collected and processed by the Consortium by means of 63 questionnaires sent to associated cheese factories, which represent the 77% of the total production. There were 1,500 PDO buffalo farms and 350,000 certified buffalos in 2014; cheese factories with the PDO logo have decreased from 111 in 2011 to 107 in 2014, with a production of 38,068 tons of Mozzarella. The turnover of the production is 300 million euros, while consumption is 500 million euros. This is an important economic entity that employs 15,000 people, satellite activities included.

49 For more recent and up-to-date studies of the production, consumption, turnover and sales of *Mozzarella di bufala campana see* the following publications by Camera di Commercio di Caserta: *Mozzarella di Bufala Campana DOP. Studio di Mercato*, Caserta, 2008 e della Camera di Commercio di Salerno, *Mozzarella di Bufala Campana DOP. Studio di Mercato*, Salerno, 2010.

In total, 75.1% of PDO buffalo Mozzarella is sold in Italy while 24.9% reaches foreign markets. In Italy, the biggest consumers are in the PDO area, the Southern regions and Islands with 30%, and 28% in Nielsen 1 area (Piemonte, Liguria, Valle d'Aosta and Lombardia), with an upward consumption trend in the South and a decreasing trend in the other areas. Despite this, the 2% increase of buffalo Mozzarella in 2014 confirms that the economic crisis did not affect this niche market. A great deal of the credit goes to the PDO logo, which continues to promote product typicality at both a national and a foreign level.

Sales channels follow modern distribution systems: big distribution chains that allow distribution costs to be decreased while guaranteeing that the market will be reached are preferred. In fact, the biggest percentage of the production, slightly above 37%, reaches consumers' table through supermarkets and hypermarkets, while low percentage shares go through traditional retail channels, direct sale, discount stores and Ho.Re.Ca. (the distribution channel of shops such as hotels, restaurants and cafes).

Graf. 2. Sales of PDO Buffalo Mozzarella by distribution channel

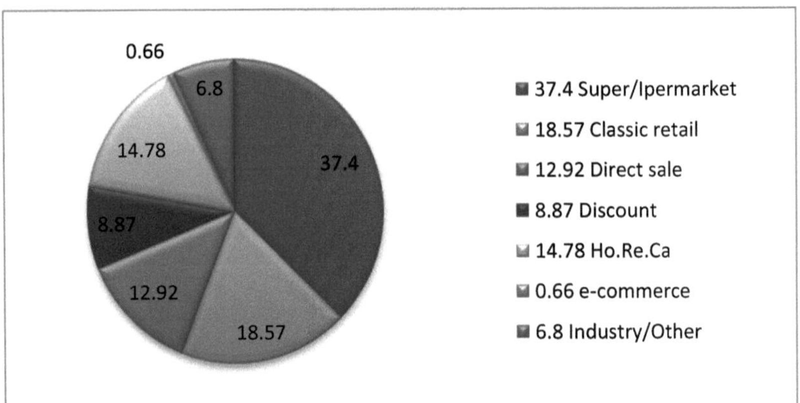

Source: Consorzio per la Tutela del Formaggio Mozzarella di bufala campana di Caserta.

As regards the external trade, France is the prime European importing country for PDO buffalo Mozzarella. It is followed by Spain and the UK, while Germany registers a decrease, even though it is still the second largest importing country after France. Decreases are also registered in the USA, in Japan and in Canada, while there is an increase in Eastern Europe and in the Arab Emirates.

According to Chamber of Commerce surveys updated to 2013, in the province of Salerno there are 1,636 buffalo farms and 226 dairy industries, with 1,626 workers. In the province of Caserta there are 2,548 breeding farms and 271 dairy industries with 1,653 workers. Only 1/5 of 500 cheese factories have subscribed to the Consortium. This means that, in order to ensure the future success of buffalo Mozzarella, it is important to act against cow's milk and mixed products competition or else buffalo Mozzarella will remain confined to a narrow segment of excellence-food products.

Tab. 2. Sector data of PDO Buffalo Mozzarella (2011–2014)

	2011	2012	2013	2014
Certified cheese factories	111	110	105	**107**
DOP production (in tons)	37,472	37,122	37,308	**38,068**
Production sector gross income (in millions of euro)	320	300	300	**300**
Consumer sector gross income (in millions of euro)	500	500	500	**500**
Number of DOP buffalo breeding farms		1,500	1,400	**1,500**
Total number of DOP certified farmed buffalo heads		340,000	310,000	**350,000**
Total employed labor force including the spin-off sector		15,000	15,000	**15,000**

Source: Consorzio per la Tutela del Formaggio Mozzarella di bufala campana di Caserta.

Conclusions

During the 1950s, because of land reclamation it seemed that the buffalo were destined to disappear. However, since the war both the number of buffalo and production of buffalo Mozzarella have increased significantly. Today, buffalo dairy companies play an important role in the economy of Southern Italy, thanks to the taste and quality of buffalo Mozzarella, which have made it a winning product that has won a growing niche of consumers.

The economic importance of this product needs protection and requires professionalism and sensitivity from producers, since many of them have not joined the Consortium. The market, however, is also affected by the problem that many producers do not use the protected name and do not follow the specifications. There is no corporate brand, and it is the dominant consortium brand that makes it difficult to build customer loyalty.

There are no reliable data for domestic consumption, but it is certain that the consumption of buffalo Mozzarella is increasing at the expense of cow's milk Mozzarella, and is mainly located in the South although there was an increase in Central Italy among middle-income families; the low consumption in the North is probably due to insufficient knowledge about Mozzarella. The competition between buffalo Mozzarella and cow's milk Mozzarella is mainly due to the lower price of the latter. It also has the advantage of being more adaptable to modern sales techniques.

The small-scale industry structure sets limits to the expansion of the market. The future of buffalo Mozzarella could depend on the distribution channel (for instance, Ho.Re.Ca. and food tourism), despite the problems surrounding the transportation of the product. After analyzing potential production, in order to expand the market for buffalo Mozzarella it will be necessary to develop modern sales strategies and encourage producers to protect the uniqueness of this typical product.

Part III

Forms of Enterprise in Italy in the Nineteenth and Twentieth Centuries

The Modernization of the Dairy Industry in Italy in the Late Nineteenth Century

Social Dairies

Manuel Vaquero Piñeiro
University of Perugia

1. Introduction

In the context of the formation of the industrial structure of post-unification Italy, production cooperatives played a key role in the birth of the agri-food sector. For nearly half a century the specialized literature and government policy were dedicated to making it possible to increase production, improve product quality, and expand national and international marketing channels. While historiographical interest has largely centered on the consolidation of the entrepreneurial state, and private enterprise in its diverse forms,[1] the late nineteenth century saw the first consolidation of industries derived from agriculture and animal husbandry. The cooperative movement was credited with an indisputable role in this development.[2] It was no coincidence that, after 1861, this was the path chosen above all by management teams of a Catholic conservative background, who viewed solidarity based solutions as the most appropriate strategy to improve the living conditions of peasant farmers, while at the same time avoiding the dangers associated with the abandonment of the countryside and the rise of the working class population in the cities.[3]

In this way, in the final decades of the nineteenth century, and in the wake of an intense propaganda campaign, the development of the Italian

---

1 Amatori F., Colli A., *Impresa e industria in Italia. Dall'Unità a oggi*, Venezia, Marsilio, 1999.

2 On the history of the cooperative movement in Italy see Fornasari M., Zamagni V., *Il movimento cooperativo in Italia. Un profilo storico-economico*, Firenze, Vallecchi, 1997; Zangheri R., Galasso G., Castronovo V., *Storia del movimento cooperativo in Italia 1886–1986*, Torino, Einaudi 1987; Degl'Innocenti M., *Storia della cooperazione in Italia. La lega nazionale delle cooperative, 1886–1925*, Roma, Editori Riuniti, 1977.

3 Battilani P., "I mille volti della cooperazione italiana. Obiettivi e risultati di una nuova forma di impresa dalle origini alla seconda guerra mondiale", in Mazzoli E., Zamagni S. (eds.), *Verso una nuova teoria economica della cooperazione*, Bologna, il Mulino, 2005, p. 105.

agricultural industry was entrusted to the formation of a dense network of cooperatives and social production organizations, which were entrusted with much of the task of modernizing a number of industrial sectors of strategic importance. This was the case with cocoon dryers,[4] wine cooperatives,[5] and, as will be analyzed in this chapter, cooperative dairies. It is also true, however, that the case of cooperative dairies,[6] and, more generally, the formation of cooperative structures within the dairy products sector, features specific characteristics.

The measures taken to encourage the creation of cooperative dairies contributed significantly to the formation of a geographically divided sector. In order to strengthen the market for agricultural products, in the last decades of the nineteenth century, cooperative dairies constituted an effective response to the problems arising from an economic structure characterized by small landholders who could, however, to their credit, count on a consolidated experience of the joint management of resources.[7] Therefore, when it came to adapting to the needs of national and European markets, the most coherent response to the context was to adapt existing structures to the industrialization that was in full swing. This was a true case of a natural industry, typical of Italy, which involved lengthy processes related to the social capacity to construct local systems of production.[8] The cooperative dairies also undoubtedly had many points of contact with the rural cooperative banks and their Catholic ethos.[9] The paths taken by both these institutions, the dairies and the banks, can above all be examined with regard to the identification of the channels of financing that allowed the alpine cooperative dairies to embrace industrialization.

However, while the Alpine region featured a dense network of small social dairies with local roots, the irrigated plains of Lombardy and Emilia witnessed the birth of the first large industrial dairies which, while they adopted the cooperative business model, in reality responded to the objectives of the new class of private entrepreneurs. This defined an industrial dualism in the dairy sector responding to two very different social contexts, which was capable of stimulating, over the last three decades of the nineteenth century, a sharp contrast of ideas, as demonstrated by a large volume of printed texts.

---

4 Bof F., "Antefatti, origine e sviluppo iniziale degli essiccatoi cooperativi bozzoli in Friuli", in *Storia economica*, X, 1–2 (2007), pp. 5–51.

5 Da Passano M., *La vite e il vino: storia e diritto (secoli XI-XIX)*, Roma, Carocci, 2000.

6 On the some what diverse development of the chocolate industry see Chiapparino F., *L'industria del cioccolato in Italia, Germania e Svizzera: consumi, mercati e imprese tra '800 e prima guerra mondiale*, Bologna, il Mulino, 1997.

7 Bozza A., "Alle origini della cooperazione trentina", in Fabbri F. (ed.), *Il movimento cooperativo nella storia d'Italia 1854–1975*, Milano, Feltrinelli, 1979, pp. 347–58.

8 Giannetti R., Vasta M., *Storia dell'impresa industriale italiana*, Bologna, il Mulino, 2005, p. 110.

9 Pecorari P., *Le banche popolari nella storia d'Italia*, Venezia, Istituto veneto di scienze, lettere ed arti, 1999.

2. The social dairies in other countries

The importance of the modernization of the dairy sector, through the involvement of farmers in the creation of cooperatives, was echoed in the pages of the agricultural almanacs, popular publications which enjoyed great success in post-unification Italy.[10] In the context of the growing expansion of theoretical and practical knowledge in agriculture, the agrarian almanacs established themselves as publication with specific characteristics, intended to meet the needs of a wide range of potential readers. While books and newspapers constituted a cultural barrier that it was impossible for the less educated sectors of the peasant population to overcome, the more easily understood almanacs proved an appropriate vehicle to promote the idea of a form of agriculture that would follow in the footsteps of more advanced countries. If the scientific-academic debate remained the preserve of a small elite for obvious reasons, at a lower level a wide range of texts and publications of various kinds[11] contributed to the spread of innovations in Italy that were already allowing other countries to achieve important goals in their economic development.

Thc 1871 Agrarian Almanac (*Almanacco agrario*) devoted a long article to the progress made in England, Switzerland and other European countries in the livestock sector, and the intensive production of butter, cheese, fresh meat, leather, etc. This growth guaranteed significant commercial trade.[12] In 1868, Switzerland alone had exported more than 28 million kg of cheese, and 26 million kg of butter, for a total of one hundred million lire. In comparison with this flourishing example, the Italian situation appeared decidedly less encouraging. The quantity of cheese sold abroad barely reached 2 million kg, a figure that came nowhere near imports of six million kg in 1868. As for butter, the ratio between imports and exports favored the latter: 452,834 kg and 842,638 kg respectively. However, better sales of Italian butter abroad did not serve to offset the poor performance of cheese exports, which was considered a shameful trend in a country that has a second Switzerland in

10 Vaquero Piñeiro M., "Gli almanacchi agrari in Italia (XVIII-XX secolo)", in *Storia economica*, XVIII, 1 (2015), pp. 59–97.

11 Wilhelm G., *Che cosa sono le latterie sociali?: una parola agli economi rurali dei paesi alpini austriaci*, Rovereto (TN), tip. V. Sottochiesa, 1874; Masino A., *Latterie sociali*, Torino, tip. Camilla e Bertolero [1875]; Bassi E., *Le latterie sociali in Italia. Loro utilità, origine, diffusione. Con progetto di statuto e prospetto di tutte le latterie Italiane. Studio*, Udine, tip. Cooperativa, 1891; De Gobbis F., *Le latterie sociali: studio amministrativo*, Modena, tip. Bassi & Debri, 1898; Bassi E., *Le latterie sociali in Italia*, manuale pratico, Milano, Tip., G. Agnelli, 1900; Bonsignori G., *Utilità e facilità d'impianto delle latterie sociali*, Brescia, tip. Queriniana, 1900.

12 *Almanacco agrario del prof. Gaetano Cantoni*, Milano, presso Gaetano Brigola editore, 1871, pp. 187–203. See Cantoni G., *L'industria del latte. Conferenze tenute a Portici in occasione della esposizione nazionale di caseificio*, Milano, Dumolard, 1878. The theme of the social dairies is covered with a more moral tone in the almanac *Il buon contadino. Almanacco del periodico l'agricoltore*, Lucca, tipografia di B. Canovetti [1872].

the Alpine region, a second Belgium in the plains of the Po, and which has, along the central and southern Apennines, in terms of climate, terrain and waterways, perennial good natural pasture. Such a disappointing situation was mainly due to the scarcity of fodder and dairy animals.

After setting out the reasons that were preventing the Italian dairy sector from making progress, by learning how to make best use of the abundant resources available, above all in Alpine areas, the *Agrarian Almanac* stated that the only solution was to pool the dairy livestock which was divided between many farmers, with the goal of making a third, or even less than half, of that which they would earn in the hands of a single operator. This was the case in Switzerland, which could boast more than 2,075 dairies,[13] or the USA, where, between 1851 and 1867, 803 associated dairies and creameries had been established,[14] for a total of 199,519 head of cattle, including 169,812 in the state of New York alone.[15] Having overcome initial criticism, within a few years the social dairies founded in the USA became an essential institution in the reduction of the risk of milk counterfeiting, and therefore ended up encouraging an increase in the consumption of a food product that the population of the city began to consider safe and of superior quality.[16] In this way, through cooperation and an increase in demand, livestock owners saw a steady growth in their incomes.

3. Public actions and private initiatives

The Italy of the 1870s saw the appearance of a number of welcome initiatives, capable of countering the atavistic individualism of the peasant farmers, who were accustomed to consuming what little milk they had at their disposal in the family. In reality there were a number of different problems to be faced. Apart from encouraging associative forms of production, it appeared essential, in order to increase the volume of exports, to learn to make the best use of milk, in order to produce cheese of good enough quality to win over the more selective foreign markets. To this end, and thanks to scientific and

---

13 Reggiani E., *La produzione del latte e le latterie sociali cooperative*, Milano, U. Hoepli, 1908, pp. 288–9. On the role of the Hoepli manuals in the spread of technical and scientific knowledge see De Mauro T., "Il caso Hoepli", in A. Assirelli (ed.), *Un secolo di manuali Hoepli 1875–1971*, Milano, Hoepli 1971, pp. 11–30.

14 On the 19th-century cooperative movement in the USA in the citrus fruits sector see Laszlo G., *Storia degli agrumi. Usi, culture e valori dei frutti più amati del mondo*, Roma, Donzelli, 2006, p. 88 [English edn, *Citrus. The Golden Apple*, Chicago, Chicago University Press, 2006].

15 Reggiani E., *La produzione del latte, op. cit.*, pp. 291–2.

16 Keillor S.J., *Cooperative Commonwealth: Co-ops in Rural Minnesota, 1859–1939*, St. Paul, State Minnesota Historical Society Press, 2000. On the role played by Scandinavian immigrants in the rapid spread of social dairies in the USA see Toninelli P.A., *Storia d'impresa*, Bologna, il Mulino, 2006, pp. 131–4.

mechanical progress, which definitively put an end to widespread ignorance about the physical and chemical properties of milk,[17] it became necessary to establish theoretical and practical dairy schools and test stations.[18] In any case, the real underlying problem was not so much the training of Italian cheese-makers, who were skilled and competent, as educating those called upon to undertake a management role in the dairies.

In 1872, government policy led to a decisive turning point in the history of the Italian dairy industry. On 25 April, the minister for agriculture Stefano Castagnola published a circular intended not only to describe the robust situation of cooperative dairies in other countries, such as Switzerland, Britain, the USA and Austria, but also to lead the dairies active in Italy towards definitive technical and commercial modernization.[19] The instrument chosen to achieve this aim was a systematic propaganda campaign involving awards and competitions. At the same time congresses were held in Milan (1874 and 1878), along with the national dairy exhibition in Portici (1877).[20] These were followed by further congresses in Udine (1886) and Treviso (1888).

In the light of this sequence of initiatives, it can be observed that the birth of the cooperative dairies occurred at the stage when Italian agriculture was suffering the effects of the much-discussed agrarian depression at the end of the nineteenth century.[21] It is, however, true that a variety of opinions exist regarding the true nature of the agrarian crisis, from the long-established view that recalls the negative consequences of the collapse of prices and profits,[22] to the more recent theories that interpret the situation as a crisis not of poverty but of renewal.[23] Leaving aside the dichotomous terms of the debate, what is beyond discussion is that the entire national agricultural sector had to adapt to radical changes in the international markets in the last decades of the nineteenth century. In the policies of the Historical Left party it appeared

---

17 Besana C., *Caseificio*. Part I. *Scienza del caseificio*. Part II. *Tecnologia del caseificio*, Torino, UTET, 1908.

18 The experimental dairy in Lodi began operations in 1860, and the Reggio Emilia School of Animal Husbandry and Dairy opened in 1874. See Galbani A., "Il caseificio lombardo e la Stazione sperimentale di caseificio di Lodi (1860–1880)", in *Storia in Lombardia*, XIV, 3 (1995), pp. 5–31; Cavazzoli L., "La stazione sperimentale di caseificio di Lodi: 1871–1900", in Decleva E., Lacaita C.G., Ventura A. (eds.), *Innovazione e modernizzazione in Italia fra Otto e Novecento*, Milano, Angeli, 1995, pp. 123–66.

19 Archivio Centrale dello Stato (ACS), Direzione Generale Agricoltura (DGA), I Versamento, b. 169, fasc. 523.

20 *L'esposizione nazionale di caseificio in Portici nel 1877 e l'industria del latte*, Annali di Agricoltura, 20, 1879, Milano, Ministero di agricoltura, industria e commercio, 1879.

21 Frascani P., *Le crisi economiche in Italia. Dall'Ottocento a oggi*, Roma-Bari, Laterza, 2012, pp. 21–36.

22 Segreto L., "Storia d'Italia e storia dell'industria", in Amatori F., Bigazzi D., Giannetti R., Segreto L. (eds.), *Storia d'Italia. Annali* 15. *L'industria*, Torino, Einaudi, 1999, pp. 7–86 (p. 12).

23 Ciocca P., *Ricchi per sempre? Una storia economica d'Italia (1746–2005)*, Torino, Bollati Boringhieri 2007, pp. 111–13.

essential to allow local production to go beyond the traditional local markets to achieve better commercial opportunities outside Italy. Under the pressure of foreign competition, the growth in exports became the pivot on which to balance the reorganization of the sector, starting with the cooperative dairies that possessed, on their own account, a consolidated wealth of knowledge and relationships that enabled them to adapt to the changing economic framework of reference.

However, while the objectives to be achieved between the 1870s and the 1880s were well-defined, in an era that saw the pressure to innovate mount in the Italian dairy sector, the overall situation appeared scarcely encouraging. The agrarian calendars were not alone in handing down outdated technical and commercial ideas in the sector. The Agricultural Survey also presented regional situations that did not invite optimism. In the Veneto the best news concerned the Belluno area, where the social dairies worked miracles despite being faced with numerous difficulties.[24] In the plains, however, the situation was very different, because the peasant farmers continued to support the tendency to fatten cattle to obtain meat and agricultural strength. Signs of progress, overall, were few and far between.

Figure 1. Dairy of Casalpusterlengo

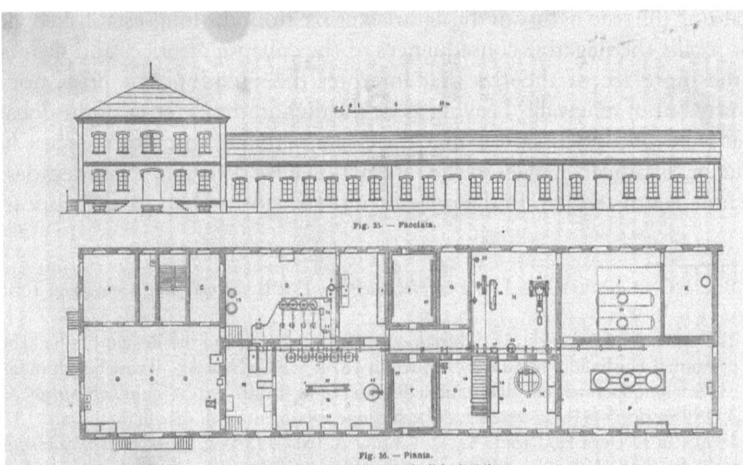

Source: Besana C., *Caseificio*, *op. cit.*, p. 130.

Tentative signs of change were widely reported at the 1874 Milan conference. The social dairy in Gemonio[25] was a particular subject of

[24] *Atti della giunta per la inchiesta agraria e sulle condizioni della classe agricola*. Vol. IV. *Relazione del commissario Emilio Morpurgosulla XI circoscrizione (provincie di Verona, Vicenza, Padova, Rovigo, Venezia, Treviso, Belluno e Udine)*, Roma, Forzani e C. tipografi del Senato, 1884.

[25] *Atti e memorie del primo congresso per l'incremento del caseificio tenutosi in Milano nel marzo 1874 presso la R. scuola superiore di agricoltura e cenni sulle esposizioni dei prodotti del latte,*

discussion. Active since 1872, it owned a building designed specifically for cheese production by the engineer Fortunato Valaperta, who was also the mayor of the town. The two-storey building featured a kitchen (or *casone*) on the lower floor for the packaging of cheeses, a room for the production of butter, a room for storing milk, an ice house, a salting room (*casirola* or *salatoio*) for salting cheeses and a storeroom (or *casera*) for maturing cheese.[26] The upper floor housed the hall used for members' meetings and the cheese-makers' accommodation. Once the resistance of small farmers, who saw the cooperative factor as a disruptive force, had been overcome, the milk supplied by the members formed a capital in terms of cheese and butter belonging to "community". The cash income was divided between the partners on a pro rata and pro socio basis. While the economic benefits, according to the speakers at the conference, appeared undeniable, the most useful benefits observed were in moral terms, as thanks to this type of partnership the farmer developed "esteem and mutual respect for partners; confidence in his peers; and the constant practice of honesty and scrupulous obedience to established norm, which was mandatory for all", essential aspects to raise the him from the state "of the brute" to "a higher state as a feeling and thinking being". Between December 1872 and January 1874 the Gemonio dairy produced 1,534.25 hectolitres of milk, 3,887.1 kg of butter, 7,645.2 kg of cheese and 4,555.6 kg of ricotta cheese. Given the positive results and the possibilities that opened up for marketing dairy products, testing began for the production of Stracchino cheese. The gross income obtained in the first fourteen months of operation amounted to 26,780.68 lire, with expenses of 3,578.26 lire. As a result, net income was 23,202.5 lire, or an average monthly income of 1,657.32 lire.

A qualifying aspect is that, in terms of production, social dairies had the undeniable advantage of encouraging an increase in sales prices, and therefore in the incomes of milk-producing farmers. In some municipalities in the Novara area (Baveno, Ornavasso, Cuzzago, Mergozzo, Pallanza), after the establishment of the cooperative dairies the price of butter rose from 1.4 to 2 lire per kilo, and that of cheese by 60 cents to 1.50 lire, bringing considerable benefit to many small producers, who until then had not derived any benefit from the low quantities of milk in their possession. At the same time, the use of modern equipment and skilled workers allowed for the production of cheese and butter of higher quality, something that would have been unthinkable if the time available had been spent handling a few liters of milk every day. Therefore, the dairies not only appeared to be "a real providence for the valleys", but were also promoted in the plains, as at the

---

Milano, tip. Lombarda, 1974, pp. 203–63. Other reports were presented at the conference in Milan concerning the dairies in Bracchio, Casale Corte Cerro, Montebuglio and Somerano. For records of dairies founded in the early 1870s, with drawings of buildings, regulations, minutes of meetings and statistical tables of the milk delivered by members see ACS, DGA, I Versamento, b. 169, fasc. 523.

26 On dairy premises and equipment see Besana C., *Caseificio*, *op. cit.*, pp. 123–37.

end of the nineteenth century they were the only alternative to making the transition from the traditional craft stage to the industrial level.[27]

4. Between tradition and innovation

However, the concept of the social dairy masked an associative reality that was quite varied, the result of the existence of a plurality of ways to pool raw materials and proceeds in cheese production. The simplest of these was based on what was known as the "rotating family method". Although it could be permanent, it was not unusual for this arrangement to be seasonal, particularly in the summer on the alpine pastures when shepherds shared milk to make cheese. In this way they could collect the 200–400 liters of milk needed for the production of a form of hard cooked cheese.[28] By contrast, during the winter months a few families joined together in associations lending each other in turn the milk needed to produce homemade dairy products.[29] In any case it resulted from a complex associative framework, as winter dairies were near barns, while summer dairies were near the pastures. It must also be said that the duration of the winter dairies depended on the varying duration of the summer pastures.

To complicate the existence of rotating dairies still further, their smooth operation depended on verbal agreements. In the absence of a statute, this appeared a fragile arrangement as it depended solely on the bond of trust between families. Another big drawback of "rotating family" (*turnarie familiari*) dairies stemmed from the fact that the milk was usually processed in unfit and unclean premises, because it was typically intended for the kitchen of each member, who often lacked the appropriate tools.[30] As it was intended for the production of dairy products for domestic consumption,[31] the manufacturing process was completely devoid of uniformity, and the dairyman was usually a figure who lacked the necessary professional skills. Although similar simple forms of cooperative dairies might eventually produce cheese and butter of quality, they did not have an effective commercial dimension, as was the case in Osoppo (Friuli), where in 1880 there were twelve rotating dairies, managed exclusively by women. Any gains from the

27 Another institution which should have made a significant contribution to the development of the dairy industry was the social or consortium warehouses, *Atti del congresso e del concorso provinciale di latterie*, Udine 10 – 17 maggio 1886, Udine, tip. di Giuseppe Seitz, 1886, pp. 39–41.

28 ACS, DGA, I Versamento, b. 169, fasc. 524.

29 *Atti del congresso e del concorso provinciale di latterie*, *op. cit.*, pp. 5–6, 222–3.

30 Reggiani E., *La produzione del latte*, *op. cit.*, pp. 298–9.

31 Statistically, the per capita consumption of cheese in Italyat the end of the 19th century was circa 2–4 kg, Zamagni V., "L'evoluzione dei consumi fra tradizione e innovazione", in Capatti A., De Bernardi A., Varni A. (eds.), *Storia d'Italia. Annali* 13. *L'alimentazione*, Torino, Einaudi, 1998, pp. 169–204 (p. 176).

sale of surplus were used for the purchase of household items, demonstrating the lack of any investment in equipment and technologically advanced premises. In any case, and despite the presence of a very elementary type of organization, the "rotating family" dairies had, in the opinion of many, the great merit of signaling the first step towards the modern dairy cooperative.

Figure 2. Butter label, Dairy Levico

Source: Image courtesy of Franco Frisanco.

A second organizational approach was the "rotating social" dairy. Compared to the previous dairy model, this had the advantage that the handling of dairy products did not occur in the house of the individual member, but rather in a dedicated room rented or purchased on behalf of the society.[32] The master cheese-maker had full freedom in the choice of processing methods, but every family, on the day he worked for them, in addition to guaranteeing him compensation had to provide the necessary assistance and bring timber for the fire. The butter was sold on behalf of the society, while the cheese and ricotta cheese were reserved for the members. Although the rotational social dairy model represented a decisive improvement in milk processing, the main limitations of this system of cooperation derived from the lack of true technical and administrative management, since both the dairyman and families in turn imposed their own criteria or tastes on the manufacturing of the cheese, which, as a result, lacked the necessary uniformity of production. Furthermore, behind the apparent greater freedom of each individual milk producer there lays a deep-rooted distrust of wider autonomous social action.

[32] Reggiani E., *La produzione del latte*, *op. cit.*, pp. 299–302.

Furthermore, the dairyman had no clear position, having to constantly respond to the will of the members of the dairy and the individual families called in turn to package the cheese. In short it was a cooperative system that still presented many ambiguities, but also undoubtedly offered several features in common with the social cooperative dairies that many believed constituted the means to modernize the Italian dairy sector at the end of the nineteenth century.

While at the end of the nineteenth century the transformation of rotating social dairies into cooperative dairies was in progress in many areas of the country, the Emilia region was witness to an entirely different phenomenon. Drawing criticism from those who argued that the milk sector should remain on the margins of speculative operations and individual profits, on the plains of Emilia, and also in the rotating dairies of Lombardy which were numerous until the end of the 1860s, they were replaced by private centers belonging to small industrialists who, on their own account, proceeded with the purchase of milk and the sale of finished products.[33] It was only in the early twentieth century, thanks to peripatetic professors of agriculture, that the social cooperative dairy movement was revived in the Emilia region.

What were in effect entrepreneurial investment transactions were not always viewed in the same positive light as pure cooperation, being seen as promoting the sale of milk by peasant farmers who, attracted by easy cash gains, deprived themselves of a staple food for the sustenance of their families.[34] A certain amount of criticism of dairies of an entrepreneurial character, such as that founded in Soligo in 1883, was also aimed at the anonymous cooperative societies which, in return for the "mirage of dividends", involved the presence of a number of figures who had neither barns nor cattle. Some were against this development, since instead of bringing true benefit to small landowners in reality it tempted them from the path of true cooperation.

When, therefore, the Italian dairy sector began to emerge from the narrow margins of family production and embarked on the road to industrialization,[35] this created the conditions for a debate that generated a contrast between those who encouraged the initiatives aimed at market-driven production and those who, by contrast, supported a cooperative model with a moral profile, assigning the sale of products with the role of supporting mutual assistance.[36] This dividing line was part of the nature of the nascent social cooperatives dairies that, beyond the common trait of being stable aggregations of milk producers organized on the basis of a

33 *Ibid.*, pp. 302–3.

34 *Atti del congresso e del concorso provinciale di latterie*, *op. cit.*, p. 6.

35 On the contrasting process of the modernization of the Italian food industry see Chiapparino F., "Tra polverizzazione e concentrazione. L'industria alimentare dall'Unità al periodo tra le due guerre", in *L'alimentazione*, *op. cit.*, pp. 205–68.

36 Zangheri R., "Nascita e primi sviluppi", in *Storia del movimento cooperativo in Italia 1886–1986*, *op. cit.*, pp. 5–218 (pp. 48–9).

statute, offered a variety of models, depending on the purpose and the aim of the partnership. Effectively, there were three possible groupings: those that applied the cooperative principle in full, in that only those who supplied milk could be partners; those moving away from the cooperative model, in so far as the partners were capitalists, whether they produced milk or not; and finally, those in which the cooperative principle was purely associated with the industrial dimension, as both capitalists and the suppliers of milk could arbitrarily become partners.

A productive sector therefore emerged composed of diversified corporate structures that, in simple terms, saw a distinction between the dairies of the irrigated plain of Lombardy and Emilia and the Alpine dairies, with the former moving towards the definitive industrialization of the sector,[37] while the latter remained anchored to a model tied to the principle of solidarity. Besides being a dichotomy that began to take shape in the course of the modern age,[38] at the end of the nineteenth century, a key moment in the theoretical development of this distinction can be traced back to the proposals approved by the dairy congress held in Udine in 1886.[39] According to participants at the meeting it was necessary to encourage the unions of milk producers that, unlike with the cooperatives, their primary objective should be not so much sales in the interest of the milk-producing partners as "putting more and better food on the table of the poor farmer". In the social cooperative dairies each partner participated in the formation of capital, which was indispensable for the establishment of the premises and the acquisition of equipment. Following this they would equally enjoy the economic benefits derived from the sale of dairy producc.

Such an approach, according to which the main aims to be achieved were mutualist, alongside the cohesion of the family, is perfectly explained when examining the identity of the individuals who founded the dairies in the Friuli. The founders included a range of subjects such as teachers, doctors, soldiers, priests and other figures, for whom cooperation was not supposed to aspire to the development of forms of production aimed at mere monetary gain. The involvement of the Catholic Church in the emergence of the cooperative movement is emblematic of the situation. In the Veneto,[40] it can be seen that the clergy involved in cooperation between milk producers justified these transactions not only because they improved production, but also for social reasons, as they were perfectly suited "to

37 There were 272 rotating dairies in the province of Reggio Emilia in 1869. Twenty years later they had all given way to private structures, Reggiani E., *La produzione del latte*, *op. cit.*, p. 302.

38 Mocarelli L. "Allevamento e produzione lattiero-casearia nella Lombardia dell'età moderna", in *Il futuro del latte in Val Padana*, S. Angelo Lodigiano 13 marzo 2015 (forthcoming).

39 *Atti del congresso e del concorso provinciale di latterie*, *op. cit.*, p. 7.

40 Tescione M., "Veneto: i cattolici per la cooperazione nella seconda metà dell'Ottocento", in *Annali dell'Istituto Alcide Cervi*, 5 (1983), pp. 327–50.

the mentality of the smallholder farmer, attached to his piece of land, and accustomed to frugality and domestic budgeting. He is a stranger to any speculative tradition, with little interest in involving himself with the laws of the bourgeois market".[41] Other sources were also in line with the Catholic vision of economic modernization,[42] although they did foresee that the development of the dairy industry at the end of the 1880s in the districts of Cadore and Agordino was related to the increase in remuneration, which presented small producers with the certainty of a higher income.[43] In 1888, encouraged by the agrarian meeting in Agordo, seventeen cooperatives in the area established a cooperative limited company for dairies in the local area.[44] Within a few years the company had succeeded in pooling the dairies of 1,193 families in the Agordo area, with 2,319 dairy cows, reaching an annual production of 1,810,075 liters of milk.[45]

Under the provisions of the statutes the management of the cooperative dairies was the responsibility of a board of directors, made up of a group of partners responsible for monitoring the technical and economic performance of the company. In order to make the dairies "stores or warehouses for food savings" it appeared necessary that each one organize itself in a centralized manner, with management and administration organized by the shareholders. Only then would it be possible to accomplish a thorough inspection of the work being done and bookkeeping, an essential requirement to put the social cooperative dairies on a sound legal and operational footing.

5. Industrial dualism

The gradual establishment of cooperative dairies in the 1880s led to the birth of the newspaper *The Italian Dairy Illustrated*, which offered the sector practical, scientific and commercial information. It is worth noting this publication, as it was the first Italian newspaper to tackle the problems of the dairy sector. Published in Reggio Emilia in 1882, and edited by Giacomo Maffei,[46] the paper took on the role of promoting the benefits of cooperative dairies for small milk producers. It featured continuous reminders of the

41 D Rosa G., *Tempo religioso e tempo storico: saggi e note di storia sociale e religiosa dal Medioevo all'Età contemporanea*, Roma, Edizioni di storia e letteratura, 1987, pp. 190–1.

42 Franzina E., "Tra Otto e Novecento", in Lanaro S. (ed.), *Storia d'Italia. Le regioni dall'Unità a oggi. Il Veneto*, Torino, Einaudi, 1984, pp. 763–861 (pp. 786–813).

43 Amantia A., *Produzione agraria e condizioni di vita nel Bellunese tra Ottocento e Novecento*, in *La montagna veneta in Età contemporanea. Storia e ambiente. Uomini e risorse*, Lazzarini A., Vendramini F. (eds.), Roma, Edizioni di storia e letteratura, 1991, pp. 69–94 (p. 77).

44 Reggiani E., *La produzione del latte, op. cit.*, p. 306.

45 *Ibid., La produzione del latte, op. cit.*, pp. 312–13.

46 Giacomo Maffei was the owner of a factory warehouse that sold technical and chemical products to the dairy industry. Therefore, the newspaper was designed not only to inform but evidently also to encourage the sale of machinery.

need to give the production of butter and cheese in Italy a precise business structure, and from this perspective cooperation appeared the most suitable solution. In order to promote the emergence of cooperative dairies, the paper provided abundant information on the statutes, economic relations and general life of dairies, focusing particularly on efforts to overcome the many existing limits, starting with the resistance that many producers showed to the use of machines and mechanical tools.[47] The theme of technological modernization was addressed several times by the newspaper, which also included numerous advertising supplements extolling the benefits of owning a modern centrifuge, skimmer or kneader. Better yet, if possible, to use facilities that were fully steam powered, as was the case in Germany.[48]

The attention of the newspaper was also drawn to the negative consequences for the development of the sector of the widespread practice of the adulteration and counterfeiting of milk. Given the importance of making the dairy industry a competitive production sector, there was a pressing need to enact laws that would give assurance to consumers on the quality and nutritional properties of milk and dairy products.[49] Indeed, the enactment of laws to combat fraud in the food sector fuelled lively debates in Parliament, as this was a matter which presented a number of implications in terms of both health and economic aspects.[50] In the specific case of milk and butter the anti-adulteration law was passed in 1894, while discussions on cheese continued into the first decades of the twentieth century. With respect to milk, where the main issue was to eliminate the addition of water, in the case of cheeses legislative action appeared more complicated, due to a lack of precise information on the characteristics of cheeses produced in Italy. It was, in truth, an issue involving health protection and the fight against commercial fraud, fields that were still in their infancy.

47 *Il caseificio italiano illustrato*, 1 febbraio 1889, fasc. 5, p. 33.

48 *Il caseificio italiano illustrato*, 31 maggio 1889, fasc. 22, pp. 173–4.

49 *Il caseificio italiano illustrato*, 28 giugno 1889, fasc. 26, 193–5; 1 novembre 1889, fasc. 44, pp. 337–9.

50 Lonni A., "Dall'alterazione all'adulterazione: le sofisticazioni alimentari nella società industriale", *op. cit.*, pp. 531–84: 566–7.

Figure 3. Kitchen in the dairy, the school of San Michele all'Adige

Source: Image courtesy of Fondazione Edmund Mach, S. Michele all'Adige (TN), Italy.

In this context, and in order to encourage exports, it was essential to provide Italian dairies with a trademark, the only way to ensure that produce would be uniform and identifiable.[51] At the end of the 1880s *The Italian Dairy Illustrated* published news from London and Amsterdam about the spread of margarine, a recent innovation of the dairy industry in the north of Europe that might damage sales of Italian butter.[52] As a result, Italian butter had to be protected, through the formation of clear categories and the reduction of discretionary margins. In this respect it was essential to establish partnerships between cooperative dairies, as was already the case in the Netherlands, Belgium, Switzerland, England, Canada, the USA and other countries. At the first congress of dairy cooperatives held in Reggio Emilia in 1907 it was proposed to establish a single federation for the whole of Italy, with the aim of regulating and coordinating the activities of the individual dairies.[53]

---

51 *Il caseificio italiano illustrato*, 27 settembre 1889, fasc. 39, pp. 297–9.

52 *Il caseificio italiano illustrato*, 20 dicembre 1889, fasc. 51, p. 394. The first margarine factory in Italy was established by a French company in Milan in 1874. It was not a great success, however. See *Il caseificio italiano illustrato*, 24 agosto 1888, fasc. 34, pp. 269–71. For Europe see Merta S., "Margarine in Competition with Butter in Germany (1872–1933). The Example of Van den Bergh's Margarine Factory in Kleve", in Oddy D.J., Drouard A., *The Food Industries of Europe in the Nineteenth and Twentieth Centuries*, Farnham, Ashgate, 2013, pp. 119–32.

53 Reggiani E., *La produzione del latte*, *op. cit.*, p. 310.

As this was a sector fragmented into many local contexts,[54] collecting accurate statistical data on the number of Italian cooperative dairies at the turn of the century proved a scarcely credible operation. However, the overall impression was of an industry that was experiencing a marked transformation, intended to establish new production structures. In any case, while doubts remain about the reliability of the quantitative data, in 1898 there were 208 active social cooperative dairies in Italy,[55] of which 124 were concentrated in the province of Belluno, followed at a considerable distance by Udine (21), Treviso (9) and Vicenza (5). In second place behind the region of Veneto, which at the end of the nineteenth century was the most advanced in the world in terms of dairy cooperatives, was Lombardy, with 31 cooperatives. In other regions such as Emilia the phenomenon of social cooperative dairies appeared to have made little impact, as indeed was the case in the entire southern part of the country where there was only one facility in Sardinia and the southern provinces.[56] The available data for the beginning of the twentieth century shows remarkable growth. The Friuli alone accounted for 195 dairy cooperatives, which did not include "the small and still primitive dairies" that continued to have a pronounced family-artisan profile.[57]

As evidence that this was a sector undergoing rapid transformation, in 1907 legally constituted social dairies of a cooperative nature numbered around 700,[58] while by 1920 the number of dairies operating in Italy, by this definition including any building for the collection and processing of milk, stood at around 4,300, creating employment for more than 15,000 people.[59] Of these plants 75% were made up of small dairies, generally run by cheese-makers in Emilia and Lombardy. Just over 20% included cooperative dairies of any form, while 5% were large cooperative dairies and large industrial plants in the dairy industry.[60]

After three decades of growth and gradual modernization in the early twentieth century, the dominant theme in the dairy sector was the predominance of a plurality of small local manufacturers that avoided the major statistical surveys. As of 31 December 1902, only 114 cooperative societies were legally constituted in the Italian dairy and butter industry,

54 In 1914 the number of social, rotating and cooperative cheese cooperatives in Trento exceeded 400, effectively one or more for every village. See Dalpiaz S., *Gli uomini del latte. La produzione lattiero-casearia nella storia della cooperazione trentina*, Trento, 2013.

55 Reggiani E., *La produzione del latte*, *op. cit.*, p. 296.

56 ACS, DGA, I Versamento, b. 752, fasc. 3269.

57 According to the 1937 industrial census, 94.5% Italian cheese production was done "domestically". See *Censimento industriale 1937. L'industria della lavorazione del latte e dei prodotti*. Monografia, 3, Roma, Istituto centrale di statistica del Regno d'Italia, 1939, p. 18.

58 *L'Italia economica. Annuario statistico-economico dell'industria, del commercio, delle finanze, del lavoro*, anno II, 1908, Milano, Società editrice di Annuari, 1908, pp. 189–90.

59 Fascetti G., *Caseificio*, Milano, U. Hoepli, 1910, p. 12.

60 *Censimento industriale*, *op. cit.*, p. 107.

with paid-up capital amounting to 1,516,448 lire.[61] The most important companies were in the province of Milan (2 with 488,974 lire of paid-up capital), Cremona (7 with 304,870 lire), Pavia (4 with 233,907 lire), Parma (24 with 164,297 lire), Treviso (5 with 118,636 lire), Brescia (8 with 45,248 lire) and Belluno (26 with 34,494 lire). Therefore, even from this angle, the dualism seen in the dairy sector remained in sharp focus, with the nascent large industrial plants with capital to be found in Lombardy, while the smaller dairies were located in mountainous areas.

Conclusions

The 1870s represented a turning point for the milk sector in Italy. Following the example of countries such as the USA, France and Switzerland, ever-growing state intervention occurred aimed at increasing production and marketing capacities. Small farmers began to be encouraged to create social dairies. However, overcoming the traditional individualism of the small milk producers was not so easy. In this respect, the role played by the Catholic clergy in some regions proved very helpful. The main objective of the establishment of cooperative dairies in the alpine areas of the country was to avoid population displacement towards the cities in the valley, where industrialization would guarantee better living conditions. Nevertheless, social dairies grown up in the Alpine regions in the second half of the nineteenth century had not only an economic purpose, but also great social importance. For this reason they are considered an interesting case history for study of the development of a modern food industry. Through an intensive marketing campaign, the milk sector found it convenient to abandon the traditional production and organization criteria and adapt itself to national and international markets that were more and more demanding in terms of both the quality and the image of the products for sale.

61 *Riassunto delle notizie sulle condizioni industriali del Regno*, parte I, Roma, tipografia Nazionale di G. Bertero, 1906, p. 128.

Appendix

Table 1. Social Dairies in Italy (1889)

Province	Social Dairies	Province	Social Dairies
Belluno	124	Cuneo	1
Udine	21	Torino	1
Sondrio	19	Mantova	1
Treviso	9	Milano	1
Brescia	7	Reggio Emilia	1
Vicenza	5	Salerno	1
Como	2		

Source: Reggiani E., *La produzione di latte, op. cit.*, p. 296.

Table 2. Social Dairies in Italy (1906)

Province	Social Dairies	Province	Social Dairies
Belluno	154	Novara	45
Udine	205	Padova	1
Reggio Emilia	37	Verona	1
Parma	17	Ascoli Piceno	1
Cuneo	5	Mondovì	2
Sondrio	56	Pavia	3
Como	50	Vicenza	10
Cremona	8	Cuneo	4
Piacenza	4	Novara	6
Modena	1	Aosta	103
Bologna	2	Torino	23
Genova	6	Brescia	31
Treviso	17	Milano	3

Source: Fascetti G., *Caseificio, op. cit.*, p. 19; *Almanacco del giornale L'industria lattiera e zootecnica* (1910), Reggio-Emilia, Cooperativa fra lavoratori tipografi ed affini, 1910, pp. 9–10.

Enterprises, Trade and Industry in the Lombardy Dairy Sector

The Origins of Locatelli and Galbani (1860–1914)

Silvia A. Conca Messina
University of Milan "La Statale"

1. Introduction

Italian dairy production has long been a marginal agricultural activity, except in the lower Po valley in the north of the country, where traditionally it had a prominent role in the local economy. In the last few decades of the nineteenth century, the sector shifted to a noteworthy economic position also in other areas of the peninsula, as the demand for Italian dairy products from domestic and international markets continuously increased (abroad, particularly in England and in the Americas, as a consequence of mass intercontinental migration). National Unification (1861), innovative technological processes and scientific knowledge, new machinery, the availability of steam and electric power and new preservation and transport systems allowed cheese production to grow continuously from 34,600 metric tons in 1861 to 181,400 in 1913. In 1894, exports of Italian cheese exceeded imports. In 1893–1900, Italy's average annual export was 9,700 metric tons, while it reached an annual average of 27,500 between 1909 and 1913 (Tab. 1).[1]

1 See also Gobbetti C., "Sunto della conferenza tenuta dal prof. Carlo Besana a Pavia il giorno 1 Settembre 1897", in *Annuario della R. Stazione Sperimentale di Caseificio in Lodi*, 1897, pp. 47–57; Ministero di Agricoltura Industria e Commercio (henceforward MAIC), Direzione Generale della Statistica, *Statistica industriale. Riassunto delle notizie sulle condizioni industriali del Regno*, Parte I, Roma, Tip. Naz. G. Bertero e C., 1906, p. 128; Besana C., "Il commercio dei prodotti del latte tra l'Italia e l'estero nell'anno 1910", in *Annuario della R. Stazione Sperimentale di Caseificio in Lodi*, 1910, pp. 44–8; Bachi R., *L'Italia economica nell'anno 1913, Annuario della vita commerciale, industriale, agraria, bancaria, finanziaria e della politica economica*, Anno V, Città di Castello, Casa editrice S. Lapi, 1914, p. 5; Porri V., *L'evoluzione economica italiana nell'ultimo cinquantennio*, Roma, Stab. Tip. G. Colombo, 1926, p. 157.

Table 1. Production and exports of cheese. Italy, 1861–1913 (in metric tons)

	Italy	
Cheese	**Production**	**Exports**
1861	34,600	1,400
1871	49,600	1,800
1881	78,300	2,800
1891	94,100	5,600
1901	115,500	10,900
1911	185,000	27,900
1913	181,400	32,800

Source: ISTAT, *L'Italia in 150 anni. Sommario di statistiche storiche 1861–2010*, 2011, Table 13.23, p. 660 and Table 16.10, pp. 740–1.

Lombardy had a long empirical tradition in dairy production, part of which was already being exported to international markets. During the last three decades of the nineteenth century farming and livestock further increased in the region and the expansion of breeding contributed to the transformation of the entire dairy chain. Lombardy confirmed itself as Italy's largest producer and in the 1890s accounted for around 1/3 of all cheese production and exports (especially Grana and Gorgonzola).[2] Former small family businesses transformed themselves into larger enterprises and industries, which controlled the purchase of milk and cheese from the farms of the plain, founded new plants, provided the ageing process for cheese and sold the produce on domestic and international markets. The advent of new manufacturing systems allowed for the pooling of large quantities of milk: large-scale mechanized cheese production became possible. Names such as Locatelli, Galbani, Invernizzi (all from Valsassina, the high hills near Lake Como, which specialized in ageing cheese) or Polenghi Lombardo (from the Lodi area) are some of the most important enterprises which emerged during this period, whose brands have lasted until today. In 1950 these companies employed nearly 40% of all workers in the dairy industry.[3]

This chapter will focus first on the main directions and character of the transformation of the dairy sector in nineteenth- to early twentieth-century Lombardy. Then it will trace the origin of the family firms, particularly Galbani and Locatelli, which shifted dairy production from a craft to an industry, examining the local leading role they played as merchant-

2 MAIC, Direzione Generale della Statistica, *Statistica industriale*, 1906, p. 130; Porri V., *L'evoluzione economica italiana nell'ultimo cinquantennio*, 1926, p. 155; Romani M., *Un secolo di vita agricola in Lombardia (1861–1961)*, Milano, Giuffrè, pp. 108–9.

3 Bertagnoni G., "L'industrializzazione del latte alimentare in Italia: criticità e ritardi visti attraverso un case study locale", in *Storia e futuro. Rivista di storia e storiografia online*, No. 39, Nov. 2015. For a general survey of the Italian agri-food industry of the period, see Chiapparino F., "Tra polverizzazione e concentrazione. L'industria alimentare dall'Unità al periodo tra le due guerre", in *Storia d'Italia. Annali 13. L'alimentazione*, Torino, Einaudi, 1998.

entrepreneurs in the collection, ageing and trade of cheese. The study will particularly focus on their business strategies, analyzing innovation in dairy processing systems and the development of commercial relations and new markets, which seem to have been particularly important in promoting the growth of production.

2. New trends and conditions favorable to growth

Until the political unification of Italy (1861), conclusively accomplished in 1870 with the annexation of Rome, the dairy production was widely spread throughout the Lombardy countryside. Various types of cheese, strong and soft, were locally produced in the plain and in the mountain valleys. Nevertheless, the main area of production and trade, with significant exports of goods to external markets, was the low plain between the rivers Ticino and Adda, around Milan and the small towns of Pavia and Lodi. Since the early modern age, the area has been characterized by an advanced agricultural system widely acknowledged as one of the most progressive in the European context, where large plots of land were rented out to agrarian entrepreneurs and used for cereal and grass cultivation, cattle breeding and milk-cheese production (Grana). One of the most renowned soft cheeses, called Gorgonzola, was manufactured by seasonal migrant herdsmen (bergamini) in the plain around Gorgonzola during the autumn and winter, where they came down to feed their cattle with the excess local grass and hay.[4]

A sizeable amount of the local production of Grana and Gorgonzola was bought by merchants from Codogno (a long-standing major trade center of the low plain) and Milan; it was then transported and aged in cool stores (*casere*) located in Codogno, in Corsico (a small village near Milan) and along the San Gottardo hamlet in Milan; almost all of it was finally sold in local markets or sent (Grana) to other Italian regions, and a small part of it was sent to foreign countries, especially within the Austrian Empire, by which Lombardy was ruled until independence and the unification of Italy.[5] The internal and international demand for butter and cheese followed a general positive trend throughout the Restoration age (1815–61), but there were temporary market contractions due to the interruption of trade during the independence wars (1848, 1859) and their consequent depressive effects or to the rise of duties introduced by the State or the Church in Central and Southern Italy (1854, 1856) combined with the difficulties connected to the crisis of silk in the mid-1850s. On the whole, the figures for cheese production and exports did not increase very greatly: in the middle of the

4 Massara F., "Dello stracchino di Gorgonzola", 19 August 1866, in *Giornale di agricoltura*, No. 8, 9, 10, 11, 1867.

5 Campiglio A., "Commercio e industria", in Litta Modignani L., Bassi C., Re A. (eds.), *Milano e il suo territorio*, Milano, Pirola, 1844, tome II, pp. 99–101.

century, the low plain produced around 15,000 metric tons of cheese (only 9–12% of which was soft cheese), while production from the region as a whole was around 25,000 metric tons and exports (Grana) varied to some degree, which can be estimated very approximately at 900 to 1,400 metric tons per annum in the early 1840s and 1,200 metric tons average per annum in the 1850s.[6]

It was only after 1870 that the effects of several political and economic changes, which were under way during the century, started to be felt and paved the way for a rapid growth in trade combined with increasing modernization and diversification of Lombardy dairy production: national unification, the transport revolution, scientific discoveries and technological improvements, the agrarian crisis – all contributed to generating a progressive trend and positive economic circumstances for the sector.

First of all, national unification, together with improvements in transport and communication, opened up a wide new market for the dairy industry. Access to the national market was now free of duties and became easier through the new railway network, which was extended both within Lombardy and the north (e.g. Milan-Lecco, 1873) and between Northern areas and Central-Southern Italy (railways in Italy extended from 2,500 km in 1861 to 8,500 km in 1881). Moreover, commercial treaties were signed with France and Great Britain (1863) and Austria-Hungary (1865 and 1867), which meant that for some years there was a reduction in customs barriers, facilitating the launch of the Italian export of agrarian produce.[7] Thanks to improvements in international railways and ships, transport costs via land and water to the rest of continental Europe and England dropped, and the same occurred in prices for the Atlantic route and transportation to the Americas. At a European level, an increasing amount of Italian dairy products was sent in particular to Austria, Germany, Great Britain, France and Switzerland. London became one of the most important marketplaces for selling Italian agri-food specialities (primarily Gorgonzola), also due to the growing migrant community living there. Over the Atlantic, the USA above all and, to a lesser degree, Argentina and Brazil became emerging destinations for Italian produce, following the millions of migrants and the changes in the way of life of the expanding multi-ethnic towns and metropolises of the American continent.

In Lombardy, the increasing availability of water in Lomellina (Pavia) and around Cremona (thanks to the private building of expensive canals

6 Levati S., "'Cibo sano, comodo a conservarsi e al trasporto, di squisito gusto': il commercio del Parmigiano nello Stato di Milano tra Sette e Ottocento", in Battilani P., Bigatti G. (eds.), *Oro bianco. Il settore lattiero-caseario in Val Padana tra Otto e* Novecento, Lodi, Giona, 2002, p. 76, No. 26; Besana C., *Tra agricoltura e industria. Il settore caseario nella Lombardia dell'800*, Milano, Vita e pensiero, 2012, pp. 64–72; Campiglio A., "Commercio e industria", p. 100, No. 1.

7 Romani M., *Un secolo di vita agricola in Lombardia*, pp. 44–5.

throughout the century),[8] combined with growing profits from the markets, led to an expansion of grazing land, livestock (also for meat) and milk production in new areas beyond the traditional region between the Rivers Ticino and Adda. The rising trend involved all the dairy provinces and significantly gathered pace in the last three decades of the period, when cattle were also better selected and an increasing number of cows of more suitable breeds were imported from Switzerland.[9] An additional decisive reason for this economic transformation was that the dairy sector offered more stable returns than silk and wine (silkworms and vines were hit by serious diseases respectively from the 1850s and the 1870s) or than cereals (due to competition from American grains). The globalization of markets and the agrarian crisis, which started in the 1870s, boosted this trend, reducing the international level of agrarian prices and particularly those of rice, wheat and silk.

In the meantime, scientific discoveries in biology (microbiology) and chemistry allowed pioneering milk processing methods; larger milk factories (often cooperatives) were founded, with huge investments in innovative technologies and mechanisms, such as the recently invented skimmer machines and modern boilers, burners and mechanical devices which enabled a more hygienic system of production to operate. The production of milk and butter went through the roof. Many small dairies (*casoni*) producing Grana closed down or were forced to innovate to remain competitive – as they were often made up of dirty hovels and managed by rough hands-on masters with an empirical method who were not able to guarantee consistent quality or to reduce high wastage from defective production. Indeed, their backwardness in methods and technology was strongly criticized by specialized journalism, which considered them responsible for Grana's loss of market share relative to its competitor, Parmesan. The production of Gorgonzola and Grana, traditionally limited to a restricted zone, spread to other areas of Lombardy or the northern Italian provinces. At the end of the 1870s Lombardy began to produce or imitate Swiss, French or southern Italian cheeses, often with the cooperation of foreign or southern Italian skilled workers: for example Emmental, soft cheeses, provolone and Caciocavallo. At the turn of the nineteenth century,

8 Besana C., *Tra agricoltura e industria*, pp. 94–6; Campiglio A., "Commercio e industria", p. 125.

9 The number of bovines increased from 626,058 in 1869 to 840,324 in 1881 in Lombardy; in Italy they went from 3,455,635 to 4,783,232: see Besana C., "Formaggi-carni salate-legumi-frutti in conserve", in *Esposizione industriale italiana del 1881 in Milano, Relazioni dei giurati*, Milano, Hoepli, 1883, p. 30. On Lombardy cattle breeding see Fumi G., "L'esportazione di bestiame dalla Svizzera e l'allevamento bovino in Lombardia (secoli XVIII-XIX)", in Piola Caselli F. (ed.), *Regioni alpine e sviluppo economico. Dualismi e processi d'integrazione (secc. XVIII-XX)*, Milano, Angeli, 2003, pp. 153–88; *Id.*, "Iniziative per il progresso zootecnico nell'Italia settentrionale tra Otto e Novecento", in Maggiore T., Mariani L. (eds.), *Seminari carne. Filiera zootecnica, valore alimentare*, Sant'Angelo Lodigiano, Museo lombardo di storia dell'agricoltura, 2014, pp. 98–108.

the amount of regional cheese production was estimated at nearly 45,000 metric tons, and that of butter at around 19,500 metric tons.[10]

The business was growing and so were the dairy trade and the fortunes of Lombardy merchants and manufacturers, who were now able to extend their networks throughout the peninsula, connecting products and producers to national and international markets. At the beginning of the 1880s, among the wealthiest firms we can cite those of Gallone and Guscetti from Milan, Bignami, Ferrari, and Polenghi Lombardo from Codogno.[11] Polenghi Lombardo, first constituted in 1871, exemplified the model for rational dairy production: owning a large plant equipped with modern machinery, using skimmers and steam power and many stores in Lombardy (Codogno, Lodi, S. Fiorano, Cremona) and London, in the early 1880s the enterprise was able to process large quantities of milk daily produced by around 1,000 cows, and to sell 21,000 wheels of Grana, 10,000 of Gorgonzola, and 790 metric tons of butter (650 of which was exported).[12] In the last few decades of the nineteenth century, new firms and a "new dairy industry" were emerging, dealing with an enlarged market, innovative technologies and products.

3. Italian traditional cheese and international markets: Locatelli (1860–1914)

Besides a large number of small and medium sized local producers and sellers, a few merchant-industrialists progressively emerged and accumulated large estates, managing milk and cheese production, ageing and trade and extending their businesses throughout the peninsula and abroad. Family firms such as Locatelli and Galbani, with strong commercial attitudes and access to credit, able to sell in both national and international markets, exploited the increasing demand and its opportunities: first they collected Lombardy and Italian traditional cheese from the plains to the hills and provided ageing stores, and later they became industrialists, acquiring or building factories also in other Italian regions and producing new types of cheese which were successful in meeting consumers' tastes.

---

10 Tedeschi P., Stranieri S., "L'evoluzione del settore lattiero-caseario lombardo dall'Ottocento al Duemila", in Archetti G., Baronio A. (eds.), *La civiltà del latte. Fonti, simboli e prodotti dal Tardoantico al* Novecento, Brescia, Fondazione Civiltà Bresciana, 2011, pp. 708–9. On the distribution of production among the various provinces and municipalities in the 1890s, see MAIC, Direzione Generale della Statistica, *Statistica industriale. Lombardia*, Roma, Bertero, 1900.

11 Besana C., "Formaggi-carni salate-legumi-frutti in conserve", *passim*.

12 *Ibid.*, p. 38; Tedeschi P., Stranieri S., "L'evoluzione del settore lattiero-caseario lombardo", *op. cit.*, p. 716, No. 37. For Polenghi Lombardo see the chapter by Gianpiero Fumi in this volume.

Both the Locatelli and the Galbani families hailed from Ballabio Inferiore, a small mountain village near Lecco (on lake Como) in Valsassina.[13] The area had a traditional local breed of cattle and small-scale production of butter and soft cheese (Stracchini and Robiole), sold on the local market,[14] and had a territorial vocation for the ageing of cheese, as this benefited from natural caves generating cool air, which were particularly suitable for the ageing process. From the 1870s, after the completion of the Milan-Lecco railway, a significant gradual expansion and enhancement of cheese manufacturing took place in the area.[15] Stracchini and Gorgonzola were in increasing demand from British and American markets, and some Lombardy firms opened a branch in London, supporting the expansion of production and the ageing of Gorgonzola in Valsassina.[16] Several small dairy merchants or herdsmen with surnames such as Locatelli, Galbani and Invernizzi were listed in the valley villages' registers (e.g. in Balabio Superiore, Balabio Inferiore, Baiedo, Barzio),[17] but among them, at the end of the nineteenth century, one of the emerging major local merchant firms with higher income and taxation was that of Mattia Locatelli. Davide Galbani was certainly one of the most important local enterprises too, but it nevertheless lagged behind.[18] How did these firms grow?

Locatelli's merchant firm was founded by Mattia (the son of Giovanni Domenico) in 1860[19] and at first specialized in ageing and trading Gorgonzola. Mattia introduced new industrial cheese ageing systems and must have had great commercial skills, exploiting the growth of the market and opening an office in London in 1885. When he died, in December

13 Galbani and Locatelli were also relatives, as Angiola Zappelli, Davide Galbani's wife, was the sister of Alma Caterina, Giovanni Locatelli's wife, see Colli A., "Egidio Galbani", in *Dizionario Biografico degli Italiani*, Enciclopedia Treccani, Vol. 51, 1998. On the complex entrepreneurial activity and system of this area, which deserves an in-depth analysis, see Besana C., "Tra monte e piano. Allevamento transumante ed attività casearie nell'area lecchese tra Ottocento e Novecento", in Fontana G.L., Leonardi A., Trezzi L., *Mobilità imprenditoriale e del lavoro nelle Alpi in età moderna e contemporanea*, Milano, CUESP, 1998, pp. 55–74.

14 Archivio di Stato di Como (ASCo), *Camera di Commercio di Como*, drafts for the triennial Report of the Chamber of Commerce of Como to the Government, referring to 1854–5–6, f. 213.

15 ASCo, *Camera di Commercio di Lecco*, Reports of the Chamber of Commerce of Lecco to the Italian Ministry of Agriculture, Industry and Trade, Mar.-April 1884, Jan.-Feb. 1885, May-June 1885; Sept.-Oct. 1885; Nov.-Dec. 1885; Oct. 1886, f. 50.

16 E.g. Paolo and Pietro Polenghi from Codogno, Carlo Bodega from Maggianico-Lecco, SACo, *Atti dei Notai*, Act 6 December 1886, Baruffaldi, f. 7593.

17 ASCo, *Camera di Commercio di Lecco*, Registers of Chamber Tax, f. 83 (1891), f. 138 and f. 140 (1904), f. 159 (1910).

18 The taxable income amounted to: 6,700 lire for Locatelli in 1891; 7,000 lire for Locatelli and 3,750 for Galbani in 1904; 11,750 lire for Locatelli and 5,000 for Galbani in 1910. See ASCo, *Camera di Commercio di Lecco*, Registers of Chamber's Tax, f. 83 (1891), f. 138 and f. 140 (1904), f. 159 (1910).

19 Mantegazza A., "Locatelli", in *Dizionario biografico degli italiani*, Roma, Istituto della Enciclopedia Italiana, 2005, vol. 65.

1886, he had recently built up a house of residence, with an adjoining cheese ageing store, in Ballabio Inferiore and had accumulated considerable real estate in Valsassina (Ballabio Inferiore, Ballabio Superiore, Pasturo, Cremeno), in Lecco, and nearby in Castello, Germanedo, Suello, Pusiano, Cesana Brianza and Bosisio, for a total value of 77,700 lire.[20] The whole legacy was officially valued at 90,000 lire, 67,400 of which was actually passed down the business, as 22,600 lire had to be transferred between 1888 and 1890 to the daughters as part of their legal inheritance.[21] It was not such a large fortune compared to, for example, the millionaire estates of the contemporary cotton industrialists, but neither was it negligible, considering that it probably represented only a part of the total assets. In 1891 the taxable income for Locatelli was calculated at 6,700 lire, not much less than that of Keller, the indisputable leading firm in Lombardy spinning silk, based in Milan but operating widely in the area, whose income from the Lecco province was estimated at 10,000 lire.[22] Confirmation of this positive evaluation comes from the consistency of the last family dowry to be conferred (11,500 lire in 1890) and from the respectable livelihoods of the daughters' husbands, two of whom were landowners, while the others were an accountant and an innkeeper.[23]

After Mattia's death, his elder son Giovanni became head of the enterprise and had the power of attorney; nevertheless the firm was held by a family partnership made up of the four sons and their mother and widow.[24] As it was based on family tradition, the founder of the business was fortunate in having many sons. Each son, once he reached a responsible age and after due training, was given the responsibility of running one of the important sections of the business, which spread over the Lecco area, plus the London office.[25] It is to be noted that Giovanni and his brother Pompeo had attended a technical school, and it was Mattia's wish that his still under-age sons Vittorio and Luigi would follow suit, as they probably did.[26] Giovanni also became the mayor of Ballabio Inferiore (1895–1901), while his brother

20 ASCo, *Atti dei Notai*, 26 January 1890, No. 3807, Baruffaldi, f. 7595.

21 ASCo, *Atti dei Notai*, 3 January 1888, No. 3528, Baruffaldi, f. 7594 and 26 January 1890, No. 3807, Baruffaldi, f. 7595.

22 ASCo, *Camera di Commercio di Lecco*, Registers of Chamber's Tax, f. 83 (1891).

23 ASCo, *Atti dei Notai*, 3 January 1888, No. 3528, Baruffaldi, f. 7594 and 26 January 1890, No. 3807, Baruffaldi, f. 7595.

24 Mattia was married to Mistica Steffanoni from Suello, ASCo, *Atti dei Notai*, 3 January 1888, No. 3528, Baruffaldi, f. 7594. ASCo, *Camera di Commercio di Lecco*, Firms Register, f. 232 and Commercial Notifications, f. 44/1.

25 This organization lasted a long time, so that the Locatelli firm long continued to be run and managed by the family. For an interesting British point of view on the strength and weakness of the firm's organization in the 1950s, see Cox, R.G.P. (head of cheese buying for Britain's Sainsbury's, visiting continental suppliers), "Italian interlude", in *J.S. Journal. House organ of J. Sainsbury Ltd*, December 1952, pp. 31–6.

26 ASCo, *Atti dei Notai*, Last will of Mattia Locatelli16 August 1884, No. 168, Bolis, f. 7606.

Vittorio was to become a first council member and later also mayor of the town (1902–8).[27]

From 1886 to 1897, under Giovanni's guidance, the brothers' company expanded further, acting as bankers to the small dairies who could not possibly afford to mature their own cheese for months, nor did they want to incur the risks. Giovanni and his wife, Alma Caterina Zappelli, who was collaborating with him, organized the production of Gorgonzola in low Lombardy and in Piedmont[28] and built up new ageing stores (*casere*) in Valsassina (in Baiedo and Maggio-Cremeno, around 1887).[29] The ageing of Gorgonzola in Valsassina underwent great development in the last part of the nineteenth century. At the beginning of the twentieth century, it was possible to observe an impressive "migration" of cheese from the low plain to the mountains, starting in May and lasting throughout the summer, which animated the entire valley, bringing occupation and wealth, whereas the production of this kind of cheese was once restricted to autumn and winter in the lowlands.[30] Giovanni organized growing exports to England, the USA and Argentina[31] and Pompeo left Lombardy for London, where he and his own family managed the British commercial branch until 1897, when the Mattia Locatelli firm came under the control of Giovanni, and the two enterprises separated.[32]

Indeed, in 1897 the Locatelli brothers decided to divide the jointly-owned firm and properties among themselves. The whole property asset was calculated at 339,624 lire and divided into four equal parts. The firm

27 *Ballabio Ieri – Viaggio nell'archivio storico del Comune … e non solo*, Ballabio, Comune di Ballabio, 2007 (digital format).

28 Archivio Storico del Senato della Repubblica, *Senatori del Regno, Segretariato generale*, f. 710, env. 1282, Umberto Locatelli, *Memoria aggiunta al ricorso all'Alta Corte di Giustizia per le sanzioni contro il fascismo di Roma*, 5 December 1945.

29 The ageing store of Maggio was placed against the mountain's rock, from which the natural air flow originated, used in the cooling of the rooms; the one in Baiedo was a cheaper building which was more distant from the Lecco railway station, but vital due to the degree of the air's coolness; the ice created in specific basins annexed to the stores was used for the transporting of cheese in the railway wagons. See ASCo, *Tribunale di Lecco, Atti e trascrizioni commerciali*, withdrawal, 7 December 1915, No. 3953, Baruffaldi, f. 7, envelop 164; ASCo, *Atti dei Notai*, 16 August 1887, No. 3422, Baruffaldi, f. 7594.

30 Cornalba G., *Il caseificio della Valsassina*, Milano, Tip. Agraria, 1912, pp. 17–19.

31 Archivio Storico del Senato della Repubblica, *Senatori del Regno, Segretariato generale*, f. 710, env. 1282, Umberto Locatelli, *Memoria aggiunta al ricorso all'Alta Corte di Giustizia per le sanzioni contro il fascismo di Roma*, 5 December 1945.

32 Pompeo (1864–?) lived in London at least from 1890 (see ASCo, *Atti dei Notai*, 26 January 1890, No. 3807, Baruffaldi, f. 7595). He married Teresa Locatelli (1869–?) in 1888, and in 1911 they were still living in London with their two sons and their daughters Mistica and Frances. At that time Pompeo was still working as a cheese merchant and, following the family tradition, the two sons were probably working in the family firm: Attilio (1890–?) was a Business Assistant, while Lucindo (1891–?) was working as a Shipping Agency Clerk; see London National Archives, *British Census 1911*, Household Record Number 313.

Mattia Locatelli with the trade credits in Lecco and Ballabio (20,300 lire), and the ageing stores in Baiedo and Cremeno, plus the real estate in Lecco, Germanedo, Ballabio Superiore and Laorca – for a total value of nearly 90,000 lire – passed to Giovanni. Pompeo inherited the British credits and goods (70,000 lire, plus 3,000 for the furniture and 12,000 lire in cash), but could no longer use the name "Mattia Locatelli". Luigi continued on his own, under the firm name "Luigi Locatelli", the cheese ageing business, as he kept a share of the ageing store in Maggio-Cremeno and the old house with storage rooms in Ballabio Inferiore, where he continued to live, marrying his niece Ines (daughter of Giovanni).[33]

Giovanni immediately involved his five sons, Ercole, Umberto, Mario, Mattia and Tranquillo, in the business. In 1897, he moved the headquarters of the firm to Lecco, where he built up the residential house on the site of the old castle and moat of Lecco and of the ancient city's fortified walls.[34] In the same year, Umberto[35] left Italy for Buenos Aires with 50 wheels of Gorgonzola and organized the second foreign branch of the enterprise. In the meantime, starting from the end of the nineteenth century, the Mattia Locatelli storehouse in London was established at Clerkenwell, firmly located in the area of Little Italy.[36] From 1897 onwards there was strong competition among the five brothers to expand international trade and affirm Italian products abroad. The brothers ceaselessly visited all the European markets, and made dozens of trips across the Atlantic, to North and South America. They were successful in creating a solid representation network and branches throughout the world, as far away as Japan, Hong Kong, Shanghai, Sidney, Melbourne, India, South and North Africa.[37] In 1904 Ercole founded the third overseas office in New York, serving for many years as director and finally as president (1931–7) of the local Italian Chamber of Commerce, established there in 1887, which organized the first permanent exhibition of Italian products in New York in 1913.[38] Since the early years of the nineteenth century, he was often involved in

33 Archivio Notarile Distrettuale di Como: Division act, 8 May 1897, No. 18556, Resinelli; Inventory 21/24 July 1904, No. 19790, Resinelli.

34 ASCo, *Tribunale di Lecco, Atti e trascrizioni commerciali*, withdrawal, 7 December 1915, No. 3953, Baruffaldi, f. 7, envelop 164.

35 Umberto would be appointed Senator in 1939.

36 In 1899, a "Mattia Locatelli" "cheese factory" was located at 12 Clerkenwell Green, London; in August 1899 a fire destroyed some of the buildings in the street, and No. 12, 13, 14, 15 and 16 principally suffered, as reported on 16 August 1899 by the *London Daily News* and the *Morning Post*. In 1941 the location of the firm was still the same, see *The London Gazette*, 3 January 1941.

37 Archivio Storico del Senato della Repubblica, *Senatori del Regno, Segretariato generale*, f. 710, env. 1282, Umberto Locatelli, *Memoria aggiunta al ricorso all'Alta Corte di Giustizia per le sanzioni contro il fascismo di Roma*, 5 December 1945.

38 *1887–2012. 125 Italy-America Chamber of Commerce*, N.Y., The Italy-America Chamber of Commerce, 2012, p. 30 and 52.

debates on many international business problems, such as tariffs, food, pharmaceuticals, transportation and legislation.[39]

Together with its commercial expansion, the productive organization of Mattia Locatelli was continuously innovated, modernized and diversified, with the aim of being able to satisfy increasing foreign demand for typical dairy products from Italy. In 1904, when Giovanni died, the firm's value was estimated at around 220,600 lire: the real estate was valued at 123,000 lire, commercial credits at 23,300 lire, mortgage credits at 39,500 lire, and the ageing store in Maggio, holding 36,480 wheels of Gorgonzola at the end of July, at 33,600 lire (Tab. 2). The Italian supply network had a wide geographical distribution; among the firms, we can cite that of the Lombard Davide Galbani (Robiole), Arrigoni Giovanni and Eugenio Cademartori (both renowned Italian producers of soft cheese nowadays called Taleggio), while other enterprises were from Palermo, Bari, Livorno, Milan, Brescia, Alessandria, Trieste, Marseilles and Grenoble.[40]

Table 2. Summary of the assets left by Giovanni Locatelli, 1904

Summary	
Assets	
1. Real estate	123,200
2. Dais at Lecco Theatre	500
3. Furniture, linen, horses, etc. in Lecco	3,465
4. Credits in Mortgages	39,458
5. Valuables and money	5,020
6. Commercial credits	23,324
7. Merchandise, horses, wagons, etc. in Maggio	33,630
Total assets	228,597
Liabilities	-8,015
NET ASSETS	**220,581**

Source: Archivio Notarile Distrettuale di Como, Inventory 21/24 July 1904, No. 19790, Resinelli.

In 1904, the Mattia Locatelli firm passed to the brothers Umberto (1878–1958), Ercole (1881–1947), Mario, Mattia and Tranquillo (the latter three were minors). Initially, Umberto, Ercole and their mother Caterina Zapelli were appointed by the power of attorney, but later, from 1910 onwards, only Umberto (married to Gattinoni Giovanna) and his mother maintained the power, probably because at this point Ercole was permanently living in New York.[41] Mattia came of age in 1906 and starting from 1912 was put in charge of managing the branch in Buenos Aires. But unfortunately the

39 Martin Sampierre, *The Legacy of Antonio Martino*, London, etc., Cornwall, 1988, p. 101.

40 Archivio Notarile Distrettuale di Como, Inventory 21/24 July 1904, No. 19790, Resinelli. financed Eugenio Cademartori from 1885 onwards through a long-term mortgage of 4,000 lire, see the act of 6 July 1885, No. 3148, Baruffaldi in ASCo, *Atti dei Notai.*

41 ASCo, *Camera di Commercio di Lecco*, Commercial Notifications, f. 49/1; *Camera di Commercio di Como*, Notifications, f. 109.

business lasted for a short time, as in 1914 Mattia took his legal part of the family enterprise and in April 1915 suddenly left his home in Argentina and went missing. Nevertheless, in those years Umberto added the ageing of Parmesan to the activities carried out by the enterprise, as it was one of the cheeses most in demand from the market. He acquired the new building of Villa Ospizio, near the railway station, in the north-west of Reggio Emilia (now incorporated into the city), and in 1913 built up new structures and building extensions.[42] It seems that he also extended the network of cheese suppliers in High Piedmont (for Gorgonzola, Sbrinz, Emmental) and Lazio (for Pecorino), creating the most important cornerstones of the enterprise.[43] Before the First World War, Locatelli was already becoming the leading industrial dairy firm in Italy and the USA's largest importer of Italian cheese. At the time, with a final more realistic evaluation than usual, the whole of the firm's assets and liabilities were estimated at 1,623,000 lire, 964,000 of which were held in the merchandise in stock and 159,000 in the firm's branch in Buenos Aires.[44]

4. Luxury cheese and innovative taste: Galbani (1880–1914)

As mentioned above, the best-selling cheeses in Lombardy had long been Grana and Gorgonzola, but other goods were coming into production, imitating French soft cheeses and re-inventing traditional local production. One of the first local firms which aimed to conquer the markets with new branded cheeses was that of Galbani.

Since its beginning in the 1880s,[45] the strategy of Davide Galbani (1821–1903) was to transform traditional dairy products into desirable industrial luxury goods, labeled for their soft and sweet taste, reminiscent of natural and traditional food processing. Galbani's enterprise, hailing from Ballabio Inferiore like Locatelli, was only worth half the value of the Locatelli firm until the First World War.[46] But the firm and its products soon became well-renowned and appreciated. Since its founding, the name of Galbani was closely associated with Robiole, a small kind of particularly fat and

42 ASCo, *Tribunale di Lecco, Atti e trascrizioni commerciali*, withdrawal, 7 December 1915, No. 3953, Baruffaldi, f. 7, envelop 164.

43 Archivio Storico del Senato della Repubblica, *Senatori del Regno, Segretariato generale*, f. 710, env. 1282, Umberto Locatelli, *Memoria aggiunta al ricorso all'Alta Corte di Giustizia per le sanzioni contro il fascismo di Roma*, 5 December 1945. In 1920 he would acquire several plants in these areas.

44 ASCo, *Tribunale di Lecco, Atti e trascrizioni commerciali*, withdrawal, 7 December 1915, No. 3953, Baruffaldi, f. 7, envelop 164 (Assets at 31 December 1914).

45 Colli A., *Egidio Galbani*; ASCo, *Camera di commercio di Lecco*, Registers of Chamber Tax (1884), f. 73.

46 See No. 791.

buttery Stracchino, which Davide collected from a number of producers in Valsassina and then aged in specific stores located nearby in Maggianico, and sold in Lecco.[47] From 1898 onwards the firm was owned by Davide and his sons Egidio and Giuseppe, who joined the business: all three of them together owned the Galbani property and had power of attorney over it.[48] The production and sale of the various dairy products took place in Maggianico (ageing), and was already under way in Melzo, where Galbani exercised a significant part of his trade, and at the end of the nineteenth century he built a dairy factory producing processed cheese and creams in the autumn and winter.[49] The firm was cash-rich and able to buy milk and cheese in advance, paying cash upon delivery and able to sell on credit.

At the end of the nineteenth century, the son and future heir of the firm Egidio Galbani (1858–1950)[50] was well aware that in Switzerland, Holland, Germany and France the production of soft cheese for desserts was rapidly increasing. But in Italy many soft cheeses still did not meet standards of hygiene and were not visually appealing, as they seemed to resemble curd more than aged cheese. Egidio went to Emilia several times to meet a dairy technician, Giuseppe Notari, trying to find a more rational system of production and a more hygienic way of manufacturing table cheese. Thanks to the scientific discoveries of Notari, Galbani began to perfect the cheese, improving the manufacturing and assuring consistent quality throughout the year, managing the ageing process and dealing with marketing, by also presenting his goods as new types of French cheese (e.g. Italian Camembert), imports of which he aimed to replace.[51] Nevertheless, the names of the products were strictly Italian, such as Robiola Lucia and Robiola Italia, or else they had royal names (Margherita cheese). When in 1906 Egidio Galbani created a new kind of cheese, Bel Paese, it featured Italy (the map of the peninsula

---

47 Cornalba G., *Il caseificio della Valsassina*, *op. cit.*; see also Besana C., "I latticini di lusso di Davide Galbani", in *L'Industria del Latte*, a. IV, No. 12, 15 giugno 1906; Cornalba G., *Considerazioni sulla composizione chimica e sull'alto valore alimentare dei formaggi molli Galbani*, Reggio Emilia, 1915.

48 ASCo, *Camera di commercio di Lecco*, Commercial Notification of 9 December 1898, f. 232.

49 *Ibid.* See also Tosi E. 1909, *Manuale pratico del caseificio*, 2nd edn (f. ed. 1904), Casale, Tip. Cassone, 1909, pp. 554–5.

50 Colli A., "Egidio Galbani", *op. cit.* The firm probably passed to the sons Egidio and Giuseppe in 1903, after Davide's death. In 1910 Egidio became the sole owner of the firm Davide Galbani (whose headquarters was still located in Ballabio Inferiore), and had full power of attorney, but his brother Giuseppe preserved an indirect power of attorney, see ASCo, *Camera di commercio di Lecco*, Commercial Notification of 16 February 1910, f. 232. A son of Giuseppe, called Davide, already employed in his father's firm, in 1909 tried to start on his own a new dairy enterprise which he could call with the same name of the most renowned "Davide Galbani". Introini from Milan suggested this adventure to him. Even if he sent cheese to the south of Italy, in Sicily, in Libya, Morocco, etc., the small capital endowment forced him to abandon the business in 1913, ASCo, *Camera di Commercio di Lecco*, Fallimenti, f. 61, env. 288.

51 Tromellini P., *In Memoriam. VI Marzo MCMXXVII. Giuseppe Notari*, Reggio Emilia, 1937, pp. 9–12.

and the face of Abate Stoppani were reproduced on the label), connecting the added value of the new cheese to a wider and more general made-in-Italy brand, and no longer to typical local traditional production (Fig. 1). The cheese enjoyed such solid and enduring domestic and international market success that it became the object of many imitations and forgeries abroad, and represented the main product of the enterprise until the end of the First World War.[52] After winning prizes at the International Exhibition of Brussels (1897), Paris (1900) and Milan (1906),[53] Galbani was now ready to go from manufacturing luxury cheese to larger industrial dairy production.

Figure 1. Popular Italian cheeses labels (1900–1920s)

Source: Museo del Marchio Italiano <http://www.museodelmarchioitaliano.it/marchi.php>.

During the early twentieth century the availability of electric power made it easier to regulate the temperature necessary for dairy production and the ageing processes. In 1911–13, Galbani started building his historical factory in Melzo, in the low plain, where vast local milk production could supply the large-scale dairy industry. Nevertheless, in the first few years the factory's machinery did not seem to be really advanced. The building resembled more a big *casera*, but was nevertheless equipped with modern refrigeration systems which allowed soft cheeses to be continuously produced even in the hot season. The war and the loss of international market were probably replaced by government commissions, and the expansion of Galbani continued. In 1920 Galbani became a limited company (S.p.A.) with 2,000,000 lire of social capital and headquarters in Melzo, with Egidio as President and his brother Giuseppe as Vice-president, together with the Invernizzi brothers (conferring assets) and other shareholders injecting capital.[54] The

52 Mandressi F., *L'industria casearia italiana dall'Unità agli anni Trenta. Il caso Galbani*, degree dissertation, pp. 190–2.

53 Besana C., "I latticini di lusso di Davide Galbani", in *L'industria del Latte*, IV, 15 giugno 1906.

54 Archivio Storico della Camera di Commercio di Milano, *Tribunale di Parma*, 26 May 1920, No. 20014, Drumid, f. A336. For Galbani S.p.A. see Mandressi F., "La nascita del caseificio industriale in Lombardia", in *Annali di storia dell'impresa*, 1999, No. 10,

new company absorbed the Galbani businesses, which were already well-established and this bears witness to the constant growth and expansion of the enterprise's production.[55]

5. Conclusion

The Lombardy dairy sector played a traditional and prominent role in the local economy. During the last three decades of the nineteenth century farming and livestock increased further in the region, and Lombardy confirmed itself as Italy's largest producer and exporter. Merchant-entrepreneurs were able to exploit the new opportunities offered by national unification, the transport revolution, scientific discoveries and technological improvements, together with the growth of international markets mainly due to Italian migration within Europe and to the Americas.

Some of the former small family businesses transformed themselves into larger enterprises and industries, which controlled the purchase of milk and cheese from the farms of the plain, founded new plants, provided the ageing process for cheese, and sold the produce in domestic and international markets. Locatelli and Galbani emerged during this period, despite there being numerous others producers and firms operating in the sector, achieving significant results. In these cases, the local leading role they played acting as merchant-entrepreneurs in the collection, ageing and trade of cheese; their ability to buy milk and cheese in advance, paying cash on delivery and selling the produce on credit; the family structure, in which every member was put in charge of running one of the important sections of the business, which spread over many areas of Italy, plus the foreign offices; their constant attention to innovation in dairy processing systems and the ceaseless development of commercial relations and new markets – all these factors seem to have been particularly important in promoting their growth and in cementing their leading position in the Italian dairy sector, which was to last for many decades ahead.

pp. 586–90. On the 20th-century Galbani see Besana C., Locatelli A.M., "Italian family business and multinationals in the Eighties of the 20th Century: Invernizzi/Kraft and Galbani/Danone", in *Revue Française d'Histoire Economique. The French Economic History Review*, No. 1, 2015, pp. 106–29.

55 Archivio Storico della Camera di Commercio di Milano, *Tribunale di Parma*, 26 May 1920, No. 20014, Drumid, f. A336.

The Rise and Decline of a Large Company: Polenghi Lombardo

Gianpiero Fumi

Università Cattolica del Sacro Cuore, Milan

1. A Protagonist in the Development of a Modern Industrial Dairy (1871–1914)

For more than a century, Polenghi Lombardo played a crucial role in the enlargement of a milk district and in the development of a modern supply chain (see Tab. 1 for the changes in firm name).[1] Since the paradigmatic value of this case concerns the transformations and problems of the whole dairy sector in Italy, the reconstruction will be divided into periods.

---

1 I would like to thank for their collaboration: Archivio storico comunale [Municipal Historical Archives], Lodi; Archivio storico Intesa Sanpaolo [Intesa Sanpaolo Historical Archives], Milan; and Archivio centrale dello Stato [Italian State Central Archives], Rome. I am also grateful to Fondazione Istituto Gramsci, Roma, and Ufficio studi of Mediobanca [Mediobanca Research Department], Milan for allowing me to consult Polenghi Lombardo's budget reports, and to Fulvio Coltorti for his precious support for this research.

Abbreviations:

- ACS, IRI: Archivio centrale dello Stato, Rome, Archivio IRI, II, Archivio storico, Sezione Finanziamenti industriali, Documents for the meetings of the Board of Directors;
- ASCL, PL: Archivio storico comunale, Lodi (Milan), Minutes of the Polenghi Lombardo's Board of Directors;
- ASI-BCI, VCD: Archivio storico Intesa Sanpaolo, Milan, Banca Commerciale Italiana documentary heritage, Minutes of the Managing Committee;
- ASI-IMI, EIB: Archivio storico Intesa Sanpaolo, Rome, Istituto Mobiliare Italiano documentary heritage, Servizio Ispettorato e consulenza tecnica, Eximbank reports;
- *Relazioni e bilancio*, followed by the year: Società di Esportazione Polenghi Lombardo, *Situazione patrimoniale e bilancio delle rendite e spese del...*, or *Relazioni e bilancio...* (Annual balance sheet and report of the directors, printed).

Table 1. Changes in the company name

Year	Name, place, type
1871	Antonio Zazzera and brothers Polenghi, Codogno (general partnership)
1879	Zazzera, Polenghi, Lombardo, Cirio and C., Codogno (limited partnership)
1880	Polenghi, Lombardo, Cirio and C., Codogno (limited partnership)
1889	Polenghi Lombardo and C., Codogno (limited partnership)
1900	Società di Esportazione Polenghi Lombardo, Codogno, later Lodi (shares-based company)
1987	Fedital S.p.A. (Federconsorzi group) → Polenghi Lombardo Division
1993	Polenghi Lombardo S.p.A. (Cragnotti & Partners group)
1994	Cirio-Polenghi-De Rica S.p.A., since 1997 Cirio S.p.A. (Cragnotti & Partners group)
1999	Eurolat S.p.A. (sold to Parmalat S.p.A. in 2000)
2008	Newlat S.p.A. (Angelo Mastrolia bought part of Lodi San Grato plant and recovered the Polenghi Lombardo brand)

Polenghi Lombardo was created in 1871 in Codogno, in the south of Milan, in the middle of a region specializing in milk production, where meadow cultivation, breeding and the cheese trade had been closely connected for centuries[2] and where some important cheese merchants – sometimes real dynasties – already operated.[3] After Italian unification (1861), the development of railway communications, new links with European markets and the free trade policy opened new prospects for expanding the dairy market. Two of the partners (the brothers Pietro and Paolo Polenghi) were local landowners who intended to sell cheeses also in foreign countries.[4] The third partner (Antonio Zazzera) was a small diary entrepreneur who wanted to start a real industry, in the light of the great changes that were taking place in this sector in Europe. From a commercial

2 Romani M., *Un secolo di vita agricola in Lombardia (1861–1961)*, Milan, A. Giuffrè, 1963, pp. 109–10; Saltini, A., "Le campagne lombarde, tessera chiave del mosaico agrario europeo", in *Glocalism. Journal of culture, politics and innovation* (e-journal), 2014, No. 1–2.

3 Besana C., *Tra agricoltura e industria. Il settore caseario nella Lombardia dell'Ottocento*, Milan, Vita e pensiero, 2012.

4 Arioli N., Cattaneo F., "La nascita dell'industria lattiero-casearia nel Lodigiano", in Battilani P., Bigatti G. (eds.), *Oro bianco. Il settore lattiero-caseario in Val Padana tra Otto e Novecento*, Lodi, Giona, 2002, pp. 271–346; Fumi G., *L'economia lodigiana tra Ottocento e Novecento. Percorsi e protagonisti*, Lodi, Camera di commercio – Milan, Metamorfosi, 2009, pp. 64–79, 179–83; Cavazzoli L., "La Polenghi Lombardo e le istituzioni sperimentali e formative del Lodigiano", in *Storia in Lombardia*, 2010, No. 1, pp. 5–40; Tedeschi P., Stranieri S., "L'evoluzione del settore lattiero-caseario lombardo dall'Ottocento al Duemila", in Archetti G., Baronio A. (eds.), *La civiltà del latte. Fonti, simboli e prodotti dal Tardo antico al Novecento*, Brescia, Fondazione Civiltà Bresciana, 2011, pp. 691–757; Dotti, M., "Punte di diamante. Dal fiume di latte emergono alcune produzioni di nicchia", in Cafaro P. (ed.), *Diseguaglianze eccellenti. Ricchezza materiale e immateriale nel Lodigiano trapassato e futuro*, Milan, Angeli, 2014; Besana C., "Polenghi", in *Dizionario biografico degli italiani*, Vol. 84, Rome, Istituto della Enciclopedia italiana, 2015 (<http://www.treccani.it/biografico>).

point of view the company was immediately successful, especially in the export of butter, thanks to the assiduous work on the London market, to the trademarks registered in the UK[5] and to the rewards obtained during national and international expositions.[6] In 1879, new partners injected fresh capital and, in the case of a Piedmontese merchant, Francesco Cirio, also a wide network of contacts. But it was clear that from an industrial point of view the company's strategy was only partially new. As a matter of fact, they used steam churns, but milk was still processed in small diaries scattered around the provinces of Lodi and Cremona. And even if the production was partially diversified – manufacturing also less mature cheeses that allow a quicker return on capital and overcome the predominance of wholesalers – the main aim of the firm was to produce butter.[7] So most of revenues came from butter and to a lesser extent from traditional products – the Grana and the Gorgonzola – while the production of Emmental, an imitation of the Swiss type, remained complementary.

The company started becoming more industrial at the end of the nineteenth century, when a plant to produce butter using selected lactobacilli and Danish machines and technicians was built. At the same time a modern plant opened in Lodi. It was connected to the railway line and employed around a hundred workers. In 1900 the company became a shares-based company (Società di Esportazione Polenghi Lombardo and C.), with new partners, coming also from banks, to finance its ambitious projects. The first years witnessed the contrast between the mainly commercial strategy of the founders, the Polenghi brothers, and the more industrial line of the other administrators.[8] In the following years the company strengthened its production by creating or acquiring other companies, an operation easy enough because of the fragmentation in the dairy sector. Including both new and already-existing factories, Polenghi Lombardo controlled up to five industrial plants, four minor plants and seven small diaries in the provinces

5 In 1884 two Polenghi Lombardo's trademarks of the "Finest Selected Lombardy Butter" were registered.

6 Such as the rewards obtained at the 1867 Universal Exposition in Paris, at the 1881 Italian Industrial Exposition in Milan, and at the 1884 Italian General Exposition in Turin. In 1878, at the time of another Exposition in Paris, the activity of the company was described as follows: "exhibitors produce the parmigiano using their steam system. They manufacture every year from 7,000 to 8,000 cheeses, that are marketed after one to four years of ripening. Exhibitors send every year almost 8,000 parmigiano cheeses, 700,000 kilograms of butter and 10,000 gorgonzola cheeses, annually produced" (*Esposizione universale del 1878 in Parigi. Sezione italiana. Catalogo generale*, Rome, Tip. Barbera, 1878, p. 254).

7 Using the remaining milk for a new product, the condensed milk, that was rewarded in several competitions.

8 Arioli N., Cattaneo F., "La nascita dell'industria lattiero-casearia", *op. cit.*, p. 331. Pietro Polenghi (1852–1932) was head of the Società di Esportazione Polenghi Lombardo for 22 years. His brother Paolo (1859–1925) was Vice-president of the company and head of the Polenghi Brothers in London. In the British capital he was also head of the Italian Chamber of Commerce for 30 years.

of Lodi, Cremona and Piacenza. The raw material came from 600 farmers owning 15,000 heads of cattle.

Given the spread of the centrifugal cream separator and the increase in the milk processed, there was a huge quantity of skimmed milk to re-use. But now, technology allowed several solutions: using it for human feeding as fresh or fermented drinks, transforming it into condensed milk, drying it to obtain powdered milk, producing low-fat cheeses and cheeses with margarine, using it to make bread or to feed animals. The company was one of the first to discern the opportunities offered by technology and by industrial research to re-use remainders of dairy processing through the extraction from the whey of a protein (the casein) and of a type of sugar (lactose). In 1904 the company started producing powdered milk by evaporation using the Hatmaker process, and in 1911 Polenghi bought the exclusive rights on Trutfood patent to condense and transform milk into powder.[9] As for the whey, traditional remainder of dairy processing, most of diaries also activated pig breeding. But to use the enormous masses of whey it would have been necessary to create excessive concentrations of pigs. Hence, Polenghi created a big chemical factory to extract and refine lactose and other by-products for both industrial and food uses. Large-scale production, diversification and flexibility gave to Polenghi a capability superior to the other producers', and allowed the company to conquer new market share.

2. A Thousand Dairy Products (1914–1928)

During the First World War the demand for milk to feed both civilians and soldiers greatly increased. Condensed milk and powdered milk, evaporated and pasteurized, allowed meeting the huge supplies demanded by big cities, under conditions decided with the authorities that managed supplies. Then there were the enormous supplies of cheese to the army, and of sterilized and condensed milk to military and civilian hospitals. No company, however big, could supply the quantities decided in the contracts with the Ministry of War. For example, in 1915 Polenghi and other industries committed themselves to provide the army with 30,000 quintals of cheese per month for four months, also turning to sub-suppliers.[10]

In the post-war period ownership changed. Since it was not a cooperative, the company could not impose on suppliers the strict regulations adopted by cooperatives to assure fresh and clean milk and preserve the quality of products.[11] Nevertheless, the company and the local agricultural world were

9 *Società di Esportazione Polenghi Lombardo, sede in Lodi, Milano (Italia)*, Milan, s.n. [1930?].

10 Fumi G., "Una grande impresa cooperativa nell'industria alimentare: la Latteria Soresinese (1900–1940)", in Battilani P., Bigatti G. (eds.), *Oro bianco, op. cit.*, pp. 410–11.

11 Concerning the increased possibility of regulating the behavior of members in dairy cooperatives see, for example, Henriksen I., Lampe M., Sharp P., *The Role of Technology*

closely linked. Firstly, for a long time Polenghi succeeded in absorbing an increasing quantity of milk locally produced by diversifying its uses. Moreover, the company also encouraged the development of high quality animal husbandry by spreading good breeding and milk collection practices.[12] But Polenghi was linked to the territories even regarding property structure and control of the company. Since its creation, shareholders had included the landlords, farmers and agrarian bodies of Lodi. In 1920 the local agriculture institutions, maybe fearing that the property of Polenghi would change, took control of the company,[13] presumably for some years.[14] Under the new guidance of Angelo Ferrari, Polenghi experienced a decade of great industrial and commercial investments.[15] Between 1918 and 1925 the capital of the company increased from 4 million to 25 million lire, to compensate for the devaluation of the currency and mainly to support the commercial and industrial expansion of the business. This increase in capital allowed Polenghi Lombardo to buy other milk factories in Lombardy and in Emilia-Romagna, to acquire new areas of milk supplies and to eliminate some competitors. These factories would work at full pace, while some minor diaries were closed. These

---

and Institutions for Growth: Danish Creameries in the Late Nineteenth Century, European Historical Economics Society, 2011 (EHES Working papers in Economic history, 1; <www.ehes.org>). In Lombardy the Latteria Soresinese, a big cooperative, established in 1902 some norms for members on the cleaning of stables, and the cleaning of cattle and their feeding with "healthy and high quality" feeds. As regards milk delivery, it was forbidden to mix milk with that of heifers which had recently gave birth to a calf or which were ill. There were also norms on drum and bucket cleaning and their sealing with the producer's name. The people responsible for the Latteria could always inspect the stables (Fumi G., "Una grande impresa cooperativa", *op. cit.*, pp. 389–91).

12 For example, the company printed and distributed the leaflet *L'allevamento razionale dei vitelli: notizie e suggerimenti pratici*, Lodi, Dell'Avo, 1900.

13 One of the promoters of this operation was the general director and future president of the Federazione italiana dei consorzi agrari (Italian Federation of Agrarian Consortia, henceforward Federconsorzi), Emilio Morandi, an engineer who would be head of Polenghi from the early 1930s. See Fontana S., *La Federconsorzi fra Stato liberale e fascismo*, Rome-Bari, Laterza, 1995; Pezzati M., "Un ingegnere al vertice della Federazione dei consorzi agrari: Emilio Morandi", in *Studi piacentini*, 1996, No. 20, pp. 11–30.

14 Control of Polenghi would return to private shareholders in the 1930s. In 1948 Angelo Polenghi, the president, and Angelo Ferrari, the managing director, owned almost 33% of the shares. "If we add to this block the shares owned by Polenghi's and Ferrari's relatives and by a group of friends, we exceed 50% of the capital of the company, assuring to the people already mentioned the control of the company" (ASI-IMI, EIB, folder 126, "Relazione contabile sulla Polenghi Lombardo, Lodi", of Aldo Cicoletti, September 1948).

15 Angelo Ferrari (1889–1964) was Polenghi Lombardo's general director and managing director for almost thirty years. Moreover, he was part of the board of directors as one of the main shareholders. He managed a number of companies controlled or participated in by Polenghi in Italy and big industrial and banking companies of Lodi. He ran the Federazione degli industriali del latte (National Federation of Milk industry) and the Federazione delle industrie agricole e alimentari (National Federation of agricultural and food industry), with 10,000 affiliated companies, and was a member of the Corporative Colonial Committee for Zootechnical Products. Because of these roles, in 1939 he became a member of the Chamber of Fasci and corporations. He was also adviser of Confindustria and Assonime. After the war he ran Polenghi again, up to 1950, when Federconsorzi changed all the administrators.

decisions triggered arguments among the board of directors. Angelo Polenghi in particular, the son of one of the founders and one of the main shareholders, criticized this "improvised expansionism" and resigned.[16]

As a matter of fact the industrial structure and the commercial activity developed impetuously. The increase in the quantity of milked processed – up to 2,400 hectolitres per day in 1927 – was followed by technical improvements of each production site that made the whole company considerably more flexible. The plants reached their maximum for the manufacturing of both Grana[17] and milk to drink. Other milk to produce cheese was found in the Veneto by buying a milk factory in Bressanvido (Vicenza) in 1928. As regards milk to drink, the company also kept supplying Rome, Bologne and La Spezia after the war. Polenghi Lombardo created a company to supply Piacenza with fresh milk and started managing, together with other producers and traders, the Centrale del latte of Milano, which could process more than 2,000 milk hectolitres per day.[18] In the following phase of cheese market stagnation Polenghi tried even more firmly to enter the milk factories of Rome, Genoa and other cities.

Since this continuous increase in the demand for milk from distant areas, in 1922 a big Krause plant was installed in Lodi. This plant, designed in Germany, could process up to 1,000 hectolitres of milk per day to produce powdered milk.[19] In this period the company of Lodi distinguished itself for its great productive versatility, thanks also to the scientific and technical skills acquired by its management, aware of the continuous advances in dairy science and technology. Besides buying patents and having frequent relations with research bodies and institutes,[20] Polenghi also greatly innovated a both processes and products through a dedicated Research Department, as attested by the trademarks registered and the new productions started in this period (Tab. 5 in Appendix). For example, the range of products for infants came to include very pure powdered milk for breastfeeding babies, sour powdered milk for summer feeding, and milk flour for weaning, to correct and integrate artificial feeding, lactobacilli and yogurt.[21]

---

16 ASCL, PL, 3 May 1924.

17 In the first four months of 1926, Polenghi Lombardo's warehouse in Parma would have sold almost 16,000 Grana Reggiano cheeses, that in Codogno 22,000 Grana cheeses.

18 See Granata M., *Municipio e mercato. La Centrale del latte a Milano*, Milan, Skira – Granarolo S.p.A., 2005.

19 At the end of 1924 the plant in Lodi processed more than 900 milk hectolitres per day; almost 700 of them were powdered, partly to be exported to Germany (ASCL, PL, 17 November 1924).

20 See, for example, R. Università degli studi di Milano – Stazione sperimentale di zootecnia, *Studi e ricerche*, Milan, S.a. Stampa periodica italiana, 1938. Polenghi gave an annual contribution to two institutions in Lodi, the Istituto sperimentale lattiero-caseario (Experimental Institute for dairy products) and the Stazione sperimentale per la praticoltura (Experimental Station for fodder crops).

21 Del Bello T., Dossena F. (eds.), *Lodi … "industre, coraggiosa, ricca di passato, ricchissima di avvenire"*, Milan, STEM, 1937, pp. 50 and 52–7.

The problem of remainders was dealt with by developing the production of cheeses with margarine, "artificial" butter, and by-products such as the casein for industry. Next to the already-mentioned factory for lactose, Polenghi created a department to process caseins using patented methods. Since 1923 the company had begun to produce "zoolite", a plastic material that could have multiple uses, especially in the button industry.[22] In this attempt to fully exploit the remainders, before the war Polenghi Lombardo had also started fattening up chickens rationally. After the war an anonymous company was set up; it became the most important Italian exporter of eggs and poultry. Finally, Polenghi Lombardo bought a big cold meat factory that could mechanically process the flesh of animals fed with serum, so completing the milk exploitation cycle (Fig. 1).

But the dairy market, especially that of cheeses, was subject to considerable variations and prices continued to fall. Besides the creation of new products, good sales organization and strong corporate communication were necessary.[23] The presence of Polenghi Lombardo in national and international expositions and in publicity can easily be attested. Its commercial organization ranged from big warehouses for ripening cheeses to agencies and representations in the main Italian and foreign cities (Figs 2–4). For the first time in the dairy sector, the company sold directly to retailers without any intermediation by wholesalers, using its own vehicles.[24] This type of development led to good economic results up until 1928, so that Polenghi Lombardo could pay high dividends.

22 Polenghi Lombardo's trademark "zoolite" was registered in 1925. Another plastic material from casein was erinoid (galalith).

23 Between 1924 and 1927 Polenghi Lombardo registered the trademarks of some "new" soft cheeses using names that echoed the geographical origin of products and several types of powdered milk.

24 Ferrari P., *L'industria del latte*, Piacenza, Camera di commercio, industria, artigianato [1971?], p. 164.

Figure 1. Complete processing cycle of milk at Polenghi Lombardo, with all products and by-products (c. 1930)

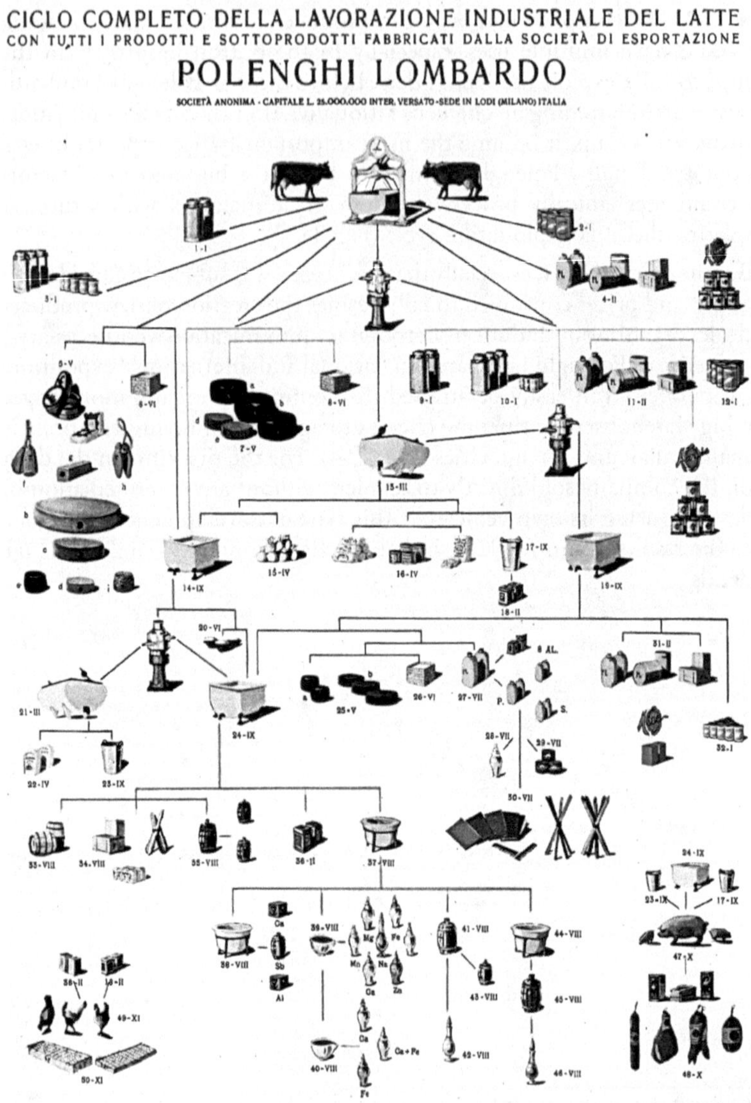

Source: Società di Esportazione Polenghi Lombardo, sede in Lodi, Milano (Italia) …, Milan, s.n., [1930?].

Figure 2. Polenghi Lombardo's production and sales organization (c. 1930)

Source: ibid.

Figures 3 and 4. Polenghi Lombardo's sales organization in European and Mediterranean countries and in the rest of the world (c. 1930)

Source: *ibid.*

3. The Years of Recession (1929–1939)

In a period in which the factories were fully efficient, high losses in 1929 marked the beginning of a long crisis. These losses depended on the reduction of products sold abroad (Tab. 2), without any compensation from the interior market because of the fall of prices that devaluated the huge stocks of cheeses under maturation.

Table 2. Polenghi Lombardo's sales in 1931–1932

	1931			1932		
	Foreign countries	**Italy**	**Total**	**Foreign countries**	**Italy**	**Total**
	Quintals			Quintals		
Cheeses	7,734	42,986	50,720	4,730	56,871	61,601
Powdered milk	2,632	7,679	10,311	802	8,757	9,559
Cold meats	876	9,682	10,558	733	8,297	9,030
Other products	268	151,844	152,112	192	126,185	126,377
	Percentages			Percentages		
Cheeses	15%	85%	100%	8%	92%	100%
Powdered milk	26%	74%	100%	8%	92%	100%
Cold meats	8%	92%	100%	8%	92%	100%
Other products	0%	100%	100%	0%	100%	100%

Source: data processed from ACS, IRI, meeting of 29 December 1933, report on the Società di esportazione Polenghi Lombardo.

It was the first crisis of the profitability of capital since the foundation of the company. Analysts did not consider these losses as worrying because of Polenghi Lombardo's turnover of capital, even if in the 1920s the company had exceeded in activities and in assets. The company had short-term and mainly payable credits with customers and short-term debts (credit facilities from several banks, especially in Lodi and Milan, and loans from the main shareholders, administrators included). In any case, the company had to reduce its capital and costs: employees' pay and milk factory staff were reduced, leases on dairies were not renewed, the amount of milk processed in less efficient milk factories was reduced, and the production of Grana was limited, while the manufacturing of goods that were quicker to make (soft cheeses, powdered milk, milk to drink) was increased and pig breeding was limited to milk factories, intensifying the activity of cold meat factories. Moreover, an inspection service to supervise the activity of agencies was created, some warehouses and agencies were eliminated, Grana Reggiano was sold only in Parma, and Polenghi Brothers in London was closed. To improve the management control system and to reduce costs, a careful study was conducted and, as a result, mechanized general accounting was introduced, together with other changes aimed at simplifying administration. The long recession forced administrators to engage in an enervating attempt to persuade banks to renew the credits, believing too early that Polenghi was overcoming the "very serious crisis" it had suffered.[25] Moreover, some of the banks that had given credit facilities to Polenghi failed.[26] Relations with banks were not easy still in the mid-1930s, as demonstrated by the refusal of some bank managers to again give credit facilities to this big company, although this refusal started to be internally questioned.[27] Fortunately, in 1933 Polenghi obtained from the Istituto per la ricostruzione industriale (IRI), a new State institution for industrial credit, a twenty-year financing of 10 million lire, that allowed it to consolidate its floating debts towards banks.[28]

From an industrial perspective, the increase in milk consumption accentuated the problem of skimmed milk, especially around big cities. In these areas milk consumption varied considerably according to season and to day. For example, in Milan the daily consumption of whole milk fluctuated from a maximum of 2,200–2,300 hectolitres to a minimum of 1,600–700; as a consequence, the organization that supplied the Centrale del latte of Milano could get to manage daily over 600 hectolitres of milk. This milk was skimmed in some collection centers using a centrifuge, the cream was sold to butter-makers, and the skimmed milk was distributed to the members of the consortium. In collaboration with research centers, Polenghi tried to feed cattle and pigs using skimmed and powdered milk.[29] There were also some innovations concerning casein, since a method to transform this substance into a by-product similar to natural wool was discovered. The textile casein, produced using machines made by firms in Lodi, was sold by Snia Viscosa as "Lanital" or synthetic wool.[30]

During the 1930s, the dairy industry tried to overcome commercial difficulties by means of agreements aimed at limiting competition. Also,

---

25 See, for example, the report of the Banca Commerciale Italiana's Managing Committee, meeting of 4 May 1932: ASI-BCI, VCD, Vol. 17.

26 Such as the Banca Commerciale Agricola Piacentina [Agricultural Commercial Bank of Piacenza], whose credit facility had allowed Polenghi to pay its suppliers, that failed in 1932.

27 In 1935–6 the branch in Lodi of Banca Commerciale Italiana criticized the top managers' decision not to give an additional credit facility to Polenghi, their best client in Lodi. After long discussions the additional facility was given (ASI-BCI, VCD, Vol. 28, meeting of 6 December 1935; *ibid.*, Vol. 29, 10 and 17 January and 28 1936). This fact highlights the relevance of Ferrari in the Italian industrial sector and for the Fascist regime, and the intention of the bank's top managers to keep taking decisions autonomously.

28 For documents, see ACS, IRI, meeting of 29 December 1933.

29 Under the guidance of its technical director, an Agriculture graduate, Polenghi invented a flour from powdered skimmed milk, margaric acid and potato flour and called it "vitelia"; this flour underwent trials in the State University of Milan.

30 The process to extract this textile fibre was discovered and patented by the inventor and entrepreneur Antonio Ferretti. A company in Lodi specializing in equipment and plants for the dairy sector (M. Sordi) helped Polenghi to find a method of extracting the textile casein, and provided Snia Viscosa with the machines needed. However, this new way of processing skimmed milk produced a whey more problematic to use for feeding pigs, because of the large quantity of sulphuric acid it contained.

Polenghi joined consortia that entailed producing and selling lactose, powdered milk and zoolite, partially in common. Finally, because of the company's and its CEO's public importance,[31] it could not to be involved in colonialism. Polenghi invested jointly with other companies in the Horn of Africa, especially in the Compagnia etiopica industrie alimentari (Ethiopic company for food industries), aimed at exploiting Ethiopian zootechnical resources to supply that country with milk and to export meat and extracts to Italy.[32] Moreover, in 1939 Ferrari and Polenghi contributed towards creating an Italian-Albanian company to produce Pecorino, with eight dairies in the provinces of Santa Quaranta, Argirocastro and Corizza. Ferrari had already tried to exploit Albanian zootechnics and ovine dairies;[33] one of the reasons for this was that Italian consumers appreciated strong-flavored cheeses, such as pecorino, which Polenghi produced in several dairies in Sardinia, and Gorgonzola. These investments in colonies were lost after the peace treaty.

4. New Ownership (1939–1951)

The Second World War upset markets. As was written in the 1939–40 budget report, the milk industry entered "a stormy and uncertain period, during which the prices of products, raw materials and stocks for consumption varied unpredictably and continuously, and also being supplied with stocks and other ingredients became difficult. Normal exports reduced, became uncertain, and for some years during European hostilities also exchanges with Italian clients became difficult, in terms of both execution and payment". Before the war Polenghi had had more than 150 representatives abroad; during the conflict, relations with most countries were interrupted. As regards the interior market, the Ministry of Corporations controlled prices and industrialists considered these prices as not profitable.[34] But the conflict also offered the company opportunities in terms of foodstuff supplies to the armed forces, milk to drink to Genoa, La Spezia, Florence, Rome and Reggio Emilia, and exports to Germany. Moreover, the company was chosen for the forced accumulation of butter and for the controlled slaughter of pigs. According to "the categorical imperative of autarchy", the plant to produce pure lactic acid was again put into operation in one of the factories, a new

31 See No. 15, *ibid.*

32 The Compagnia etiopica industrie alimentary [Ethiopic Company for alimentary industries], of whom Angelo Ferrari was the president, exploited 1,800 hectares of soil to breed cattle and pigs and to set up dairies and cold meat factories. Moreover, it created some milk centrals near the main cities (Ferrari P., *L'industria del latte*, *op. cit.*, pp. 402–5).

33 In 1920 Angelo Ferrari and Polenghi Lombardo had participated in the creation of a company to drain land and to breed milk livestock in Albania, counting on 38 thousand hectares of surface. The country's political events prevented the company's program from being realized (*ibid.*).

34 *Relazione e bilancio 1939–1940.*

department for extracts and stock cubes from milk proteins and blood was created, and the equipment to produce plastic materials from milk proteins to replace metallic materials in the manufacturing of handles, cutlery, electric devices, etc. was enabled.[35] However, while the war was continuing, work in factories diminished because less milk was produced and this milk was mainly used for direct consumption. All during the war the company realized modest profits, because of the strict price and raw material regulations and the huge increase in operating expenses. These expenses included also money for welfare activities, which the company bore according to the instructions from the regime (canteens, shops for employees, sea camp, recreational organizations and houses for workers, pensions). In 1943, the company had 1,359 employees, more than three-quarters of them blue-collar. Aerial bombings damaged some milk factories and plants, but machines and goods were not always damaged because they had been removed in time.[36]

After the Second World War, the already mentioned Federconsorzi, the major Italian cooperative, took over individual agricultural consortia and assumed control of Polenghi Lombardo.[37] Also, the governing bodies of the company were transformed, ceasing to represent more locally rooted ownership. In fact, interlocking directorates for the years 1952–83 show that only a few of the new Polenghi administrators had a role in other firms of the territory. Polenghi Lombardo became part of a national group of companies specializing in the various branches of agriculture, ruled by Federconsorzi. Even if the conquering (or reconquering) of Polenghi by the agricultural organizations led to protests from the members and to criticisms in newspapers, it is necessary to remember that after the war the company was almost failing. Goods still had to be sold at pre-established prices and stocks of products had to be consigned at fixed prices. Since mid-1947 the prices of cheeses, meat colds and other products had collapsed because of the deflationary monetary policy and goods imported from the USA,[38] but the cost of the milk did not decrease immediately, given the resistance of farmers. Moreover, the company did not enjoy the same tax benefits as cooperatives. To find money and reduce the loan interest burden on Polenghi Lombardo, the capital of the company was increased from 30 to 180 and then to 270 million lire. At the same time, the Board of Directors obtained an important loan from the USA through the Export-Import Bank.[39]

In order to more easily access bank credit some activities were transformed into autonomous companies, in particular Polenghi Laziale, which produced

35 *Relazione e bilancio 1940–1941.*

36 *Relazione e bilancio 1944–1945.*

37 According to a later bank source, Polenghi Lombardo was "in practice an emanation of Federconsorzi" that owned 98% of its shares (ASI-BCI, VCD, Vol. 192, meeting of 3 December 1964). Thanks to Claudio Besana for having indicated this document to me.

38 *Relazione e bilancio 1947–1948.*

39 Papers documenting this operation can be found in ASI-IMI, EIB and in other series of this archive.

Pecorino in around ten dairies in Lazio and Sardinia. The Companhia brasileira de laticínos Polenghi was set up, and thanks to Italian technologies and technicians created a huge factory in San Paolo and several milk factories in the rest of the country.[40] But in 1950–1 results were still negative (1 billion lire losses) due to a year of industrial and commercial crisis in the sector, and despite the important industrial equipment, the high quality products and the diffuse commercial organization. The price of milk remained higher than that of dairy products, which kept decreasing. This reduction had negative repercussions on stocks of long-ripening cheeses. Powdered milk sales were spoilt by huge imports of low-fat powder, at extremely low prices. Exporting cheeses was not competitive given the high costs of the raw material in Italy[41] and the resistance of countries such as France, Great Britain and the USA to liberalize exchanges.

5. Pressure from the European Community (1952–1978)

While it ranked among the first milk producers at international level, with more than 70 million hectolitres of cow's milk in 1951, Italy was the twenty-sixth country in the world as regards per capita consumption of milk and of dairy products. Almost 65% of milk was thereby turned into different types of cheeses, hence the relevance of companies like Polenghi to develop Italian agriculture. Since the beginning of the 1950s new financial resources had enabled the making of new investments to modernize plants and to meet trends in consumption. A new pavilion for milk multiple processing was built, processes previously carried out in peripheral milk factories were concentrated in the main factory, in Lodi, a plant to pasteurize cream, and a new department for melted cheeses and another for homogenized sterilized milk were created. This long-life milk (marketed as "Stella") was a novelty for the Italian market, where it brought about a revolution.[42] After the whole type, a lower-fat milk was invented for consumers who wanted a low-fat regimen. Polenghi Lombardo constantly tried to develop its sales organization, being present in each regional capital and sufficiently important cities through branches, warehouses, agencies, and a partial renewal of the company's vehicle fleet.

Between the 1950s and the 1960s, Polenghi launched other products such as Galatine sweets, shaped as milk bars, and new types of cheeses, soft

40 Ferrari P., *L'industria del latte*, *op. cit.*, pp. 163–4, 404–5.

41 "A comparison between milk prices on their market of origin in December 1950 shows, as a matter of fact, that we had to pay more than 4,300 lire per hectolitres, in Denmark 3,100, in New Zealand 1,800, in Australia 1,700, in Argentina 2,500. As if that were not enough, the governments of these countries had implemented and are studying continuous and appropriate measures to increase exports, also by means of rewards" (*Relazione e bilancio 1950–1951*).

42 *Relazione e bilancio 1952–1953*; *Relazione e bilancio 1954*.

and spreadable. Besides the high quality butter, consumers appreciated mascarpone cream (a dessert) and a new type of long-life cream for domestic consumption and the confectionery industry. Thereby, Polenghi did not lose any segment of the market, from milk to drink to butter, from creams to fresh, soft, semi-hard and hard cheeses, to diet and infant products. For sixteen years Polenghi obtained good financial results.

The situation changed in 1968–9, when the company suffered significant losses (of 797 and 865 million lire) and had to draw on reserves and on its capital, which was subsequently restored to the previous amount and then doubled (from 1.5 to 3 billion lire, afterwards reduced to 2.5). The crisis depended on the interests on mortgages that had been taken out in the previous years, on the cost of short-term credit, and on the cost of labor, which had increased considerably since 1969 because of contract renewals. Another negative factor was the reduced profitability of Polenghi butter, a high prestige product marketed as "Optimus". On the one hand, since 1968 the European Community policy had encouraged butter exports from over-supplied countries to countries that lacked it, such as Italy. On the other, the company suffered the consequences of the high price of milk as raw material for dairy products due to the creation of consortia of drinking milk producers; the creation of these consortia had been encouraged by national and European contributions. In Italy, consumption of fresh and long-life liquid milk was considerably lower than in other European countries, although on the increase.[43] Because of this trend, the quantity of milk for manufacturing was decreasing, leading to serious consequences for cheese production. Since 1968 Polenghi had limited dairy production to some of its factories (Lodi, Bressanvido and Bologna, the latter only for a few years), and considered it not profitable to keep manufacturing in Sardinia.[44] The company continued to increase drinking milk, but in Italy less than half of its production was hygienically treated. Polenghi, one of the first companies to use modern refrigerated stainless steel equipment to collect and transport milk, had to face increasing demand for sterilized and pasteurized milk. Besides the traditional glass bottle, in 1969 Polenghi was the first company in Italy to use Brik-Pak containers for UHT milk, as already experimented with abroad. Milk was sterilized through highly productive automatic plants that could operate for many hours a day. Packaging was carried out by means of machines that permitted aseptical operation and the introduction of new sizes and shapes, such as the 1/5 liter tetrahedron for cream and new milk and cacao beverages.

---

43 In 1970 the average milk consumption per capita in Italy was 64 kg, in France 108 kg and in Germany, Denmark and Holland 125–35 kg.

44 *Relazioni e bilancio 1968*. For the competition in fresh milk market between Polenghi and cooperative enterprises in Bologna see Battilani P., Bertagnoni G., *L'utilizzo del social network nel marketing: le imprese cooperative negli anni Sessanta*, in Società italiana degli storici economici, *Innovare nella storia economica: temi, metodi, fonti*, Prato, Fondazione Istituto Internazionale di Storia Economica "F. Datini", 2016, pp. 166 ff. (<www.istitutodatini.it/collane/htm/varie9.htm>).

The situation was far more delicate for Grana and other long-ripening cheeses, whose production had extended outside their traditional area, southern Lombardy and Emilia. The Grana over-production crisis seemed to stop temporarily because the AIMA, the Italian State agency for interventions in the agricultural market, bought large quantities of this product. But in the following years the national Grana market again suffered over-production, and the increase in consumption and the creation of the first consortia to protect typical cheeses (Parmigiano-Reggiano and Grana Padano) could not contain this trend.[45] In the baby food sector, Polenghi had good industrial and qualitative results, but there was very strong competition from multinational companies. The range of these products, the fruit of Polenghi's studies and research, was enlarged, adding a new type of powdered milk, followed by the same milk in liquid form, sterilized and homogenized, ready to be used, besides new dietetic products.

Despite positive trends in consumption, there were tensions in the markets for table cheeses (Gorgonzola, Provolone, Stracchini, Italici, etc.), soft and spreadable cheeses, spun-paste cheeses (Mozzarella and other similar products) and Pecorino cheeses. In this climate, the dairy organization had to undertake hard tasks. Since the end of the 1960s the Italian commercial interchange had reached extremely low levels of passivity, and the zootechnical sector exhibited the same trend. Increasing competition inside the EEC depended also on an European policy aimed at developing production, keeping prices high, through intervention and defense mechanisms. This policy favored countries with lower milk and cheese production costs.[46] In Italy, fresh, concentrated and powdered milk imports boomed (for example, between 1969 and 1972 imported liquid milk quintupled). Also, the pressure from foreign cheeses and butter was increasing. Conversely, Italian exports remained modest and steady. This trend depended on high production costs, on delays in returning import duties, and on health requirements from the USA and Canada, two of the most important markets for Italian dairy products because of their large emigrant communities.[47]

6. New Investments, New Owners, Decline (1979–1992)

Polenghi Lombardo decided to use imported milk for a part of its production, but since 1979 the company had chosen to buy milk for its two Italian factories only from national suppliers. The plant in the Veneto was

45 Ferrari P., *L'industria del latte*, *op. cit.*, pp. 280–1.

46 According to Polenghi Lombardo's administrators, the price of industrial milk (in Italian lire) in the second half of 1979 was: Lombardy 326; Denmark 263; Germany 233; Holland 230; France 229; Belgium 221; UK 199; Ireland 185 (*Relazioni e bilancio 1979*).

47 In foreign countries companies were starting to produce imitations of typical Italian cheeses, at significantly lower prices (*Relazioni e bilancio 1972*).

completely renovated and developed, also thanks to EEC and Veneto region funds, with from 350 to 1,000 milk hectolitres processed per day. This factory, intended exclusively to process milk for dairy products, manufactured new types of soft cheeses and dairy specialities, besides melted pasteurized cheeses and milk sweets (Galatine), which were produced in a highly mechanized production department. These sweets, which could have different flavors, were greatly appreciated by consumers as concentrated food and attracted interest also from foreign groups (from the Soviet Union, Yugoslavia, China and Canada), which wanted to reach co-production agreements or to exchange know-how.[48] After a short number of years Polenghi butter production returned to growth, also thanks to the EEC's "Christmas butter" operations.[49] The brand-new plant near Lodi (called Lodi 2) opened at the end of 1979. This factory, which had been built using EEC funds and by taking out a twenty-year mortgage, was intended for milk to drink, creams and mascarpone. New products and new packages fulfilled public programs to increase milk consumption.[50] Drinking milk and cheeses accounted for, respectively, 36 and 44% of overall company revenue (Tab. 3).

Table 3. Polenghi Lombardo's total net turnover in 1982

Division	Billion lire	%
Milk to drink	70.2	35.9%
Butter, cream, condensed milk	21.8	11.1%
Cheeses	87.1	44.5%
Cold meats	10.4	5.3%
Dietetic products and Galatine	5.6	2.9%
Others	0.6	0.3%
Total	*195.7*	*100.0%*

Source: *Relazioni e bilancio 1982.*

Despite a considerable increase in national milk production (almost 100 million hectolitres of cow's milk in 1982), there was considerable pressure from milks and cheeses from other European countries. Imports, in terms of liquid milk, exceeded 60 million quintals in 1982, for more than 2,200 billion lire.[51] Besides these general difficulties experienced by the overall sector, there were some internal problems. When Polenghi exceeded the historic level of 150 million liters of milk processed, increasing both sales and revenue, its financial situation again worsened. The increase in industrial and commercial activity entailed higher expenses, but the

48 *Relazioni e bilancio 1982.*

49 Given the butter market situation inside the EEC, characterized by high stocks and costs for storing, the European Commission financed butter sales at reduced price for direct consumption during the Christmas period. The "Christmas butter" operation was repeated for several years.

50 *Relazioni e bilancio 1981.*

51 *Relazioni e bilancio 1982.*

company's resources were limited and, as a consequence, Polenghi had to turn to banks; the heavy bank charges would add to depreciations in the currency and to losses for the previous financial years. The situation became highly critical in 1981–2, with losses of almost 8 and 5 billion lire.[52] Results were negative also in the following years, despite capital increases operated by Federconsorzi, the increase in revenue, and the decisions of management. Besides cost reductions, there were considerable investments in manufacturing, marketing and organization of the company. The entire production, including cheese manufacturing, was centered in the new factory, focusing mainly on soft and "invented" cheeses that generated the highest profit margins.[53] Among new products was the long-life "Mascherpone di Lodi" in heat-sealed trays, intended for supermarkets, canteens and catering. The other largest production was of long-life and fresh pasteurized milk. The company kept manufacturing dietetic products and the whole baby food line was renewed. A new technically advanced creamery for butter was set up. This creamery had a department for the production of anhydrous butter, a particular type of butter for professional consumers, that is, confectionery industries, confectioneries, ice cream shops and bakery product industries.[54] The factory in Lodi San Grato became a big complex made up of four technically advanced units[55] characterized by high productive capacity, high productivity and flexible plants.

In the same period, the marketing approach was completely revolutionized. Given the changes in consumption, greater attention was paid to packaging, and to setting up a plant for products already packed and divided into portions. Moreover, the packaging of all the products was revised as part of the corporate communication. At the same time, a territorial expansion strategy was adopted. Polenghi Laziale became Polenghi Sud to facilitate its presence in Southern Italian regions.[56] Polenghi UK was created in Great Britain, the first foreign country the company had penetrated at its origin. Other branches were set up in Spain, Germany and Brazil, even

52 *Relazioni e bilancio 1981*; *Relazioni e bilancio 1982*.

53 Milk that was processed by Polenghi, considering both its own factories and third-party plants, exceeded 196 million liters, collected only from Po Valley farms. Almost 100 million of these liters were processed to make cheeses, with considerable increases in the production of Gorgonzola, Provolone, Mozzarella, Ricotta and, sometimes, Grana Padano.

54 *Relazioni e bilancio 1983*.

55 The new factory consisted of "Lodi 1" and "Lodi 3" for fresh and semi-fresh spun pastes, Mozzarella and Ricotta, "Lodi 2" for milk to drink, and "Lodi 4" for Gorgonzola and soft cheeses.

56 In 1986 the subsidiary Polenghi Laziale became Polenghi Sud, and its seat as a legal entity was transferred to Naples. The subsidiary had a factory in the province of Chieti for spun pastes and Mozzarella. The company had decided to make further investments in the south of Italy (*Relazioni e bilancio 1985*).

if it is necessary to consider whether they actually operated.[57] So far as the organization is concerned, in 1985, after a thorough diagnosis on the part of McKinsey, a complex reorganization of all the departments took place: reassignment of responsibilities, and adoption of management by objectives and of more modern accounting and information systems.

Nevertheless, Polenghi Lombardo returned again considerably in the red. In 1988, the historic Lombard company became a branch of Fedital, the financial holding for the food sector created by Federconsorzi. Federconsorzi committed itself to covering the losses and to underwriting a capital increase. Moreover, a recovery plan through disinvestments was launched, leading it to close and sell the factory in the Veneto, to centralize the whole production in Lodi, to close some warehouses, and to sell other business units. Polenghi Lombardo Division drastically reduced its hard cheese market share.

Over the previous years, several concentration operations had taken place in the Italian dairy industry; of these, the IFIL-Danone-Galbani operation had been the most striking.[58] In the 1990s Polenghi Lombardo went through a series of sales and business combinations hard to justify in terms of industrial alliances and strategies. First of all, after the failure of Federconsorzi, Fedital debts could not be covered any more. In the last months of 1992 Fedital was sold to Cragnotti & Partners group for 30 billion lire, a price very much lower than the starting price, 106 billion lire.[59] Controlled by a finance company based in Holland, the Cragnotti group penetrated the Italian market by acquiring the Lombard company, that again took the name Polenghi Lombardo S.p.A., and other companies. At that time Polenghi ranked still sixth in the dairy sector in terms of revenues and eighth as to number of employees (470 units).[60] Thanks to a new recovery plan the budget returned active after years of negative results. The revenue was constantly decreasing, but the new ownership seemed to have ambitious projects and merged Polenghi with Cirio-Bertolli-De Rica S.p.A. (which grouped together three recently privatized companies of the public group Iri). The merger led to the creation of an Italian food giant called Cirio-Polenghi-De Rica S.p.A., then Cirio S.p.A., which produced and marketed tinned tomatoes, wine, milk and foodstuffs in general. The group was a leader in the dairy sector, including Polenghi Lombardo, the Venetian Ala, the milk

57 The foreign subsidiaries were: Polenghi Iberica S.A., Spain; Polenghi UK Ltd, Great Britain; Polenghi Lombardo do Brasil, San Paolo. In 1989 all these firms were still inactive or running at a loss.

58 See Besana C., Locatelli A.M., "Italian family business and multinationals in the Eighties of the 20th century: Invernizzi/Kraft and Galbani/Danone", in *Revue française d'histoire économique*, 2015, No. 1, pp. 116–29.

59 See Commissione parlamentare di inchiesta sul dissesto della Federazione italiana dei consorzi agrari [Italian Parliamentary Committee for investigating the collapse of the Federconsorzi], *Relazione finale (relatore: senatore Melchiorre Cirami), approvata nella seduta del 28 febbraio 2001*, in Senato della Repubblica – Camera dei deputati, *Atti parlamentari*, XIII legislature, doc. XXIII, No. 61.

60 *Annuario dell'industria alimentare in Italia*, 1994, pp. 91–2.

factories of Rome, Ancona and Naples, and other companies. The group was also the first national operator in the fresh milk sector, with a 19% market share.[61]

Cirio dairy unit was hived off in 1999 as Eurolat and after the failure of Cirio it was sold to Parmalat, a group that was considerably expanding in Italy and in foreign markets.[62] The sale to Parmalat had important effects on the production structure. The manufacturing of UHT milk was transferred to the factory near Parma and the production of butter to Reggio Emilia, while the factory in Lodi maintained only the manufacturing of milk in glass bottles and of all the other dairy products, with its staff reduced to 280 employees. A company that had permitted the Italian dairy sector to be successful at the European level continued its agony.

Conclusions

During the late nineteenth and twentieth centuries a modern supply chain developed in Northern Italy, made up of different phases – milk production, industrial processing, storing and ripening of cheeses, by-product processing, national and international distribution. Moreover, specialized producers of machines and chemical products for the supply chain appeared. Polenghi Lombardo has been a major player in this evolution of the dairy industry for more than a century. Although it was not a cooperative, the company was closely linked to agricultural interests and organizations, at a local and then at a national level. The enormous quantity of milk it worked and the invention of new products and uses of this raw material stimulated the growth of animal husbandry and intensive agriculture in several Italian regions. An open-minded property and management operated large technological and commercial investments, transforming Polenghi Lombardo into a big corporation, with a multi-plant and highly diversified asset and commercial branches extended all over the world. The company also started industrial activities abroad. But this story also has a paradigmatic value with respect to the difficulties faced by the dairy industry in Italy. Despite new investments in productive and marketing innovation, following the 1960s it suffered stronger international competition and the negative impact of the Common Agricultural Policy. These issues led to the merger of Polenghi Lombardo with other Italian groups, whose strategies largely erased its heritage.

---

61 These pieces of information were inferred from the resolution of the Autorità Garante della concorrenza e del mercato [Italian Competition Authority], No. 2222 (C1571), <www.agcm.it>.

62 For Eurolat see Coppe R. (ed.), *Guida alla cultura, al turismo e all'economia della provincia di Lodi*, Fidenza, Mattioli 1885, 2001, pp. 246–51. On these events see Felice E., "Il settore lattiero-caseario in Italia dal dopoguerra al Duemila", in Bertagnoni G. (ed.), *Una storia di qualità. Il Gruppo Granarolo fra valori etici e logiche di mercato*, Bologne, il Mulino, 2004, pp. 77–80.

Table 4. Polenghi Lombardo position between the top 10 dairy companies in Italy by share capital and active asset (1913–1983) (thousand lire)

	Company	Year of foundation	Town	Share capital	Active asset
	1913				
1	**Società di esportazione Polenghi Lombardo**	1900	Codogno	3,000	9,457
2	Società romana per il formaggio pecorino	1907	Rome	3,000	7,125
3	Forgrana	1906	Reggio Emilia	1,200	2,936
4	Latteria di Locate Triulzi	1901	Milan	600	2,010
	1921				
1	**Società di esportazione Polenghi Lombardo**	1900	Lodi	10,000	25,458
2	Società romana per il formaggio pecorino	1907	Rome	8,000	13,102
3	Industria del latte	1921	Milan	1,000	9,726
4	Latteria di Locate Triulzi	1901	Locate	3,500	7,115
5	Gianelli Maino	1911	Milan	2,250	5,059
6	S.I.L.P.A.	1918	Monza	4,000	4,964
7	Latteria moderna	1923	Turin	2,000	4,833
8	F.lli Corsi	1919	Milan	2,500	3,488
9	Egidio Galbani	1920	Melzo	2,000	2,564
10	Luigi Invernizzi fu Rocco	1921	Novara	1,500	1,874
	1927				
1	**Società di esportazione Polenghi Lombardo**	1900	Lodi	25,000	58,798
2	Latte condensato lombardo	1901	Milan	12,500	28,458
3	Egidio Galbani	1920	Milan	12,000	19,817
4	Società romana per il formaggio pecorino	1907	Rome	8,000	19,661
5	Società agricola industriale salernitana	1918	Salerno	5,100	13,307
6	Lang & Schutz	1925	Turin	10,000	11,707
7	Industria del latte	1921	Milan	3,100	9,726
8	Latteria di Casalbuttano	1917	Casabuttano	482	8,744
9	F.lli Corsi	1919	Milan	2,500	7,578
10	Angelo Arrigoni caseifici e cremerie	1924	Milan	2,000	5,226
	1936				
1	**Società di esportazione Polenghi Lombardo**	1900	Lodi	15,000	51,000
2	Egidio Galbani	1920	Melzo	18,000	37,880
3	Nestlé	1924	Novara	10,000	23,527
4	Consorzio laziale produttori latte	1917	Rome	600	22,896
5	Generale centrali latte	1929	Milan	12,000	17,726
6	Latteria Soresinese	1900	Soresina	2,317	17,139

	Company	Year of foundation	Town	Share capital	Active asset
7	Società agricola industriale meridionale	1918	Salerno	10,000	17,064
8	Società romana per il formaggio pecorino	1907	Rome	8,000	15,370
9	Centrale del latte della Grande Genova	1932	Genoa	945	12,666
10	Emilio Bianchi Società d'esportazione	1928	Milan	6,350	11,442
	1952				
1	**Società di esportazione Polenghi Lombardo**	1900	Lodi	45,000	6,193,161
2	Egidio Galbani	1920	Melzo	468,000	4,097,610
3	Società agricola industriale meridionale	1918	Rome	480,000	3,417,011
4	Nestlé	1924	Milan	341,000	1,824,934
5	Centrale del latte di Torino	1950	Turin	375,000	1,366,266
6	La Prealpina	1947	Milan	330,000	1,202,354
7	Angelo Arrigoni caseifici e cremerie	1921	Milan	30,000	1,091,477
8	LIR Latterie Industriali Riunite	1935	Robbio	4,000	1,003,745
9	Centrale del latte della Grande Genova	1932	Genoa	65,000	891,591
10	Società romana per il formaggio pecorino	1907	Rome	60,000	764,202
	1960				
1	Egidio Galbani	1920	Melzo	1,500,000	17,338,756
2	**Società di esportazione Polenghi Lombardo**	1900	Lodi	1,485,000	12,390,539
3	Locatelli (previously LIR Latterie Industriali Riunite)	1935	Milan	1,000,000	8,293,885
4	Nestlé	1924	Milan	558,000	4,499,597
5	La Prealpina	1947	Milan	720,000	3,091,207
6	Centrale del latte di Torino	1950	Turin	375,000	2,560,031
7	Invernizzi	1953	Melzo	300,000	2,169,162
8	Consorzio laziale latte e industrie agricole	1917	Rome	80,000	2,017,565
9	Società agricola industriale meridionale	1918	Pontecagnano	253,000	1,715,916
10	Società romana industria e commercio prodotti dell'agricoltura	1907	Rome	60,000	1,654,168
	1972 (1971 for asset)				
1	Egidio Galbani	1920	Melzo	4,500,000	51,862,063
2	Locatelli	1935	Milan	4,000,000	22,095,407
3	Alimont (previously Pavesi Biscottini di Novara)	1953	Milan	2,221,000	21,170,210

	Company	Year of foundation	Town	Share capital	Active asset
4	**Società di esportazione Polenghi Lombardo**	1900	Lodi	2,599,000	19,667,326
5	Nestlé	1924	Milan	1,200,000	13,479,664
6	Invernizzi	1953	Melzo	1,500,000	9,239,693
7	Tanara (afterwards Italgel)	1960	Parma	342,000	8,527,363
8	Centrale del latte di Torino	1950	Turin	375,000	6,277,004
9	Kraft	1964	Milan	475,000	5,939,581
10	Prealpi industria casearia	1968	Varese	1,000,000	4,287,024
	1983 (1982 for asset)				
1	Egidio Galbani	1920	Melzo	27,000,000	507,650,408
2	Alivar (previously Pavesi Biscottini di Novara, afterwards Alimont)	1953	Novara	30,000,000	280,342,704
3	Parmalat	1973	Collecchio	20,000,000	227,819,337
4	Italgel (previously Tanara)	1960	Parma	30,000,000	199,536,118
5	**Società di esportazione Polenghi Lombardo**	1900	Lodi	32,874,188	166,284,367
6	Nestlé italiana	1924	Milan	7,426,100	126,559,426
7	Locatelli	1935	Milan	4,000,000	107,602,548
8	Invernizzi	1953	Melzo	8,000,000	100,072,808
9	Approvvigionamento Latte Alimentare ALA	1950	Fossalta di Portogruaro	4,200,000	69,015,493
10	Kraft	1964	Milan	4,000,000	62,142,308

Source: data processed from IMITA.db (<http://imitadb.unisi.it>). This site recollects data from the yearbook *Società italiane per azioni*, edited from 1908 by Credito Italiano and Associazione fra le società italiane per azioni (Assonime). The source includes all the listed Italian joint stock companies and those with a specific threshold (1 million lire up to 1940, but ½ million in 1913; 10 million in 1953, 25 million in 1956, 50 million in 1961, 100 million since 1964).

Table 5. Trademarks patented by the Polenghi Lombardo (1900–1965)

Period 1900–1944

Butter, milk cream	Cheese	Milk, powdered milk, milk drinks
Burro extra centrifugato (1911)	Fior d'alpe (1922)	Mirum (1924)
Margo-Palma burro artificiale (1919)	Cremino Fanfulla (1924)	Verolattosiero (1933)
Stella burro artificiale (1919)	Laudensis (1926)	Vitelia [per il bestiame] (1933)
Burro extra di crema pastorizzata (1927)	Valdarda (1926)	Mirella (1937)
Burro di panna dolce (1935)	Morlacco (1934)	Lattalfrutto (1938)
Burro Optimus (1935)	Padus (1934)	Domus (1939)
Burro naturale di Milano (1943)	Primavera (1934)	Corrida (1941)
Concordia burro di crema pastorizzata (1943)	Fior di Sole (1935)	**Extracts**
Stella burro artificiale (1943)	Bel Fiore (1936)	Fosfolactol (1910)
	Croce del Sud (1936)	Zoolite (1934)
	Formaggio pecorino Sardinia (1936)	Bonbrod (1942)
	Nuovofiore (1936)	**Dietetics and baby foods, other products**
	Alpina (1937)	Miranda (1924)
	Silvana (1939)	Mire (1924)
	Silvano (1939)	
	Do-Re-Mi (1941)	
	Mozzarella Carrozzella (1941)	
	Formaggio Amoretto (1943)	
	Formaggio Unio (1943)	
	Primavera (1943)	
	Stracchino di Milano doppia panna (1943)	

Period 1945–1965

Butter, milk cream	Cheese	Milk, powdered milk, milk drinks
Burro finissimo di cremeria (1946)	Astro (1948)	Caola (1953)
	Valledoro (1948)	Caolatte (1953)
Galacrema (1958)	Blumenthal (1949)	Caomil (1953)
Burro finissimo da tavola (1959)	Caciotta Pratolina (1952)	Coc-Lat (1953)
	Caciotta Vallebella (1952)	Milcao (1953)
Conalga (1960)	Fior d'Amore (1952)	Spumetto (1953)
Burro Val di Savena (1961)	Fior di Campo (1952)	Stella (1953)
	Fior di Ginestra (1952)	Mir (1954)
	Formaggino del Quadretto (1952)	Stabilatte (1954)
	Formaggino Folletto (1952)	Stelatte (1954)
	Susi (1952)	Ster (1954)
	Vitto (1952)	Stop (1954)
	Italverde (1953)	Stelat (1956)
	Selva (1955)	Stemag (1959)
	Cremola (1956)	**Extracts**
	Giunchiglia (1956)	Bos (1950)
	Robioletta di Lodi (1956)	Gala (1950)
	Creme Berg (1957)	Laus (1950)
	Due Torri (1957)	**Dietetics and baby foods, other products**
	Ekorn (1957)	
	Viking (1957)	Sierovit (1950)
	Bonus (1958)	Crema Jovo (1952)
	Pol (1958)	Galatine (1956)
	Silvapanna (1958)	Miralat (1957)
	Bel Prato (1959)	Miravit (1957)
	Parmes (1959)	
	Babyzola (1960)	
	Cremalpe (1960)	
	Emmental Italia (1960)	
	Granetto (1960)	
	Scugnizzo (1960)	
	Tamburino (1960)	
	Topolino (1960)	
	Creme Blu (1962)	
	Gorgocreme (1965)	
	Polenghi Laziale:	
	Monte Mario Brand (1955)	
	Appia (1955)	
	Sardinia Brand [pecorino romano] (1955)	
	Pecorino Romano Cheese Tirso (1955)	

Source: data from the dataset *Marchi di fabbrica* (<http://dati.acs.beniculturali.it/mm/local>) concerning the documentation collected in Archivio centrale dello Stato, Rome, Ufficio italiano brevetti e marchi, Archivio generale, Modelli 1874–1964. The year indicates the date of the product registration by the company.

The Multinationals and the Italian Agri-food Industries

The Dairy Sector (1974–1993)

Andrea M. Locatelli

Università Cattolica del Sacro Cuore, Milan

1. Introduction: notes on the evolution of the food industry and dairy sector in the 1970s and 1980s

During the 1970s, productivity and profits within the food manufacturing industry increased significantly after the implementation of wide-ranging structural changes within the context of a period of transformation characterized by the fact that production was being overtaking by the consumer base as the driving force within the industry.[1] The increasing size of companies and the appearance of foreign competitors were just two of the elements in this new landscape, which was also characterized by an increase in the number of producers specializing in local and traditional products. A host of new phenomena were materializing within the manufacturing industry as it began to develop strategies involving structural concentration on one hand and dispersion on the other, while in certain areas products were gaining a strong identification with their place of origin and some producers were developing "flexible specialization" techniques.[2] A growing level of investment flooded into the sector throughout the 1980s, allowing it to achieve a higher level of restructuring and modernization than was possible for other industries. At the same time, the food industry's manufacturing capacity was growing while the opposite was happening in other sectors, where production methods were undergoing a period of rationalization.

---

1 Chiapparino F., Covino R., *Consumi e industria alimentare in Italia dall'Unità ad oggi*, ed. Giada, 2002; Nomisma, *Politica agraria e sistema agro-alimentare: scopi ed impieghi degli strumenti di intervento nell'agricoltura italiana. Rapporto 1993 sull'agricoltura italiana*, Bologna, il Mulino, 1993, pp. 4–14; on the relationship between industrial models and the role of businessmen, see Amatori F., Brioschi F., "Le grandi imprese private: famiglie e coalizioni", in F. Barca, *Storia del capitalismo italiano dal dopoguerra a oggi*, Roma, Donzelli, 1990, pp. 118–53.

2 In the agriculture-industry the value added per employee was 24.7 million while the whole sector had 19.5. See also Sicca, L., *Lo straniero nel piatto: internazionalizzazione o colonizzazione del sistema alimentare italiano*, Milano, 2003, pp. 15–24.

The food sector was the third largest industry industrial sector in the country behind machinery and textile production, and it was among manufacturers of niche products that the highest levels of concentration appeared.[3] Although the food industry grew faster than the agricultural and other manufacturing sectors, a fall in sale prices meant that it counted for a smaller proportion of overall profits, although the milk and dairy sector remained stronger than others. Employment figures within the industry fell slightly from 8.3% in 1970 to 7.7% in 1979, with Italy faring badly in the context of overall European food industry employment figures, representing a lower than average percentage of the national workforce due to its underdeveloped processed foods market and the fact that it imported more food than it exported.

By the end of the 1970s, the milk and dairy sector represented 13% of the food industry's 18 billion lire annual turnover and 9.1% of its 3,501 billion lire gross product, and employed 10.1% of its 206,954-strong workforce.[4] The preserved foods sector accounted for the largest relative portion of the industry, while investment return figures showed the dairy industry placed midway between sugar in the top spot and milk with the lowest returns. Between 1970 and 1979 the milk and dairy industry achieved an average yearly growth of 3.2%, with the category known as "other cheeses", at +9.0%, far stronger than soft cheeses and Grana, while the performance of butter and melted cheeses was well below average. A detailed analysis of individual products shows that 1974 – generally seen as a turning point in the economy as a whole – was a transformative period for consumer tastes in the cheese sector: sales of some products such as butter and hard cheeses fell between 1970 and 1974 before rising again over the following five years. Soft cheese sales followed precisely the opposite trend.[5] Distributor profit margins grew in reaction to slowly rising agricultural and industrial production costs, while dairy product sale prices reflected those in the wider economy.[6] The sector's largest and most prosperous companies began to make more use of

---

3 Food industry statistics took account of all activities involved in the processing of agricultural products, together with the category of "food and related industries". Throughout the 1970s food industry production volumes underwent a slow but steady growth, reaching a value added/sales ratio of 17.7%. The number of people employed in the sector was estimated to be around 432,800, of which 47.1% worked for businesses with at least twenty employees. The food industry accounted for 8.9% of the overall profit value of the processing industry, 7.7% in terms of occupation, and received 6.9% of gross investment. Bertelé U., *L'industria alimentare, fattori di cambiamento e necessità di innovazione*, Milano, Federalimentare 1989, pp. 31–2; *The European Dairy Industry: Consumption Changes, Vertical Relations and Firms Strategies*, eds. D. Rama, R. Peri, Milano, Angeli, 1995, pp. 114–24.

4 Sicca L., *L'industria alimentare in Italia*, Bologna, il Mulino, 1977, pp. 74–8.

5 Istat statistics comparing dairy product wholesale and consumer prices show that there was little change in these figures between 1959 and 1979. Istat, *3 Censimento generale dell'agricoltura*, Roma 1983; Istat, *Relazione generale. 6 Censimento generale dell'industria, del commercio e dell'artigianato*, Roma, 1987.

6 Profit margins in this context had been higher in the 1960s than in the following decade.

raw materials sourced from other European countries, leading agricultural products to build up a growing deficit on the export market throughout the 1970s.[7]

The production of milk and its derivatives has a relevant value for the rural sector. According to surveys conducted by the economic research institute Nomisma, almost three-quarters of Italy's cheese production in the 1980s was represented by five cheeses: Grana, Parmesan, Gorgonzola, Pecorino and Provolone. Apart from the more dispersive nature of Parmesan production, the manufacturers of these cheeses operated medium sized factories in an industry dominated by four large groups: Galbani, Locatelli-Nestlé, Invernizzi and Polenghi Lombardo.[8] During this period the dairy sector seemed to be moving in the opposite direction to that of industry as a whole: large businesses were able to guarantee a higher level of quality than smaller companies, despite the fact that a refinement in consumer tastes had forced them to offer a wider range of niche products. A "destructuralization" process at the manufacturing level began with small and medium sized businesses, whose commercial reach began to grow beyond their home provinces into a region-wide market, and was accompanied by the merging together of many producers, with some large companies becoming increasingly strong, much as was happening in the food industry as a whole.

The highly profitable food processing sector was a further element in this overall growth, with a 13.2% average increase in overall earnings. The sector continued to be one of the most stable of all industries, and was little affected by economic cycles as it enjoyed a constant growth in profits. The Italian dairy industry differed from its European counterparts in that it was characterized by a high level of product fragmentation, although this was beginning to change under the influence of a process of reorganization involving all participants in the industry which, just as in the food and agricultural sector as a whole, was based on a campaign of acquisitions, mergers and commercial agreements.[9]

The campaign of acquisitions and mergers was a sizeable phenomenon – representing 10% of all purchases registered throughout the 1980s – and mirrored similar processes taking place in the chemical and mechanical

7 The national agricultural program of 1977 took account of the pressures brought by Italy's foreign trade balance, and its main objective was to achieve a drastic reduction in real terms of the food industry deficit through a package of incentives to agricultural producers and, above all, by limiting access to the consumer market. Di Sandro G., *Agricoltura e sviluppo economico: il ruolo della politica agraria in Italia (1944–1982)*, Milano, Angeli, 2002, pp. 134–8.

8 Nomisma, *L'agricoltura italiana in un mercato meno protetto: risultati del 1987 e tendenze attuali*, Bologna, Nomisma, 1998, pp. 5–10.

9 Casati D., "Evoluzione strutturale dell'industria agroalimentare", in *L'Italia agricola*, 125, 1987, 1, pp. 30–42.

sectors.[10] This reorganization was not simply aimed at gaining control of companies; it also involved the introduction of management and technical innovations from which many dairy businesses benefited, although there were numerous exceptions throughout the sector. Access to a wider market was connected to a differentiation of the product range, which was achieved through new food production technologies, consumer services and packaging, rather than through the modernization of manufacturing plants.

The 1980s were also a period of substantial change for the entire dairy industry in line with the transformations that were taking place in the food manufacturing industry as a whole. Unlike in the private sector, institutional changes were having a marked effect on public intervention policies, affecting relations between the state and regional governments (after a referendum that had led to the abolition of the Ministry of Agriculture), although these changes were most heavily felt in relations between the European Community, the State and regional governments. The new reform aimed at the so-called "structural funds", which were introduced by the Community in 1988, contemplated planning and evaluation activities with regard to State and private involvement in industry.[11]

The dairy industry, just like the rest of the food industry, was becoming increasingly interconnected with other industry sectors: what was once a "static force" had been transformed into a strategic front line in the country's economic development. At the same time, the wide-ranging campaign of renewal and modernization of the distribution infrastructure made the industry more susceptible to influences and market penetration from foreign competitors and large business groups. Simultaneously, the large conglomerates and the collectives coexisted with a substantial number of small and medium sized businesses, which represented a higher proportion of the total than in other European countries. Most of these were not simply the remnants of an older, backward, industry, but were, rather, locally based activities that were responding to a specific demand for specific high quality products.

2. Milk production trends

With regard to the provision of the basic raw material, statistics collected in 1974 show that in Northern Italy more than 75% of the 7,000 tons of milk produced was destined for the processing industry (4,881.6 tons), while in Central Italy half of the almost 700 tons of milk produced was destined for direct human consumption (398), with a similar figure for Southern

10 Alzona G., "Acquisizioni e cessioni nella dinamica dei grandi gruppi industriali italiani", in *Antitrust e gruppi industriali*, a cura di P. Bianchi, Bologna, il Mulino, 1998, pp. 241–5; *International handbook on industrial policy*, ed. P. Bianchi, S. Labory, Edward Elgar, 2006.

11 Laschi, L., *L'Unione europea: storia, istituzioni, politiche*, Roma, Carocci, 2002, pp. 45–56.

Italy, which produced 841.4 tons, of which 485.1 were destined for human consumption and 356.3 was transformed into other products.[12]

From the mid-1970s there was a sharp rise in imports, although export figures saw no such increase. In total, 25% of agricultural imports were dictated by the need to make up for a shortfall in milk production.[13] More than 60% of the milk used in the dairy processing industry came from abroad.[14] This deficit in milk production, with a relative increase in imports, was seen as a political rather than as a commercial problem, one which was closely linked to the European Community's quota regime.[15] Agricultural union representatives repeatedly voiced the opinion that the deficit was in large part the fault of the quota system, and that elevated production costs were also to blame since they had an inflationary effect on prices.[16] Standardized milk production began to take hold in the milk and dairy industry when dairy processors acted to ensure that farmers received a fair income by producing milk derivatives which were aimed at the mass market and whose quantities and quality were controlled by the "guiding hand" of the European Community. These difficulties were compounded, from the point of view of the processing industry and particularly of dairies, by the approval of a law proposed by the Christian Democrat and Communist parties which handed over negotiations for the purchase of milk to the regional governments, resulting in companies being obliged to purchase Italian produced milk even if cheaper alternatives were available abroad.

Milk in Italy was, on average, 22% more expensive than that produced in the other eight European Community member states, whose products were

---

12 Istat, *3° censimento generale dell'agricoltura, op. cit.*

13 In 1974 Italy imported 421,206 tons of milk from EC countries, of which 357,780 tons came from Germany and 62,886 from France. In the same year, 115,933 tons of cheese were imported from the EC and 19,149 from Switzerland. 5,600,000 hectolitres were imported in 1975, in addition to a domestic production of 83 million hectolitres. According to Association of milk and cheese producers (Assolatte): "In Italy, with its 17.5 million hectares of agricultural land, around 3 million milk-producing cattle produce 83 million hectolitres of saleable milk. Holland, which has 2.1 million hectares of land and only 2.1 million animals, produces 99 million hectolitres of milk". The established dairy manufacturers, Galbani and Invernizzi, protested that "government policies had the effect of increasing the price of raw materials and lowering that of the final product". Assolatte annual report, 1974 and 1975.

14 Antonio Invernizzi spoke for all top businessmen in the sector when he said that "independent activities, including, of course, that of agriculture and husbandry, cannot allow themselves the luxury of a fair stable income immune from financial risks as well as losing a part of their profits to the growth of a market in which they do not participate". Assolatte annual report 1975.

15 See for a complete analysis Laschi G., *L'agricoltura italiana e l'integrazione europea*, Brussels, P.I.E. Peter Lang, 1999.

16 In the following years, as the PAC agreement was being modified, some agricultural unions (Fai-Cisl) asked that rather than the quotas being eliminated, they should be raised to a sustainable level.

increasingly present in the domestic market (Tab. 1).[17] Italy produced 8.6% of the European Community's annual milk production, which had reached 970 million hectolitres by 1974. According to the Association of Milk and Cheese Producers (Assolatte), "Such a percentage means that we are subject to Community regulations without having any influence upon them".[18] Italy had a deficit in milk destined for human consumption (34 million hectolitres), which was equivalent to 7.5% of the amount of Italian cow's milk destined for the processing industry. Imported products reached 47% of the country's overall dairy market, with foreign competition growing in the areas of foodstuffs like Grana cheese at the same time as Italian products experienced a phase of stagnation.

In 1976, the divergence between industrial costs and profits was so large that the government, which had already begun a campaign to promote the drinking of milk, took the exceptional step of raising value-added tax on milk and transferring the extra income directly to the producers. Assolatte was promoting awareness of the negative effects of a system where dairy producers were forced to seek bank loans in order to make the extra VAT payments to milk producers, payments which were reimbursed by the tax authorities, but usually with a delay of more than twelve months.[19]

During the same period, heated debates were taking place between Italy and Brussels on the need for an agricultural policy tailored to the Mediterranean area after Greece, Spain and Portugal had applied for Community membership. According to Assolatte, the problems surrounding the milk sector in Italy received little attention in these debates, meaning that the system of economic support for the milk-producing cow farming industry would continue to function according to the previous model which had been set up on the basis of what were essentially Dutch methodologies.[20] Italy's milk and dairy industry was particularly opposed to the way milk production had been subjected to regional authorities, causing farmers to abandon the rearing of milk-producing cows because the northern regions of the European Community were able to achieve a higher average production per capita. Some Italian milk producers began to express doubts about the European Community's policy regarding surpluses. Despite the fact that Italy was the biggest surplus importer, the European Community promoted the purchase of surplus milk production in the form of powdered milk and butter by European institutions which would then resell them at subsidized prices or export them along with cheese to non-Community countries. In 1984, this mechanism was the basis of the so-called

17 In 1970 the price of a quintal of milk in Italy was 6.485 lire, by 1978 it had risen to 19.015 lire and in 1985 it was 35.588 lire. Figures from Lombardy record the price in the period 1972–3 as being 8.960 per quintal, a figure which had tripled by 1977.

18 Assolatte annual report, 1975.

19 A small producer that processed 10 tons of milk a day paid out 10 million lire each month to the state, or 120 million a year, while foreign companies received compensation payments in as little as fifteen days. Assolatte annual report, 1976.

20 Nomisma, *Rapporto 1992 sull'agricoltura italiana*, Bologna, il Mulino, 1993, pp. 15–19.

milk quota and "additional levy" measures applied to excess production. Italy became the only country to consume a quarter of European milk surpluses (5,500,000 million tons of cheese and 2 million of powdered milk).[21] Italian producers increasingly saw the quota system as a barrier to innovation, although it was well-known that the Italian agricultural industry was well behind its European competitors in terms of technological progress. Also, in the mid-1980s, Italy was able to free itself from the obligation to spread its quota allowance among all the country's producers.[22] During the same period the milk and dairy industry, particularly in Northern Europe, obtained a number of advantages thanks to the milk quota system. Management of excess produce generally made it possible to shore up local supplies thanks to the strong position it gave them in their dealings with producers; and multinational companies benefited from the availability of milk at prices that were artificially controlled by government intervention, and were able to operate on the international market by taking advantage of powdered milk and butter surpluses that resulted from the European Community's stockpile management policy, which kept prices lower than those on the international market.[23]

Table 1. The price of milk (lire/liter – VAT not included)

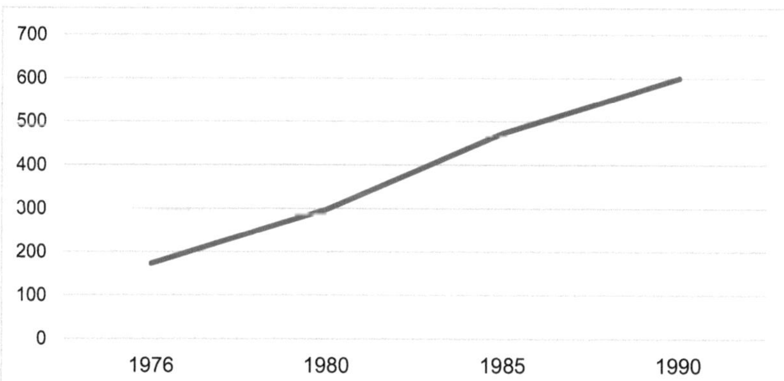

Source: Assolatte annual report 1975–1991.

[21] Vieri S., *La politica agricola comune dal Trattato di Roma alla riforma Mac Sharry*, Bologna, Edagricole, 1994, pp. 34–51.

[22] Later on a change in direction took place, caused in part by pressure from Italian producers, when the value of quotas attributed to a company became part of that company's assets (1992). Fauri F., *L'Italia e l'integrazione economica europea: 1947–2000*, Bologna, il Mulino, 2001, pp. 230–7.

[23] The Italian agricultural industry gained nothing from these monetary measures because devaluation of the lira meant that monetary compensation packages benefited animal farming industries in countries with a strong currency. See Condorelli L., *L'agricoltura fra CEE, Stato e regioni*, Bologna, il Mulino, 1973, pp. 76–9.

Table 2. Milk use (tons)

	1969	1978
Livestock	1,442,000	800,000
Food	3,263,493	4,300,000
Processing industry	4,449,238	6,274,212

Source: *Reports of Ministry of Agriculture.*

Table 3. Milk price 1985 (566 liras for a liter): real price of ECC

Italy	35,888
Denmark	26,491
Netherlands	25,697
Belgium	25,527
France	25,486
Germany	24,984
UK	23,718
Ireland	21,436

Source: *Reports of Ministry of Agriculture.*

Table 4. Italian milk import/export balance (x €1,000)

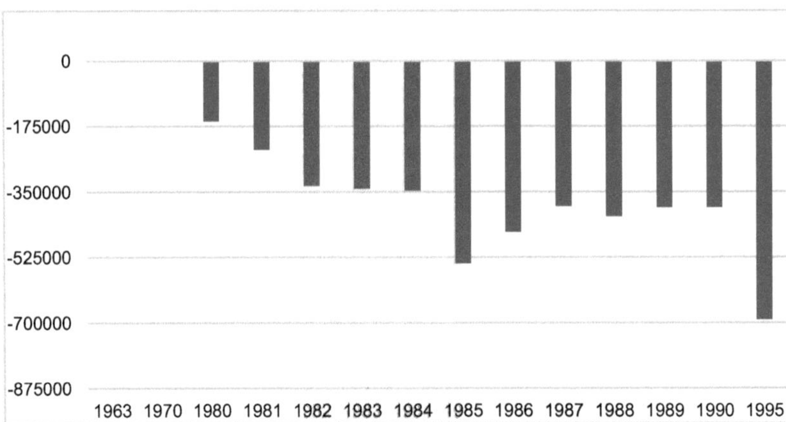

Source: Assolatte annual reports 1960–1995.

Table 5. Italian cheese import/export balance (x €1,000)

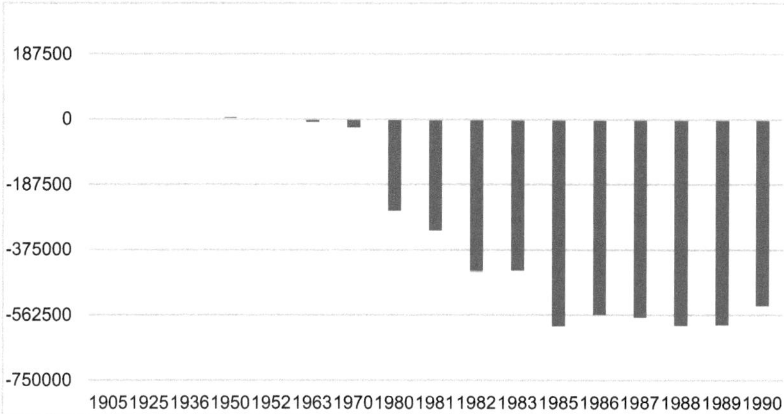

Source: Ferrari P., *L'industria del latte*, Camera di commercio di Piacenza [197–] Istat, *Fatturato, prodotto lordo e investimenti delle impres commerciali, dei trasporti e comunicazioni*, Roma 1983–1993.

In 1985, Germany was Italy's main trading partner for cheese exports, with 57,239 tons, followed by France with 44,011 tons, while 12,020 tons were imported from Holland.[24] The same period saw a rise in milk prices accompanied by a contraction in dairy product sales; this situation took place in areas which had traditionally enjoyed a consumer base that was little affected by price changes. The continuing link between the low production of milk and its derivatives and the inflated prices of these products eventually forced the authorities to intervene in order to reduce surpluses (Tab. 3).

By the middle of the 1970s Italian producers were exporting 6,791 tons of cheese to the USA, 4,502 to Switzerland, 2,735 to France and 2,552 to Germany. The sales abroad were linked to the presence of Italian immigrant communities. The UK and Canada followed, with the relatively low figures of 1,367 and 1,176 tons respectively. Of the different cheese varieties, Pecorino had the highest export volumes, with 5,584 tons, while sales of Grana Padano on foreign markets continued to fall. A large part of the cheese export market was represented by Gorgonzola (3,216 tons), while a minor position was held by melted cheeses (926 tons) and fresh cheeses (682 tons).[25] Ten years

24 Other cheese imports came from Belgium (1,722 tons), Denmark (834 tons), Ireland (100 tons) and Great Britain (7 tons).

25 In 1978 exports to the USA fell to 4,951 tons, while those to Switzerland (4,376) and France (3,744) remained steady. Growing markets, on the other hand, were those of Germany, with 3,274 tons Canada with 2,502 tons. Italian exporters found more hospitable markets for their products in Belgium (1,522) and the UK (1,234). In the same year, Pecorino continued to be the most exported Italian cheese with 5,162 tons, while exports of Grana fell dramatically (3,700). Italian producers managed to reach new

later, in 1985, Italian cheese exports amounted to around 42,000 tons; record sales in the North American market (9,577 tons) were accompanied by an increase in export volumes to Germany (9,024 tons). France imported 6,727 tons of Italian cheese, while 6,602 tons arrived in Switzerland. Between the end of the 1970s and the first half of the 1980s sales to North America began to slow down, while Italian products became increasingly successful in French and German markets, aided by an overall rise in consumption and changes in Western European culinary tastes. In 1985, foreign exports of Pecorino amounted to 13,638 tons, with 7,879 tons of Gorgonzola, which was produced entirely in Northern Italy, and 6,770 tons of Grana, which was returning to its previous popularity after a negative period between the end of the 1970s and the beginning of the 1980s.[26] Sales of "grated cheese" and Asiago were growing rapidly with 4,332 and 3,002 tons respectively. Many private operators in the dairy processing industry believed that Grana production volumes could not grow beyond a certain limit. Factory owners maintained that other derivatives faced with strong competition should be supported.[27] Furthermore, the volume of milk sales, which was less than a tenth the size of its French counterpart, was the cause of bitter clashes within the industry.

3. The appearance of multinationals

Between 1983 and 1986 foreign companies began to appear on the market with the intention of enlarging their product portfolio and increasing their geographical reach, while at the same time some medium sized producers such as Auricchio, Ferrari and the cooperatives began to strengthen their international position. In both cases, companies were taking advantage of the stabilizing factor produced by the supply of raw materials under a controlled price regime that had been elaborated by the European Community in collaboration with industry representatives. The large companies operating in the sector, which were mainly French, were able to operate in the international market with the benefit of having access to powdered milk and butter surpluses at lower prices than those available

markets to export almost 4,500 tons of Gorgonzola. Statistics from the 1979 Assolatte annual report.

26 Taken from the 1985 and 1986 Assolatte annual report.

27 The small group of large-scale dairy producers, while remaining competitors, were united in their struggle against the cooperative companies, revealing a disparity between the manufacturing and commercial conditions of Grana production and that of soft cheeses. Another area of contention was the quality of milk. Both the regional agriculture offices and milk producer associations called for milk to be purchased according to its quality. The industrialists believed that milk quality judged by the sanitary conditions of cattle should not constitute a premium over the base price because a product sold under the current legislative requirements had to have been declared healthy before being sold. See Assolatte annual report, 1975–1985.

to countries outside the European Community. Italy's agricultural industry, on the other hand, was unable to obtain any particular benefit from these European measures because devaluation of the national currency led to the animal husbandry sectors of countries with a strong currency being granted "monetary compensation payments".

Just as the multinationals and the most successful Italian dairy companies were becoming increasingly interdependent in terms both of ownership and of management, Assolatte was highlighting the gradual loss of competitiveness resulting from rising production costs and the impossibility of passing on such costs to the consumer.[28] The only consolation in all of this was that inflation was finally under control.[29] The industry's businessmen and managers saw the system of guaranteed milk prices as a negative factor, contributing to a reduction in competitiveness and the devaluation of their products. A new slogan, "No more laughing cows",[30] was coined in protest at what was seen as excessive concern on the part of national and regional governments for the financial well-being of the agricultural industry, at the cost of organizational stability in the processing industry.[31] The meat industry was experiencing similar problems. New national employment regulations in the food industry were introduced to provide extra protection for workers, with the application of automatic pay scales as well as more rigid controls on working hours and a substantial reduction in the permitted number of overtime hours. Cheese-producing companies, led by Galbani, requested that the price of raw materials be adjusted gradually and consensually.[32] Public institutions at all levels were accused by company owners of allowing "intermediary structures" to exert an unfair influence within the industry and of granting them a disproportionate amount of the resources destined for the agricultural sector as a whole. There were also requests for regulatory reforms, including guidelines for ultrafiltration technology and the sale of "low fat" cheeses. These changes came about amid differences of opinion between the industry and Health Ministries on the one hand

28 Assolatte annual report, 1977.

29 It is also worth noting that, between 1975 and 1979, the entire dairy sector faced rising energy costs together with a shrinking consumer base due to rising inflation.

30 "La vacca deve smettere di ridere". "Vacca" in Italian literally means "cow", but is also used figuratively in a similar way to "fat cat".

31 During his report to the annual assembly the Assolatte President summarized the situation as follows: "Otherwise we will be forced to bitterly consider the only alternative that the current market can offer us if we wish to survive: to become salesmen. It is no secret that we are becoming the main outlet for our competitors and are losing ground in our own market". Over the previous fifteen years, in the international market France's cheese exports had risen from 98,000 to 282,100 tons (+187%), Holland's from 173,300 to 355,700 tons (+105%), Denmark's from 67,900 to 212,600 tons (+213%) and Germany's from 51,900 to 287,800 tons, an increase of 454%. Italy's cheese exports, on the other hand, amounted to only 49,600 tons. Assolatte annual report, 1986.

32 Industrialists wished, theoretically at least, for the price of milk to be defined in full agreement with farmers.

and the Agricultural Ministry on the other, while attempts to block foreign products entering Italy if they did not conform to national health and safety requirements contravened the principle of "mutual recognition" which had been confirmed on more than one occasion by the European Union Court of Justice, most famously with the "Cassis de Dijon" ruling that allowed for the import of products manufactured under health and safety regulations different from those of the country of destination. There remained a number of problems that stood in the way of regulatory and commercial harmony with Italy's European partners; prominent among these was competition from imitation products and industrial disputes which arose from calls for increased mobility and flexibility in the workforce.[33]

By the mid-1980s, a prudent sense of optimism had settled in, and it was expected that export volumes would continue to grow. But the situation suddenly deteriorated after a rise in production volumes in the USA, New Zealand and Eastern Europe, leading to a growth in surplus stocks and a reduction in the availability of financial resources. In 1985, Europe had amassed a surplus to the value of at least 10 billion ECU with storage costs of between 2 and 3 billion ECU a year. Brussels moved to lower fixed prices in order to reduce spending, but this led to a number of other consequences, which led to "marathon talks" throughout the spring which ended up adopting measures that appeared to favor the Northern European countries that produced butter and milk in large quantities. With the regular appearance of organizational crises in the European Community, the Italian milk and dairy industry voiced the need for a "new Stresa agreement" that would carry out a wide-ranging reform of the PAC mechanism.[34] Meanwhile, the companies involved in the processing and sale of dairy products protested over provisions in the health reforms which transferred oversight of food products to the public health authority, USL, complaining that this created a less effective system and meant that many of its responsibilities overlapped with those of other institutions. In response to a scandal surrounding illegal food additives, the government gave local councils powers to close factories that contravened health and safety regulations, even though health and safety officials already existed with the very same powers (legislation No. 283/1962). Assolatte called for the Agriculture Ministry to pass reforms bringing factories in line with current regulations. Ministerial powers were to be altered to reflect the increase in agricultural producers, developments in their representative organizations and their relations with the processing

33 *Ibid.*, and the 1988 report.

34 Stresa Conference, 3–12 July 1958. Convened in July 1958 by Sicco Mansholt, Vice-president of the Commission and the Dutch head of the Common Agricultural Policy, the Conference met Commission officials, national experts and representatives of farmers' organizations. The aim was to decide how to take action to achieve the objectives of the common agricultural policy, as established by art. 33 of the EC Treaty. The most important decision taken in Stresa was to help agricultural development not through the politics of income support but with a system of guaranteed prices.

industry. The transferring of powers to local authorities had deeply affected the sector as a whole, while the approach of bureaucratic functionaries appeared to still be founded on the idea that it was necessary to protect the agricultural sector from the influence of the industrial world.

An apparent rigidity in the workforce led to a policy centered on technological investment aimed at optimization of the production cycle. Galbani, for example, opened a new meat processing factory in Melzo in 1980.[35] Factories were modernized, and another was purchased in Casalbuttano[36] Large industrial producers invested in technological innovation aimed at standardized mass production in order to face external pressures including the price of milk and increasing energy costs. The Galbani company was revising its strategies in order to position itself better internationally,[37] in the knowledge that its position on foreign markets suffered from a lack of competitiveness, built as it was on niche products inappropriate for large-scale distribution. Galbani's outlook was unanimously shared among the other companies in the sector: commercial success would depend on their ability to satisfy the market for standardized mass-produced products. Their attempts at reaching a wider market with more standardized products were frustrated by the protectionist policies of countries like the USA, Germany and Switzerland. On the domestic market there had been a dramatic reduction in the popularity of hard cheeses. A law which had been approved eighteen years previously, "Legislation No. 283 on the production and sale of cheese", only now came into force, leading to a clash between the public administration and industrialists who protested that it was an obsolete piece of legislation.

From the mid-1970s, the larger dairy companies adopted a strategy aimed at obtaining an ever-growing share of the markets in which they operated, in an attempt to create a situation where they could gain a monopoly of the domestic market for certain products. In a number of cases Italian companies saw this strategy as an effective way of fighting competition from home and abroad. In 1986, the total profits of the top forty Italian companies were less than those earned by Nestlé alone. Compared to Nestlé's 22,000 billion

35 The factory was located in Melzo (Milan).

36 Plants were in Casale Cremasco, Certosa di Pavia and Casalbuttano.

37 In the mid-1970s, Galbani's accounts valued its property at 28 billion lire, while it had a positive balance of 12 billion lire with banks and 13 billion with its clients. The company's debts with suppliers amounted to 17.3 billion lire, and it had a liquidity of 2.4 billion. 154 billion lire had been used for purchases, and employment costs were 43 billion compared to a turnover of 228 billion. In 1977 its property was worth 37 billion and it had a positive balance with banks of 13.7 billion. The company's debts with suppliers amounted to 29 billion lire. Expenditure on raw materials came to around 234 billion while employment costs were 70 billion, with a turnover of around 355 billion. Rinaldo Invernizzi died in November 1977. Galbani's real estate patrimony in 1979 was valued at 43 billion lire, while the company had a positive balance with banks of around 100 billion. Suppliers were owed 33 billion. The company's balance sheet registered around 307 billion in raw materials costs and 84 billion in employment expenses, with a turnover of 469 billion.

lire profits, Italy's highest earner was another dairy company, Galbani, with 1,288 billion lire.[38]

A transformation took place between 1983 and 1987 when foreign competitors began to opt for a strategy of direct participation in the Italian market. This period has been the subject of many academic studies, not so much because foreign multinationals were so active in Italy (Nestlé, Unilever, Kraft, Bsn-Danone, Besnier Lactalis), but because the acquisitions carried out in that period were seen as highly significant, as was the size of the market that those acquisitions brought with them. While some Italian companies pursued niche markets, foreign "big business" tended to aim at occupying a wider section of the market by offering a highly varied range of products.[39]

The domestic phase of concentration and rationalization was following by a period in which family-owned businesses, no longer able to remain competitive, were forced to cede their companies to others. Meanwhile, the public authorities, which had until then been active in supporting the milk industry, decided to cede SME into private hands. The privatization of SME, or La Società Meridionale di Elettricità [the "Southern Electric Society"], which over the years had evolved into a public holding company within the food sector, involved breaking up its component parts and selling them off to its Italian and European competitors. This followed the definitive failure on the part of Carlo De Benedetti and the CIR to create a national food industry "champion".[40]

International consultants and investment banks expressed the opinion that big Italian food companies lacked the necessary combination of brand recognition and market reach, or in other words, they tended to develop one of these aspects at the expense of the other. These shortfalls, in the context of a highly fragmentary sector, could well have been the stimulus for a restructuring process within the sector and led to a new market equilibrium. The acquisition of Buitoni by Nestlé appeared to be this first step in this transformation.

The companies which produced the most recognizable brands of all in the food industry were Sme, Galbani, Star and Peroni, but they were far less successful in terms of geographical reach. Martini & Rossi, Parmalat, Cinzano, Ferrero, Buitoni, Lavazza and Barilla, on the other hand, were able

38 In 1986 the profitability of Italian companies with a majority foreign ownership was still limited: Kraft Italia registered a turnover of 760 billion lire, Nestlé Italia 730, Hienz 560 and Bsn-Danone 550. Galizzi A., *Strategie di internazionalizzazione dell'industria alimentare e l'esperienza italiana*, Atti del Convegno "Strategie di internazionalizzazione dell'industria alimentare europea", Milano, 1988, pp. 185–91.

39 *La politica industriale tra Europa e regioni: quarto rapporto sull'industria e la politica industriale italiana*, a cura del Centro Europa Ricerche e dell'Istituto per la ricerca sociale, Bologna, il Mulino, 1990, pp. 34–93. See also Heer J., *Nestlé. Centoventicinque anni dal 1866 al 1991*, Nestlé Sa, 1991, pp. 265–88.

40 Chiapparino F., *Consumi e industria, op. cit.*, pp. 123–34.

to place their products across a much wider area but could not command the same level of brand recognition. Italian food producers were at a disadvantage compared to their European competitors, and even more so with regard to the USA, and international consultants advised them to restructure their product portfolios and to expand their market reach through a campaign of acquisitions and sales. The tactics they had been advised to follow would see them aiming at a less diversified business model which would be more concentrated on a restricted number of strategic products. The mergers were to involve large and medium sized companies rather than the smaller ones, which it was suggested should pursue a "niche policy", producing products to be marketed by larger brands or working together in cooperative associations. For their part, the main Italian brands engaged in a revision of their market position and industrial strategies.[41]

The Italian cheese sector began to attract foreign investors, leading to a complex period of restructuring and to Kraft's acquisition in 1985 of 75% of Invernizzi's shares for 96 billion lire. Invernizzi had achieved 300 billion lire in sales the previous year, with profits of 3.4 billion, and had a workforce of 2,574. In the same year, Kraft's Italian branch had registered sales of 280 billion lire, a growth of 10.5% over the previous year. Kraft's profits had amounted to 11 billion lire, including a 2.2 billion lire investment. Kraft-Italia had a factory in Zingonia near Bergamo which employed 582 workers. The company controlled 45% of the "sliced cheese" market, 30% of mayonnaise sales, and 14% of margarine on the market. The new group formed by Invernizzi and Kraft joined their products together with those of another Italian company that had been taken over by Kraft, Osella, which specialized in fresh cheeses. Kraft's directors planned a coordinated approach which also included the Philadelphia spreadable cheese that was produced in their German factories.[42] Their policy of diversification included redirecting Invernizzi's products towards foreign markets. With a higher volume of sales, Invernizzi offered more of an opportunity than had Osella to conquer a large slice of the Italian market with a wide range of products, from hard to fresh cheeses. The Italian subsidiary, by now fully integrated with its German counterpart, was able achieve a turnover of around 650 billion lire.[43] After paying 100 billion lire to Edo Invernizzi and allowing his brother Romeo to stay on as a minority shareholder, Kraft pursued an industrial policy that ensured continuity for the new subsidiary and its products.

41 Rinaldi R., *Aziende tuttofare nel futuro dell'industria degli alimenti*, "Il Sole 24 ore", 9 July 1986.

42 Kraft had seen "Operation Osella" as a chance to complement its advanced cold processing technology with a well-known brand made by a medium sized Italian family business which had always maintained good relations with milk producers.

43 Other industry operators had shown interest in Invernizzi's production process and market share, including the "the white cooperatives" network, an investment company led by Jody Vender which specialized in the acquisition of family businesses, and Yomo, a company based in Lombardy which produced soft cheeses and yoghurts.

For a large multinational, acquiring a well-known and established company was an alternative to pursuing a policy of internal growth. The efficacy of this approach depended on the size of the newly acquired market and on the ability to move forward with a unified strategy.[44] Kraft had purchased a company that, until the 1970s, had carried out a policy of product innovation and the building up of extensive market coverage.[45] At the time of its sale to Kraft, Invernizzi exemplified the "classic" management model of a large Italian dairy company.[46] Kraft's takeover of Invernizzi took place within the context of the multinational's intense rivalry with Nestlé and Unilever, which were all searching to expand their consumer markets. The rivalry between these companies was entirely based on the need to increase their volume of sales rather than on a policy of integration, an approach dictated by the low profit margins of the products that made up these markets.[47] The multinationals' business strategy began to change halfway through the 1980s when Kraft aligned itself with the tactics of its competitors by initiating an acquisition policy that was aimed at products made by small and medium sized companies that were deeply rooted in the

---

44 Philipp Morris offered Kraft $11 billion for a controlling share on the basis of a diversification strategy which was being followed in response to predictions of a dramatic fall in the cigarette market.

45 This tried and tested model had been reinforced by advertising campaigns that aimed at promoting brand visibility through new media channels. Invernizzi had storage facilities in almost every Italian province, which helped it to be less reliant on local distributors and guarantee a regular supply of its products to the market. Its preferred method was "off-truck" sales. From the 1980s ownership was entrusted to technicians who had risen through the company, with the onus on planning and control. In 1981 Romeo Invernizzi appointed Mario Valtolina to the board of directors. Valtolina had worked for the company since 1986. The new management, with the blessing of the owners, initiated a process in which operations were to be concentrated into one single site and distribution was to be reorganized by bringing together different production lines and reducing overall costs. At the same time, challenges posed by the mass distribution system moved the management to decided to abandon their production of hard cheese in favor of the more profitable soft cheeses. The company's owner, Romeo Invernizzi, was getting older, and the absence of an heir ready to take over the running of the business forced the directors to begin looking for a potential buyer, since the small group of directors running the company at the time had no intention of becoming shareholders themselves. See "La famiglia Invernizzi tratta la cessione", *La Repubblica*, 25 May 1985.

46 The family who owned the company managed their finances on the basis of accessible liquidity, including the purchase of large amounts of government bonds. Italian businesses traditionally tended to prefer a self-financing approach when it came to the day-to-day running of their companies, and saw the innovation process as an incremental system of "learning by doing". Since the post-war years the principal dairy producers had managed their production lines through an incremental improvement process, introducing new technologies and products as they appeared. On the other hand, until the end of the 1970s they had not concerned themselves with strategies for increasing profit margins through the development of traditional and locally produced products. See Colli, A., *Il capitalismo familiare*, Bologna, il Mulino, 2006, pp. 44–53.

47 Knopp M., Wilson B., *Europe's major dairy companies in an era of change*, Cambridge, Cambridge University Press, 1992, pp. 74–7, 91–6.

culinary traditions of their home territory. The most prominent of these acquisitions was that of the Modena-based company Fini.

The marketing sector was also undergoing a streamlining process. A period of innovation on the part of cheese producers in terms of brand promotion and advertising was followed, around the beginning of the 1980s, by a phase characterized by the shoring up of existing brands through a concerted effort at improving the efficiency of their sales networks. The system of mass distribution, with sales strategies based on competitive pricing and high production volumes, stood directly in opposition to the Italian dairy industry's system in which products were supplied directly to the resale sector. Another well-known brand controlled by the Nestlé group, Locatelli, had introduced "packaged cheese" (mascarpone) in 1961, and in 1987 the company built a new factory aimed at the production of "green cheese", coinciding with its abandonment of the cured meats market. Nestlé intended to bring to the Italian market its highly integrated industrial strategy based on four different areas of activity: the drinks and nutritional products line was strengthened by the addition of the Sasso olive oil brand, fresh foods were represented by Locatelli for cheese and Vismara for cured meats, Perugina produced sweets and chocolate, while pasta was made by Buitoni.[48]

This process of change in ownership and management also affected the top Italian producer of cheese and dairy products. Midway through the 1980s, Galbani was facing steadily rising production costs while at the same time sales had stagnated, meaning that the extra production costs were transferred directly to the consumer. In 1987, the company raised its capital from 27 to 90 billion lire, at a time when it presented a turnover of 115.2 billion and negative assets of 111.8 billion lire, a positive balance of 3.3 billion. Many observers have stated that Galbani found itself in a very unusual situation in 1998: its controlling shareholders were divided on which strategy to follow and the managing director, Guido Campominosi, was about to leave his position after fifteen years at the head of the company. On the other hand, Galbani was financially sound and its products had achieved wide market reach. Galbani had much to offer a multinational that wished to enter the Italian market or to strengthen its current position. In 1987 the company had a turnover of 1.46 trillion lire, comfortably ahead of its many Italian competitors. It declared profits in the same year of 110 billion lire, as against 90 billion in 1986. Its cash flow was over 150 billion and it possessed 660 billion lire in government bonds, which guaranteed interest payments of at least 60 billion lire. Added to this favorable situation was an absence of debt and an estimated credit value of 100 billion lire. Company policy was

48 It is worth noting that this process involved a number of hurdles. Dairy production lines suffered a decline in their market position throughout the decade. In 1998 the French group Besnier purchased Locatelli from Nestlé. See also "La Nestlé colosso mangia aziende", *La Repubblica – Affari & Finanza*, 29 Jan. 1986.

to prefer cash transactions and to operate a deferred payment system with its milk suppliers; they also managed their production line carefully in order to reduce storage costs to a minimum. Galbani remained the top Italian cheese producer as well as the second largest producer of cured meats. By the end of 1987 the company controlled 18% of the Mozzarella market, 51% of Crescenza sales, and 17.5% of all other cheeses. Galbani's share of the highly fragmented cured meats market was 3.7%, with a 5.5% slice of mortadella sales.[49]

In 1988, Galbani began to attract the attention of a number of French companies.[50] One of these possible buyers was Bongrain, the sixth largest food producer in France, which was especially strong in the milk and dairy sector, with internationally successful cheeses like "Caprice des Diuex" and "Tartare".[51] Bongrain had an Italian subsidiary with a turnover of around 30 billion lire, and controlled a small cured meat factory in Parma, called La Nuova Boschi. The French company obtained a 230 million ECU (350 billion lire) loan with a multi-option facility through a banking consortium that included the Banco di Napoli. Another French company, Besnier, was also interested in acquiring Galbani. This too was a family business and had built its success on the collection and distribution of fresh milk (under the Lactel brand). It boasted a 2 trillion lire turnover and some world-famous cheeses like the Camembert cheese "President".

---

49 Throughout the first half of the 1980s, Guido Campominosi preserved Galbani's traditional industrial structure, which had been in place since the early post-war years: it possessed nine factories and farms which were spread throughout the Italian peninsular, with a total of 7,000 employees. A constant supply was assured by moving excess production into foreign markets, a policy which also helped to contain manufacturing costs. The company's true strength was its commercial network which was managed directly from the company's head office. This, together with easy access to bank loans, meant that Galbani, on paper, was able to transform itself from "prey" to a group in expansion by initiating a campaign of acquisitions, although until the mid-1980s its management preferred internal development to acquisitions. Linda R., "Crescita, diversificazione e denominazione nelle strategie delle megaziende europee dell'industria alimentare e delle bevande", in *Rivista di politica agraria*, 4, 1988.

50 Journalistic sources maintained that in 1974, one year before the European parliament approved the new currency legislation, the Invernizzi family ceded 60% of Galbani to three Luxembourg companies. This was no isolated case in Italy's recent financial history, judging by court cases over a similar situation that were brought against the Ferrero and Irneri families by Lloyd Adriatico.

51 The Bongrain group was controlled by a family headed by Jean-Noel Bongrain, whose activities were divided into two areas, one of which was represented by the Bongrain group, quoted on the stock market and with a capital value of more than 700 billion lire, sales worth 1.3 trillion and 62 billion in net profits (1987). Bongrain employed 5,400 workers and 46% of its sales came from abroad, mostly from the USA, Germany and Spain. The family also controlled another food company called Soparind, which was not a stock market company and specialized in cured meat production. See "La Ifil (Agnelli) acquista il gruppo Galbani", *Il Sole 24 ore*, 13.07.1989; "L'accordo Ifil-Bsn-Galbani rafforza un settore industriale poco concentrato", *Il sole 24 ore*, 21 July 1989.

Another company looking at acquiring Italian brands was the top French food producer, Boussois Souchon Newvesel – Gervais Danone (Bsn-Danone), which produced a wide range of products (drinks, yogurt, pasta, biscuits, sauces and jams) and was allied with Ifil, a company within the Agnelli group. Until 1987 Bsn had been interested mainly in pasta factories and mineral water producers, with its purchases of Sangemini and Ferrarelle. BSN's managers explained their renewed interest in the Italian dairy industry by pointing to Italy's manufacturing technologies and the high quality of their cheese and cured meat products. Direct investment in Italy had two separate aims: firstly to facilitate penetration in what was seen as a market with complicated tastes, and secondly to make a number of "national symbols" of Italian food available to international customers. Bsn-Danone (born from a union between Danone, owned by Isaac Carasso, and Bsn, owned by Antonie Riboud) substantially wished to follow in the footsteps of Kraft-General Foods. The entrance of the multinationals Nestlé, Unilever, Bsn and Kraft-General Foods, into the Italian market was essentially the product of the homogenization of European consumer tastes and behavior that had accompanied the spreading popularity of the Italian diet. The only two big European markets where prominent groups were not yet present were those of Germany and Italy: production of beer, cured meats, cheese and frozen products was in the hands of small and medium sized businesses which were more exposed financially and constituted easy pickings for takeover bids.[52]

Before changing hands Galbani had been valued at 1.067 trillion lire, with a government bond portfolio of 685 billion, and was owed 83 billion lire by the banks. Its outstanding debt towards suppliers was 143 billion, and it had positive assets of 123 billion, with purchases worth 832 billion, an income of 1.444 trillion lire along with 70 billion in government bond returns.[53] On 19 July 1989, a joint-venture formed by Ifil and Bsn purchased Galbani for 2.241 trillion lire, and transformed it into a subsidiary of the Danone group.[54] At the time of its purchase Galbani had a turnover of 1.595 trillion lire, and employed 7,200 workers. Bsn-Danone's entrance into the Italian market had been made possible by its alliance with Ifil. The investment holding company controlled by the Agnelli family and managed by Umberto

---

52 "Galateri illustra l'operazione: così l'unione Galbani-Ifil", *Il sole 24 ore*, 21.07.1989. See also Knopp M., Wilson B., *Europe's major dairy*, *op. cit.*, pp. 99–101; Nomisma, *Il sistema agro-tecnologico verso il 2000: per una prospettiva europea. Conferenza internazionale (Bologna 19–20 settembre 1988)*, Bologna, Nomisma, 1989, pp. 23–8; Londeix O., *Le biscuit et son marché*, Rennes, Presses universitaires de Rennes, 2012, pp. 310–15.

53 Galbani S.p.A. operational budget, 1983–8.

54 The five Lichtenstein-based investors (Kraven, Lanser, Luframil, Monti and Cif) who controlled Galbani had come together at the end of the 1960s in a highly turbulent economic and currency exchange context, and had now joined together in the newly constituted Galbani finanziaria, which incorporated Egidio Galbani S.p.A., and was in turn incorporated by Ifil Partecipazioni (65% Ifil and 35% Bsn), which changed its name to Spa Egidio Galbani.

Agnelli had initiated a campaign of diversified investments which included food products.[55] In 1990, Bsn acquired a 15% share of Galbani from Ifil, in 1992 another 10%, 10% more in 1993, 5% in 1995, 5% in 1996, and in 2000 yet another 5%, and in 2001 it became a majority shareholder by purchasing another 5% share.[56]

Bsn-Ifil's takeover was accompanied, as expected, by a 10% increase in sales (+1.59 trillion lire) along with investments worth 53 billion lire aimed at raising Gorgonzola production levels (in the Corte Olona factory) destined for the European market, and a wholesale restructuring of the line of ricotta products (made in Bozzolo). Plans were also made to rebrand the company's more traditional products (Bel Paese, Santa Lucia and Certosa), and products were generally readied for placement on foreign markets. The fresh cheese product line was strengthened and a cured meats brand (Casa Romagnoli) was heavily promoted. The company also owned its own pig farms (in Bozzolo and Certosa di Pavia). The new owners expected to achieve around 151 billion lire in sales abroad. Galbani itself became a minority shareholder of Fiat S.p.A., purchasing a 7% share. The new owners had also gained control of Galbani's subsidiary companies, and were now looking at ways to improve its sales network, for which they foresaw a need for 3,500 extra employees and 2,400 new vehicles.[57]

---

55 In the 1980s the Fiat group initiated an extra-industrial diversification strategy both within the group and outside with the help of Ifil. It is worth noting that, in the 1990s, this strategy saw the Fiat group grow to include not only Fiat car makers, but also companies that operated in a variety of other sectors: publishing and advertising, insurance, financial services, estate agents, production and distribution of electricity. In the same period Ifil acted as a container company not only for the Fiat group, but also for other companies active in mass distribution sectors, the production and distribution of paper products, inspection and certification of raw materials and industrial products, and tourism and hotel industries. See Nomisma, *Rapporto sulla ristrutturazione dell'industria italiana. Laboratorio di politica industriale*, Bologna, Nomisma, 1985, pp. 23–46.

56 The operation was overseen by the Luxembourg bank Kredietbank, and the shares had been ceded by five Italian companies. The partners and administrators involved were Rinaldo (who died in 1977) and Antonio Invernizzi, together with a number of associates connected to the family who were also linked to the Banca del Gottardo, which was controlled by the Banco Ambrosiano. Costant Franssens, who was president of Galbani from 1978 and administrator and executive director of the Luxembourg bank, was the main overseer of the operation. In 1981, just before the Banco Ambrosiano crash, the aforementioned associates were replaced with other figures from Kredietbank group, which was also made a fiduciary of the *Anstalts*. Other members of Galbani's administrative board included Carl von Castelberg, who was also a trustee of the Banco Ambrosiano and a lawyer from Ticino, Fernando Garzoni, president of the Banco del Gottardo and directly involved in the management of Galbani. All these people were gradually replaced after the bankruptcy of the Banca Ambrosiano.

57 President's speech to Galbani shareholders, 1989 and 1990.

4. The dairy industry in the 1980s and 1990s

Italy still ran a high import/export deficit in the milk sector at the beginning of the 1990s. This was compounded by the industry's weak image abroad, and its inability, apart from a few exceptions, to make the most of its unique dairy traditions.[58] On the one hand, medium sized businesses and a large number of small dairies (1,200 according to industry surveys) were concentrating on the development of traditional products, and in particular of DOP cheeses. Specialized cheese products, whose popularity was growing in response to the increasingly demanding tastes of the consumer market, underwent a further development when the link between cheeses and their traditional area of production was strongly emphasized.[59] Latte Sole is an example. In the 1950s, Salvatore Puglisi Cosentino in Catania (Sicily) produced in a small company sterilized milk in glass bottles with the label "Milk the Sun". In 1963, he launched the "Latte Gala" micronized cream, packaged in tetrapaks. In the 1970s, the Sicilian firm attempted the conquest of the domestic market by acquiring companies in Central and Northern Italy while strengthening the corporate image and brand creation: the grass and the sun. The European grants had been a condition for the development. During the 1980s and 1990s, Gala started a consolidation plan with its plant in Ragusa for the production of fresh cheese and yogurt.[60]

Conversely, in 1989 almost 70% of cheese produced in Italy was represented by only five varieties: Grana Padano, Parmigiano-Reggiano, Gorgonzola, Pecorino Romano and Provolone Valpadana. This period saw a steady rise in fresh cheese sales volumes, although this was not accompanied by a similar rise in profits due to the expensive promotional campaigns that arose from intense competition between producers. It had become more difficult to gain entrance into the dairy market without a high level of experience in the manufacturing process, an efficient distribution network and a powerful brand name, which had become prerequisites in a highly competitive industry where producers needed to be able to negotiate access to mass distribution networks.

Twenty-seven per cent of the fresh cheese market was in the hands of just a few large companies under the control of Galbani-Danone and Kraft Foods Italia, with, respectively, 15% and 12% of the market. The disproportionate influence of the cost of raw materials over other elements of the production process meant that these companies needed to prioritize innovative methods for raising the value of their products so as to increase profit margins. Medium sized companies generally focused their attention

58 Assolatte, Italian milk and dairy industry report 2005, Milan 2006.

59 Istat, *Fatturato, prodotto lordo e investimenti delle imprese commerciali, dei trasporti e comunicazioni* [con almeno 20 addetti], Roma, 1983–93.

60 Since 2000 Gala Italia has changed its trademark to Latte Sole S.p.A. Moreover, Parmalat Group controls the Sicilian firm.

on one specific product range, and invested in efforts to increase their own brand recognition with the intention of reaching a wider consumer base. The smaller businesses, often built around the production of just one product, suffered from a limited distribution network and were finding it difficult to conform to new quality control regulations and technological requirements, and were therefore gradually being incorporated into larger companies.

The decisive strategy in the fresh cheese market was that of product innovation, which was designed to respond to consumer demand while at the same time presenting brands that differentiated themselves from their competitors. Rationalization of logistical costs and the broadening of distribution networks were seen as essential in a scenario of limited profit margins and fast production rates necessary to satisfy a mass market.

In the second half of the 1980s the mature soft cheese industry tended to be represented by traditional products, while "industrial" mature cheeses were generally imported from abroad. This was a somewhat fragmented market, led by Galbani and Ferrari (of Lodi), which introduced a new product called "il gran mix ricetta classica", a combination of grated cheeses including Parmigiano-Reggiano, Grana and Emmental which was packaged and sealed while fresh.[61] The challenge that now presented itself to companies wishing to improve their market strategy was that of the creation of regional operational centers with the function of presenting a unified product range and strengthening their negotiating position and their commercial performance. Within this sub-sector were a number of companies that manufactured high value products, while a group of medium sized companies specialized in the production of traditional cheeses for which the advertising budget was small but quality was kept high; this meant that they needed to achieve a respected status in their often limited catchment areas. Other companies aimed at using retail networks to achieve nation-wide coverage, and prioritized imported products which they sought to differentiate through package design and a range of different portion sizes (Ferrari and Prealpi). Companies on the rise like Auricchio and Mauri adopted the policy of a *core business*, concentrating on one specific product with the aim of placing it in shops right across the country. The strategy adopted by medium sized companies involved the development of an image connected to high quality traditional products. The entire sub-sector pursued a policy of separate product lines by offering cheaper alternative brands.

The hard cheese sector (Grana Padano and Reggiano) continued to operate according to older models in which collectives played a central role, acting as centers of aggregation for small and medium sized producers. The processing sector was in a weaker position with respect to that of resellers,

---

61 The Ferrari company was a family business led since 1960 by Giovanni Ferrari which produced butter and Grana Padano. At the beginning of the 1990s, Auricchio and Bayernland set up production lines and market strategies that made it possible for them to conquer a large slice of the market.

because wholesalers also carried out the storage and seasoning phases, making them indispensable to the dairies which did not have the infrastructure or financial means to season their own products.

Conclusions

During the 1970s, the food manufacturing industry increased after the implementation of wide-ranging structural changes while the production was being overtaken by the consumer base as the driving force within the industry. The landscape was also characterized by an increase in the number of producers specializing in local and traditional products.

By the end of the 1970s the milk and dairy sector represented 13% of the food industry. Between 1970 and 1979 the milk and dairy industry achieved an average yearly growth of 3.2%. The highly profitable food processing sector was a further element in this overall growth, with a 13.2% average increase in overall earnings.

The production of milk and its derivatives has a relevant value for the rural sector. Almost three-quarters of Italy's cheese production in the 1980s were represented by five cheeses: Grana, Parmesan, Gorgonzola, Pecorino and Provolone. Apart from the more dispersive nature of Parmesan production, the manufacturers of these cheeses operated medium sized factories in an industry dominated by four large groups: Galbani, Locatelli-Nestlé, Invernizzi and Polenghi Lombardo. During this period the large businesses were able to guarantee a higher level of quality than smaller companies, despite the fact that a refinement in consumer tastes had forced them to offer a wider range of niche products.

The dairy industry was becoming interconnected with other industry sectors: what was once a "static force" had been transformed into a strategic front line in the country's economic development. At the same time, the modernization of the distribution infrastructure made the industry more susceptible to influences and market penetration from foreign competitors and large business groups.

The 1980s were also a period of substantial change for the entire dairy industry in line with the transformations that were taking place in the food manufacturing industry as a whole. Unlike in the private sector, institutional changes were having a marked effect on public intervention policies, affecting relations between the state and regional governments (after a referendum that had led to the abolition of the Ministry of Agriculture), although these changes were most heavily felt in relations between the European Community, the State and regional governments.

The campaign of acquisitions and mergers was a sizeable phenomenon – representing 10% of all purchases registered throughout the 1980s – and mirrored similar processes taking place in the chemical and mechanical sectors. This reorganization was not simply aimed at gaining control of companies; it also involved the introduction of management and technical innovations from which many dairy businesses benefited, although there were numerous exceptions throughout the sector.

Between 1983 and 1986 foreign companies began to appear on the market with the intention of enlarging their product portfolio and increasing their geographical reach, while at the same time some medium sized producers and the cooperatives began to strengthen their international position. In both cases companies were taking advantage of the stabilizing factor produced by the supply of raw materials under a European controlled price regime. The large companies operating in the sector, which were mainly French, were able to operate in the international market with the benefit of having access to powdered milk and butter surpluses at lower prices. Italy's agricultural industry, on the other hand, was unable to obtain any particular benefit from these European measures because devaluation of the national currency led to the animal husbandry sectors of countries with a strong currency being granted "monetary compensation payments".

The large national companies are influenced by the strategies of multinationals. The causes are identified, firstly, in the complex transition between the different generations of entrepreneurs which weakens strategic power and management, especially for the historical brands. At the same time, the combination between national actors to hold off national competition fails while the production chain maintains a high degree of segmentation. The cooperative organizations cannot express an industrial project with a national dimension for local cases and identify strongly with the different cases of crisis management. The most important case is Federconsorzi and Polenghi Lombardo.

The transfer of ownership is not determined primarily by errors in the manufacturing and trade, but by a growing lack of preparation to handle the changes in a market that was impacted by rising production costs imposed by changes in the national economy. The complex competition with multinationals is determined not only by the integration of the systems, but also by consumption styles.

For a large multinational, acquiring an established company was an alternative to pursuing a policy of internal growth. The competition between these companies was entirely based on the need to increase their volume of sales rather than on a policy of integration, an approach dictated by the low profit margins of the products that made up these markets.

The marketing sector was also undergoing a streamlining process. A period of innovation on the part of cheese producers in terms of brand promotion and advertising was followed, around the beginning of the 1980s, by a phase that was characterized by the shoring up of existing brands through a concerted effort at improving the efficiency of their sales networks.

Italy still ran a high import/export deficit in the milk sector at the beginning of the 1990s. This was compounded by the industry's weak image abroad, and its inability, apart from a few exceptions, to make the most of its unique dairy traditions. Medium sized businesses were concentrating on the development of traditional products and in particular on the "DOP", while in 1989 almost 70% of cheese produced in Italy was represented by only five varieties: Grana Padano, Parmigiano-Reggiano, Gorgonzola, Pecorino Romano and Provolone Valpadana.

A large part of the fresh cheese market was in the hands of just a few large companies under the control of Galbani-Danone and Kraft Foods Italia, with, respectively, 15% and 12% of the market. The disproportionate influence of the cost of raw materials over other elements of the production process meant that these companies needed to prioritize innovative methods of production. Faced with these trends, some Italian producers develop strategies of resistance and in some cases of transformation by focusing on production niches. The specialized products make use of innovations in the production and marketing, and in a long-term perspective of the new production model of the "terroir".

The decisive strategy in the cheese market was that of product innovation, which was designed to respond to consumer demand while at the same time presenting brands that differentiated themselves from their competitors. Rationalization of logistical costs and the broadening of distribution networks were seen as essential in a scenario of limited profit margins and fast production rates necessary to satisfy a mass market.

Abstracts and Notes on Contributors

Claire Delfosse, *Cheese-making in France at the Beginning of the Twentieth Century: A Spatially Diffuse, Complex and Unique Organization?*

Abstract. In the late nineteenth century, scientific and technical innovations, the development of dairies and new trade conditions led to a process of differentiation, which modified the geography of cheese-making in France. In parallel with regional agricultural specializations and the birth of the cheese industry, successful systems for cheese production were being affirmed. Truly successful cheese production systems are affirmed. These cheese-making systems may be likened to productive systems. They combine a type of dairy business, a preferential cheese, a dairy breed and a marketing mode. Four systems can be identified for cheeses, and one for butter: the cooperatives of the Charentes area. They spread innovation and contributed to areas of expansion, which reached their peak on the eve of the Second World War, transforming local economies. However, the cheese-making systems did not cover the whole of France. They met other dairy specializations and did not disappear completely from the areas of farmhouse production of butter and cheese which had been identified from the middle of the nineteenth century.

Claire Delfosse is Professor of Geography at the University of Lyon 2. She heads the Laboratory of Rural Studies, a multidisciplinary research laboratory. Her research focuses on the geo-history of dairy products, the protection of historical food products, food quality, and food and rural-urban linkages. She has recently published a book on the history of the craft of cheese and dairy in France since the nineteenth century (*The Business of the Dairyman-cheesemaker from 1850 to the Present*, Boulogne, Editions de la Mer du Nord, 2014). Her main publications are *France the Cheesemaker (1880–1990)*, Paris, Boutique de l'histoire, 2007; *Brie Stories*, Deauville, Edn des musées, 2008; and *The Terroir: Fashion and Food*, Paris, Les Indes savantes, 2011.

Luigi Lorenzetti, *Switzerland's Dairy Industry in the Twentieth Century: Small-Scale Artisan Production Meets the Market Economy*

Abstract. At the turn of the twentieth century, the Swiss dairy industry had a strong tendency to export. However, with the First World War a new scenario emerged, and in order to ensure that food requirements were met in the country, the Federal Government promoted a "controlled" market regime that bound several players to some precise rules regarding milk and dairy production. This regime, which was based on a system of public subsidies to support the whole sector, was crucial in guaranteeing milk and dairy producers adequate production prices. After the Second World War, following a fall in milk consumption, the regime fostered a steady increase in cheese production that had to be absorbed by exports, supported by growing public subsidies. Their suppression at the end of the 1990s decreed the fall of exports of certain types of cheese and a partial concentration on the dairy sector.

Luigi Lorenzetti is a Professor at the Università della Svizzera italiana (Accademia di architettura, Mendrisio), where he conducts the Laboratorio di Storia delle Alpi. He is editor-in-chief of the review *Histoire des Alpes – Storia delle Alpi – Geschichte der Alpen.* He has published widely on economic and social history of the Alps in the modern and contemporary age. His research interests include the history of populations, the history of rural societies and the history of territory.

Claudio Besana, *The Italian Dairy Industry between 1930 and 1970. Production and Organizational Structure*

Abstract. The first part of the contribution analyses the Italian dairy industry, using data from the 1937 industrial census. This study shows a complex sector, characterized by the presence of numerous small and medium sized companies and by a strong link between milk production and cheese-making. The second part of the chapter deals with the main trends in the industry between the end of the Second World War and the beginning of a real integration of EC markets. The increase in production and in imports, as well as the difficulties of exporting, are highlighted.

Claudio Besana is Associate Professor of Economic and Business History at the University Cattolica del Sacro Cuore of Milan. His studies have focused on economic and social patterns in Italy, and especially in Lombardy, from the eighteenth to the twentieth century. At present his research activity is focused on the Italian food industry in the nineteenth and twentieth centuries. Among his recent publications are *Tra agricoltura e industria. Il settore caseario nella Lombardia dell'Ottocento* (2012) and (with A.M. Locatelli) *Italian Family Business and Multinationals in the Eighties of the twentieth century: Invernizzi/Kraft and Galbani/Danone* (2015).

Stefanella Stranieri and Paolo Tedeschi, *The Influence of European Institutions and Policies on the Italian Dairy Sector*

Abstract. The chapter shows how European institutions and policies influenced the Italian dairy sector from the 1960s up to the new millennium. It illustrates in particular the effects of the Common Agricultural Policy on the productions and price of milk, the productive systems of the dairy sector, the market and the foreign competition for the Italian dairy products.

All the decisions taken by the European institutions in fact modified the existing conditions of European dairy markets and obliged the dairy firms to adapt in order to function optimally and at the same time to maintain their fully competitiveness. As European institutions were conscious of the relevant influence of their choices and policies on the development of the dairy sector, so they also controlled different phases of dairy production and its quality labeling system. Besides, the chapter shows how a relevant European institution, such as the European Investment Bank (EIB), helped the Italian dairy sector. The EIB in fact financed the improvement of the infrastructures which linked the Italian dairy factories to the European markets. Moreover, it partially financed the renewal of some dairy plants in the Southern Italy.

Stefanella Stranieri is a lecturer in Agro-food Economics at the University of Milan DEMM. Her research and publications focus on supply chain analysis, consumer behavior and food quality standards. Email: stefanella.stranieri@unimi.it.

Paolo Tedeschi is an Associate Professor of Economic History at the University of Milan-Bicocca DEMS. His research and publications focus on the history of European integration, the economic history of rural and Alpine Lombardy (18th-20th centuries), and the history of Lombard business interest associations, trade unions and friendly societies. E-mail: paolo.tedeschi@unimib.it.

Benoit Mario Papillon, *The Canadian Dairy Sector in the Twentieth Century: From Openness to Closure and Self-Sufficiency*

Abstract. Over the last few decades, the Canadian dairy sector has been a central issue in trade negotiations for the Government of Canada (GATT/WTA level/Uruguay round, NAFTA, more recently with the European Community, and currently with Asian and other Pacific Rim countries). The first part of the chapter describes the Canadian dairy sector, which, historically, contributed significantly to Canadian exports. With the introduction of the "supply management system" nearly half a century ago, Canadian dairy exports have gradually vanished almost completely, and the Canadian price of dairy products, including milk, has become much higher than in otherwise comparable countries such as the USA or France. This system is described in the second part (production and import quotas, decision-making processes and coordination along the value chain). In many places, dairy products are basic ingredients of human diets, particularly for those of young ages. The third part of the chapter takes a look at access to dairy products in terms of pricing under the "supply management system" and at some welfare consequences for participants in dairy product transactions.

Before entering academia, Benoit Mario Papillon worked for the Canadian Government in various research groups relating to trade policy issues, financial markets and regulation reform. He holds a Ph.D. from Northwestern University (including Finance in Kellogg Management School). Besides publications on the dairy sector (Measurement of non-tariff barriers, *Contemporary Economic Policy*, *Canadian Journal of Agricultural Economics*), his published research includes writings on economic history, the theory of the firm, bankruptcy law and finance.

Hildete De Moraes Vodopives and Valter Galan, *Contemporary Challenges to the Brazilian Dairy Industry*

Abstract. As the largest country in terms of arable land, Brazil is one of the emerging dairy agri-powerhouses challenging the supremacy of Europe and North America. Nonetheless, the Brazilian dairy industry faces an array of challenges and contradictions. Ranking number one global supplier of beef and among the largest possessor of dairy herds in the world, Brazil remains a net importer of milk. Furthermore, in the context of contemporary globalization, the industry has to adapt to higher quality standards, and compete in the international market. Looking back on the twentieth century, the chapter identifies the facts impacting on the industry. These include macro-economic policies, regulation, technological advances, and the industry level of consolidation.

Ms Vodopives is a partner in Investfort Asset Management and a member of the Brazilian Financial Analysts Association – Apimec, where she served as director of corporate relations and on the board of directors. She coordinates the Harvard Strategists and the Brazil at Harvard groups, and is a member of the Harvard Crimson Compass Mentor program. Ms Vodopives holds a Masters in History from Paris Sorbonne University, an MBA in Capital Markets from PUC, and a Certificate in Special Studies in Administration and Management from Harvard University

Valter Galan is an Agronomist specialist, a partner of MilkPoint Inteligencia in Brazil. A former LATAM Manager of milk sourcing at Nestlé Brazil, he holds a Masters in Business Administration from the University of São Paulo.

Julia Lapp, *New York State Cheese: Nineteenth and Twentieth Century Transformations*

Abstract. New York state was the leading producer of cheeses for a span of approximately a hundred years, from the early nineteenth century to shortly before the First World War. Many innovations in North American cheese production occurred in the state that fostered the development of cheese production elsewhere, such as in Wisconsin and Canada. Italian cheeses have also been an important sector of New York state (and North American) cheese industries, largely arising from the large Italian immigrant population that settled in the New York city area through the Italian diaspora. This chapter provides an overview of the trajectory of cheese production from the early nineteenth century up until the mid-twentieth century, by which time Wisconsin had surpassed New York state in annual cheese production.

Julia Lapp, Ph.D. is a Nutritionist and Nutritional Anthropologist who is researching contemporary cheese-making in New York state. She teaches classes in nutrition and food systems as an Associate Professor in the Department of Health Promotion and Physical Education at Ithaca College in Ithaca, New York.

Stefano Magagnoli, *Parmigiano-Reggiano Cheese: The Industrialization of Typicality*

Abstract. Parmigiano-Reggiano cheese is one of the best-known typical products in the world, and enjoys an excellent reputation, deeply rooted in history. As far back as the fourteenth century it was mentioned by poets and writers as a symbol of opulence and good living, while in the sixteenth and seventeenth centuries it appeared in the cookbooks of the Renaissance. In those years, however, Parmigiano-Reggiano cheese was an expensive luxury product that needed controls that ensured the highest quality. A turning point occurred in the last decades of the nineteenth century, when the demand for Parmigiano strongly increased as a result of the expansion of markets. The Parmigiano brand was known worldwide, so producers were able to exploit the reputation of the product and the fame of Parma as a marketing tool. The beginning of a true communication strategy for the product took place in the early 1960s, in parallel with the expansion of markets in the economic boom.

Stefano Magagnoli is Associate Professor of Economic History at the University of Parma and Chercheur Associé of the Centre d'Etudes des Mondes Moderne et Contemporaine of the Université Montaigne – Bordeaux 3. He is among the founders of Food Lab made up within the University of Parma, and a member of the investigation team of the research project TERESMA Produits des terroirs, espaces et marchés, hier et aujourd'hui (2016–20) funded by the Région Aquitaine.

Sylvie Vabre, *Roquefort: a Large Company Creates Its Territory*

Abstract. Today, identifying food territorially appears to be "naturally" attached to craft production targeting a small market. Yet the Roquefort model, the first to be certified by the French Parliament (1925), shows that in fact there was nothing natural about it: quite the contrary. The chapter shows how the anchoring of this model in the territory through the law was in fact driven by the company's strategy to reach national and international markets. Breaking into these markets and staying in them were the reasons for the company's choices, especially with the radical shift caused by the First World War.

Sylvie Vabre is Associate Professor in modern history at Toulouse Jean Jaures University. Her Ph.D., *Le sacre du roquefort – l'émergence d'une industrie agroalimentaire fin XVIIIe siècle – 1925*, was published in March 2015. Her researches and teaching are dedicated to consumer history, business history and the history of local food productions.

Rita d'Errico, *The Production of Pecorino in The Roman Countryside from the End of the Nineteenth Century up until the 1930s*

Abstract. Pecorino cheese has been made for centuries in the Agro Romano, and was the result of a complex production chain that began with the requirement of winter pasture lands for sheep flocks from the Central Apennine mountains and ended up in the city, where urban groceries seasoned the cheese before selling it. Towards the end of the nineteenth century, Pecorino romano rose to become one of the most exported Italian cheeses thanks to a combination of factors occurring beyond the local context, such as the increase in demand represented by Italian emigrants abroad and the abilities of wholesalers and exporters. At the same time as Pecorino was cementing its position in the international market, its production base was extended to Sardinia thanks to the arrival on the island of dairy experts from Latium. While production in Sardinia left behind its pastoral origins to take on industrial characteristics, sheep farmers in Latium continued to produce Pecorino using traditional techniques; the decline of cheese production in Latium only began after the transformation of the Agro to arable land and the consequent disappearance of the migratory system.

Rita d'Errico is Associate Professor at the Department of Business studies – Roma Tre University – where she teaches Economics, Finance and Business History. Since 1995 she has been a member of the Scientific Council of the University Centre for the Study of Rome (CROMA). Her research interests focus on two specific fields: analysis of banking and credit history; and analysis of economic and social transformations in Rome and Lazio from the nineteenth to the twentieth century.

Giuseppe Doneddu, *The Production and Processing of Sheep Milk In Sardinia (1950–2015)*

Abstract. Between 1850 and 1950 Sardinia was characterized by chronic under population, and it remains largely tied to a combination of cereal and sheep farming. The intervention that followed to improve the land and agrarian management and to change the adverse natural conditions did not deliver the expected results. Despite the difficulties, the processing activities undertaken in the dairy sector allow the maintenance of a substantial trade balance. The late twentieth century opens with high expectations regarding the total eradication of malaria. However, the cereal industry enters into crisis once and for all, and the rural population is largely forced to emigrate. Only the dairy sector seems to resist, despite growing difficulties, the crisis taking place at the start of the millennium.

A full Professor of Economic History at the University of Sassari, Giuseppe Doneddu initially dedicated his research to the events occurring in Sardinia during the contemporary age, with particular reference to the mechanism of feudal management and issues of land. His interests progressively moved towards maritime economic problems, and more particularly towards fishery. Prof. Doneddu currently aims to investigate the state of insularity and economic development, the role of the institutions and the phases of production and commercialization, and to identify the human and financial resources used in a long-term perspective.

Renato Ghezzi, *The Tuscan Ovine Milk Industry, 1950–2000: A Quantitative Analysis*

Abstract. During the 1950s, Tuscan ovine farming found itself in difficulty. The situation grew critical, due to farmers moving away from the countryside and from the agricultural reform which brought about a restriction of pastures and herding, promoting the development of small farming properties in place of large sharecropping farms and favoring the growth of farming. In the second half of the 1960s a new expansionary pattern began, mostly thanks to the Sardinian immigration. Sardinian herders undertook the job of recovering and reconstructing the old colonial farmhouses to create their own farms specifically for the production of milk. During the same years, an important cooperative movement began to establish itself in the area and brought about several new elements in the field of cheese factories. The majority of Tuscan cooperatives joined together in two groups, the Lugalat and the Castor. Thanks to the successful initiative of the Castor group, in 1987, Tuscan sheep cheese was recognized as a registered name cheese brand. Nine years later it obtained the PDO trade mark.

Renato Ghezzi is Assistant Professor in Economic History at the University "Magna Graecia" of Catanzaro, where he teaches History of Economic Thought and History of Economic Analysis. His research mainly pertains to seventeenth-century Sea Trade and the modern economy of Tuscany.

Vanessa Pollastro, *From Milan to Novara: The Production and Sale of Gorgonzola in the Twentieth Century*

Abstract. The aim of this chapter is to show the changes in Gorgonzola production and marketing throughout the twentieth century. At the end of the nineteenth century, Gorgonzola production moved from the mountains of the north of Italy to the lowlands along the Po valley. The production and ripening of this cheese lasted only a few months, and it was traded in the big market of Milan. After the First World War its exports increased, thanks to its competitive price and to the Italian migration flow to Europe and America. After the Second World War, the strong competition with European products and the resulting loss of export markets led producers to search for solutions: in the 1970s they founded a consortium for the protection and defense of the product quality, and in the 1980s protection policies were introduced. Meanwhile consumer tastes were changing, and "mild" Gorgonzola achieved popularity in foreign markets. Throughout these years several changes also took place in production: many producers started to ripen their own cheeses and the wholesalers gradually lost their importance due to the introduction of large distribution.

Vanessa Pollastro (Ph.D. in Business History and Corporate Finance, University of Milan) is a Research assistant of Economic History and Labor History in the Faculties of Economics and Languages, Catholic University of Milan, and archivist at the "Mario Romani" Archive for the History of the Catholic Social Movement in Italy. Her research topics are: the history of industrial relations and trade unions, women's and business associations, and rural history.

Franca Pirolo, *Buffalo Mozzarella in Campania from Its Origins to the Twenty-first Century*

Abstract. The importance that the market of cheese products has assumed over time at both national and international level leads us to consider also the importance of fresh cheeses, despite the scarcity of documentation. Among them is Mozzarella, production of which takes place in the same areas of milk production, where the figure of the farmer often coincides with those of the producer and the seller. Buffalo Mozzarella in particular, thanks to the type of milk used and the close interrelation with the territory, has constituted one of the most important sectors of production for the compartment in Campania and in some areas of the south of Lazio.

This chapter aims to offer a historical but also an economic reconstruction of Buffalo Mozzarella. It traces its history from its origins and its evolution over time, from the first production processes to the improvements in manufacturing techniques, to the gradual and progressive expansion of the sales markets, and, finally, to the need to protect this product against fraud and adulteration. This problem induced the traditional producers to claim a trademark policy in order to identify with the name "Mozzarella di bufala campana" DOC only fresh cheese produced exclusively from whole buffalo milk and manufactured in the farms and dairies of the Campania coastal strip through the institution of a Guardian Consortium in 1993.

Franca Pirolo is Assistant Professor of Economic History and History of Economic Thought in the University of Catania. She has participated in a scientific and organizing capacity in many conferences and as a member of research groups funded by MURST, and has taken part as a speaker in seminars and conferences of national interest.

Manuel Vaquero Piñeiro, *The Modernization of the Dairy Industry in Italy in the Late Nineteenth Century: Social Dairies*

Abstract. The chapter analyses the birth and consolidation of dairy cooperatives in Italy in the period spanning the nineteenth and the twentieth centuries. From the 1870s, the Italian government began to encourage the creation of cooperatives and associations of dairy farmers in order to promote the industrial production of cheese and butter. Italy had to learn the technique of milk production in order to compete with other countries such as France and Switzerland, which attached great importance to exporting cheese. In particular, the phenomenon of dairy cooperative formation involved the Alpine areas, where social and religious factors stimulated the birth of cooperatives. At the same time, other associative models, more business-oriented, were created in Lombardy and the Emilia plains. So there is a dualism between the management system of mountain areas, based on traditional techniques, and the plain areas, exposed to international trade and competition. However, the dairy cooperatives played a prominent role in the modernization of the Italian dairy sector.

Manuel Vaquero Piñeiro has a Ph.D. in Modern and Contemporary History from the University of Cantabria (Spain). Since 2006 he has been Associate Professor of Economic History in the University of Perugia (Italy). In 2010 he received the international award Daria Borghese for studies of the history of the Renaissance in Rome. His main research interests include the socio-economic transformation of agriculture.

Silvia Antonia Conca Messina, *Enterprise, Trade and Industry in the Lombardy Dairy Sector: The Origins of Locatelli and Galbani (1860–1914)*

Abstract. Lombardy had a long empirical tradition of dairy production. In the last three decades of the nineteenth century the region confirmed itself as Italy's largest producer, and in the 1890s it accounted for around one third of all cheese production and exports (especially Grana and Gorgonzola). Names such as Locatelli, Galbani, Invernizzi and Polenghi Lombardo are among the most important enterprises which emerged during this period, whose brands have lasted until today. In 1950 these companies employed nearly 40% of all the workers in the dairy industry. The chapter focuses first on the main directions and features of the transformation of the dairy sector in nineteenth- to early twentieth-century Lombardy. It then traces the origins and the business strategies of the family firms, particularly Locatelli and Galbani, which transformed dairy production from a craft to an industry, examining the leading local role they played as merchant-entrepreneurs in the collection, ageing and trade of cheese and their efforts to develop commercial relations and new markets, which seem to have been particularly important in promoting the growth of production.

Silvia Antonia Conca Messina is Assistant Professor of Economic History at the Department of Historical Studies at the University of Milan. Her main research interests are: Italian Business History in the eighteenth to twentieth centuries (the nobility as entrepreneurs; silk, cotton, agro-industry; local credit systems) and European economic history between the early modern age and the Industrial Revolution. Her latest publications include *Cotton Enterprises: Networks and Strategies: Lombardy in the Industrial Revolution 1815–1860*, Routledge, 2016.

Gianpiero Fumi, *The Rise and Decline of a Large Company: Polenghi Lombardo*

Abstract. Polenghi Lombardo has long been a major protagonist in the development of the dairy industry in Italy. Although it was not a cooperative, since its origin in late nineteenth century it was strongly linked to Lombard and Italian agriculture, even in property and control. Large-scale production, diversification and flexibility – due to a complete exploitation cycle of milk – gave the company strengths superior to those of other producers. This allowed it to expand in Italy and win relevant market shares abroad. The chapter summarizes also the crisis the company had to face with the emergence of new ownership in the period following the Second World War (when the majority shareholding passed under the control of the main Italian agricultural cooperative group), and then the pressure of foreign competition on the dairy sector due to the European Common Market and the Common Agricultural Policy. In early 1990s, the acquisition of the company by competitors with inconsistent industrial strategies determined the gradual downsizing and actual disappearance of Polenghi Lombardo.

Gianpiero Fumi is Associate Professor at the Università Cattolica del Sacro Cuore, Milan and Rome, where he teaches Economic History and Business History. His research concerns Italian economic history, focusing especially on technical cultures and vocational education, enterprises and interest organizations, and public policies, at a regional and national scale. Recent publications: "Imprenditori culturali. L'impegno nella gestione del Teatro alla Scala" (2014); "La terra migliora l'uomo. Le colonie agricole per la gioventù 'irregolare' nell'Italia del secolo XIX" (2015); "L'esercito lavorante. L'industria lombarda e la guerra 1914–1918" (2015).

Andrea Maria Locatelli, *The Multinationals and the Italian Agri-food Industries: The Dairy Sector (1974–1993)*

Abstract. The chapter discusses the historical evolution of the dairy-cheese chain between the 1970s and the 1990s. During the 1970s, productivity and profits increased after the implementation of wide-ranging structural changes: the production was being overtaken by the consumer base as the driving force within the industry. The increasing size of companies, the appearance of foreign competitors and the increase in producers specializing in local and traditional products were the new elements. A growing level of investment flooded into the sector throughout the 1980s. The dairy sector was dominated by four large groups: Galbani, Locatelli-Nestlé, Invernizzi and Polenghi Lombardo. They guaranteed a higher level of quality than smaller companies, despite the fact that a refinement in consumer tastes had forced them to offer a wider range of niche products. The national large companies are influenced by the strategies of multinationals: the complex transition between the different generations of entrepreneurs. The combination of national actors fails while the production chain maintains a high degree of segmentation. The cooperative organizations cannot express an industrial project with a national dimension. The campaign of acquisitions and mergers was a sizeable phenomenon. The institutional changes were having a marked effect on public intervention policies, affecting relations between the state and regional governments (after a referendum that had led to the abolition of the Ministry of Agriculture), although these changes were most heavily felt in relations between the European Community, the State and regional governments.

Andrea Maria Locatelli is Associate Professor of Economic History in the Faculty of Economics, Università Cattolica del Sacro Cuore, Milan (Italy). His research topics are fiscal policy, rural history, business history, history of financial systems, and the European integration process. Among his publications are: "A new common knowledge in agronomics: the network of the European Agrarian Reviews and Congresses during the first half of the 19th century", in Aprile S., Cassina C., Darriulat P., Leboutte R. (eds.), *Europe de papier. Projets européens au XIXe siècle*, Villeneuve-d'Ascq, Presses Universitaires du Septentrion, 2015, pp. 187–204; "Le trasformazioni del caseificio Lombardo tra Ottocento e Novecento. Il peculiare percorso della regione di montagna", in *Des manufactures aux fabriques. Les transformations industrielles, XVIIIe-XXe siècles*, Zurich, Chronos Verlag, 2015, pp. 125–48; "Italian family business and multinationals in the Eighties of the 20th century: Invernizzi/Kraft and Galbani/Danone", in *Revue Française d'histoire Économique*, 3, 2015, 1, pp. 116–29; and "Entre réforme fiscale et développment économique: les cadastres en Lombardie aux XVIIIe et XIXe siècles", in Bourillon F., Vivier N. (eds.), *La mesure cadastrale: estimer la valeur du foncier*, Rennes, Presses Universitaires de Rennes, 2012, pp. 19–41.

Index*

*This index does not include the names of authors mentioned in the bibliography and those inserted in the figures.

V

W

Y

Z

European Food Issues

For several decades now, our attention has been drawn to expanding agricultural output and the proliferation of powerful food companies. At the same time, in the process of European integration, the adoption of the Codex Alimentarius (1963), the Food Law (2002), and the recognition of PGIs for many products have contributed to the creation of a common European "food space". Today, these systems of supply and distribution have between them given Europeans quite varied dietary possibilities. This situation stems from various developments, linking the economic to the technical and amounting to a long-term trend. Cultural issues bear upon this, whether culinary transmission from generation to generation or the increasingly diverse catering sector, and political decisions also contribute through the establishment of standards and regulations. Hence, traditions and ruptures, innovations and continuities are permanently unsettling the European diet. Using original sources, doctoral theses, conference papers, monographs and testimonies, this series examines historical developments at the national scale and also, more generally, in a transnational perspective. The series hopes to make a significant contribution to understanding the processes of food innovation, which are powerful factors of difference and identity in contemporary Europe.

Series Titles

No. 1 – Antonella CAMPANINI, Peter SCHOLLIERS et Jean-Pierre WILLIOT (dir.), *Manger en Europe. Patrimoines, échanges, identités*, 2011.

No. 2 – Daniëlle DE VOOGHT, *The King Invites. Performing Power at a Courtly Dining Table*, 2012.

No. 3 – Josep L. BARONA, *From Hunger to Malnutrition. The Political Economy of Scientific Knowledge in Europe, 1918-1960*, 2012.

No. 4 – Giovanni CECCARELLI, Alberto GRANDI & Stefano MAGAGNOLI (eds./dir.), *Typicality in History. Tradition, Innovation, and Terroir / La typicité dans l'histoire. Tradition, innovation et terroir*, 2013.

No. 5 – Nicolas MARTY, *L'invention de l'eau embouteillée. Qualités, normes et marchés de l'eau en bouteille en Europe, XIXe-XXe siècles*, 2013.

No. 6 – Viktoria VON HOFFMANN, *Goûter le monde. Une histoire culturelle du goût à l'époque moderne*, 2013.

No. 7 – Jean-Pierre WILLIOT (dir.), *Du feu originel aux nouvelles cuissons. Pratiques, techniques, rôles sociaux*, 2015.

No. 8 – Olivier DE MARET, *Of Migrants and Meanings. Italians and Their Food Businesses in Brussels, 1876–1914*, 2016.

No. 9 – Maureen CHINYERE DURU, *Diaspora, Food and Identity. Nigerian Migrants in Belgium*, 2017.

No. 10 – Jean-Pierre WILLIOT (dir.), *La restauration ferroviaire entre représentations et consommations. Consommateurs, images et marchés / Railway Catering Between Imaginary and Consumption. Consumers, Images and Markets*, 2017.

No. 11 – Claudio BESANA, Rita D'ERRICO and Renato GHEZZI (eds.), *Cheese Manufacturing in the Twentieth Century The Italian Experience in an International Context*, 2017.

www.peterlang.com